Developments in Soil Science 8

SOIL ORGANIC MATTER

Developments in Soil Science 8

SOIL ORGANIC MATTER

Edited by

M. SCHNITZER
Soil Research Institute
Agriculture Canada
Ottawa, Ont., Canada

and

S.U. KHAN
Chemistry and Biology Research Institute
Agriculture Canada
Ottawa, Ont., Canada

ELSEVIER SCIENTIFIC PUBLISHING COMPANY
Amsterdam Oxford New York 1978

ELSEVIER SCIENTIFIC PUBLISHING COMPANY
335 Jan van Galenstraat
P.O. Box 211, Amsterdam, The Netherlands

Distributors for the United States and Canada:

ELSEVIER NORTH-HOLLAND INC.
52, Vanderbilt Avenue
New York, N.Y. 10017

ISBN: 0-444-41610-2 (vol. 8)
ISBN: 0-444-40882-7 (series)

Printed in The Netherlands

List of Contributors

V.O. BIEDERBECK
Research Station, Agriculture Canada
Swift Current, Sask., Canada

C.A. CAMPBELL
Research Station, Agriculture Canada
Swift Current, Sask., Canada

S.U. KHAN
Chemistry and Biology Research Institute
Agriculture Canada
Ottawa, Ont., Canada

C.G. KOWALENKO
Soil Research Institute, Agriculture Canada
Central Experimental Farm
Ottawa, Ont., Canada

L.E. LOWE
The University of British Columbia
Department of Soil Science
Vancouver, B.C., Canada

M. SCHNITZER
Soil Research Institute, Agriculture Canada
Central Experimental Farm
Ottawa, Ont., Canada

PREFACE

Soil organic matter, a key component of soils, affects many reactions that occur in these systems. In spite of this, soil organic matter remains a neglected field in soil science and receives but scant attention in soil science courses. One of the purposes of this book is to remedy this situation and to provide researchers, teachers and students with an up-to-date account of the current state of knowledge in this field.

The first three chapters of the book deal with the principal components of soil organic matter, that is, humic substances, carbohydrates and organic nitrogen-, phosphorus- and sulfur-containing compounds. In Chapter 4 reactions between soil organic matter and pesticides are discussed, whereas Chapters 5 and 6 are concerned with the more practical aspects of soil organic matter. The author of each chapter is an active researcher in the field about which he is writing. We were hoping that the direct involvement that each author has with his subject would result in a more adequate and relevant book.

Hopefully, the book will be of interest not only to soil scientists and agronomists but also to oceanographers, water scientists, geochemists, environmentalists, biologists and chemists who are concerned with the role of organic matter in terrestrial and aquatic systems.

Ottawa, April 1977   M. Schnitzer
                                                                          S.U. Khan

CONTENTS

Chapter 1

HUMIC SUBSTANCES: CHEMISTRY AND REACTIONS

M. SCHNITZER

INTRODUCTION

Humic substances, the major organic constituents of soils and sediments are widely distributed over the earth's surface, occurring in almost all terrestrial and aquatic environments. According to recent estimates of Bohn (1976), the mass of soil organic C ($30.0 \cdot 10^{14}$ kg) more than equals those of other surface C reservoirs combined (atmospheric CO_2 = $7.0 \cdot 10^{14}$ kg, biomass C = $4.8 \cdot 10^{14}$ kg, fresh water C = $2.5 \cdot 10^{14}$ kg, and marine C = $5.0-8.0 \cdot 10^{14}$ kg). Because between approximately 60—70% of the total soil-C occurs in humic materials (Griffith and Schnitzer, 1975a), the role of humic substances in the C cycle as a major source of CO_2 and as a C reservoir that is sensitive to changes in climate and atmospheric CO_2 concentrations has certainly been underestimated. According to Bohn (1976), the decay of soil organic matter provides the largest CO_2 input into the atmosphere. It is true that deeper C deposits in the form of marine organic detritus, coal and petroleum, deep sea solute C and C in sediments are much larger, but these are physically separated from active interchange with surface C reservoirs (Bohn, 1976).

Humic substances arise from the chemical and biological degradation of plant and animal residues and from synthetic activities of microorganisms. The products so formed tend to associate into complex chemical structures that are more stable than the starting materials. Important characteristics of humic substances are their ability to form water-soluble and water-insoluble complexes with metal ions and hydrous oxides and to interact with clay minerals and organic compounds such as alkanes, fatty acids, dialkyl phthalates, pesticides, etc. Of special concern is the formation of water-soluble complexes of fulvic acids (FA's) with toxic metals and organics which can increase the concentrations of these constituents in soil solutions and in natural waters to levels that are far in excess of their normal solubilities.

Chemical investigations on humic substances go back more than 200 years (Kononova, 1966; Schnitzer and Khan, 1972). The capacity of humic substances to adsorb water and plant nutrients was one of the first observations. Humic substances were thought to arise from the prolonged rotting of animal and plant bodies. Since that time several thousand scientific papers have been written on humic materials, yet much remains to be learned about their origin, synthesis, chemical structure and reactions and their functions in terrestrial and aquatic environments.

Soils and sediments contain a large variety of organic materials that can be grouped into humic and non-humic substances. The latter include those whose physical and chemical characteristics are still recognizable, such as carbohydrates, proteins, peptides, amino acids, fats, waxes, and low-molecular weight organic acids. Most of these compounds are attacked relatively readily by microorganisms and have usually only a short life span in soils and sediments. By contrast, humic substances exhibit no longer specific physical and chemical characteristics (such as a sharp melting point, exact refractive index and elementary composition, definite IR spectrum, etc.) normally associated with well-defined organic compounds. Humic substances are dark-coloured, acidic, predominantly aromatic, hydrophilic, chemically complex, polyelectrolyte-like materials that range in molecular weights from a few hundred to several thousand. These materials are usually partitioned into the following three main fractions: (a) humic acid (HA), which is soluble in dilute alkali but is precipitated on acidification of the alkaline extract; (b) fulvic acid (FA), which is that humic fraction which remains in solution when the alkaline extract is acidified; that is, it is soluble in both dilute alkali and acid; (c) humin, which is that humic fraction that cannot be extracted from the soil or sediment by dilute base or acid. From analytical data published in the literature (Schnitzer and Khan, 1972) it appears that structurally the three humic fractions are similar, but that they differ in molecular weight, ultimate analysis and functional group content, with FA having a lower molecular weight, containing more oxygen but less carbon and nitrogen, and having a higher content of oxygen-containing functional groups (CO_2H, OH, C = O) per unit weight than the other two humic fractions. The chemical structure and properties of the humin fraction appear to be similar to those of HA. The insolubility of humin seems to arise from it being firmly adsorbed on or bonded to inorganic soil and sediment constituents. The observed resistance to microbial degradation of humic materials appears to a significant extent also to be due to the formation of stable metal and/or clay-organic complexes.

SYNTHESIS OF HUMIC SUBSTANCES

The synthesis of humic substances has been the subject of much speculation. Felbeck (1971) lists the following four hypotheses for the formation of these materials.

(a) The plant alteration hypothesis. Fractions of plant tissues which are resistant to microbial degradation, such as lignified tissues, are altered only superficially in the soil to form humic substances. The nature of the humic substance formed is strongly influenced by the nature of the original plant

material. During the first stages of humification high-molecular weight HA's and humins are formed. These are subsequently degraded into FA's and ultimately to CO_2 and H_2O.

(b) The chemical polymerization hypothesis. Plant materials are degraded by microbes to small molecules which are then used by microbes as carbon and energy sources. The microbes synthesize phenols and amino acids, which are secreted into the surrounding environment where they are oxidized and polymerized to humic substances. The nature of the original plant material has no effect on the type of humic substance that is formed.

(c) The cell autolysis hypothesis. Humic substances are products of the autolysis of plant and microbial cells after their death. The resulting cellular debris (sugars, amino acids, phenols, and other aromatic compounds) condenses and polymerizes via free radicals.

(d) The microbial synthesis hypothesis. Microbes use plant tissue as carbon and energy sources to synthesize intercellularly high-molecular weight humic materials. After the microbes die, these substances are released into the soil. Thus, high-molecular weight substances represent the first stages of humification, followed by extracellular microbial degradation to HA, FA and ultimately to CO_2 and H_2O.

It is difficult to decide at this time which hypothesis is the most valid one. It is likely that all four processes occur simultaneously, although under certain conditions one or the other could dominate. However, what all four hypotheses suggest is that the more complex, high-molecular weight humic materials are formed first and that these are then degraded, most likely oxidatively, into lower molecular weight materials. Thus, the sequence of events appears to be HA → FA.

EXTRACTION OF HUMIC SUBSTANCES

The organic matter content of soils may range from less than 0.1% in desert soils to close to 100.0% in organic soils. In inorganic soils, organic and inorganic components are so closely associated that it is necessary to first separate the two before either component can be studied in greater detail. Thus, extraction of the organic matter is generally the first major operation that needs to be done. The most efficient and most widely used extractant for humic substances from soils is dilute aqueous NaOH (either 0.1 N or 0.5 N) solution. While the use of alkaline solutions has been criticized, there seems to be little evidence to show that dilute alkali under an atmosphere of N_2 damages or modifies the chemical structure and properties of humic materials. Thus, a HA extracted with 0.5% NaOH solution had similar light absorbance characteristics as the same HA extracted with 1% NaF solution (Scheffer and Welte, 1950; Welte, 1952). Other workers (Rydalevskaya and Skorokhod, 1951) found no substantial differences in elementary composition and CO_2H content between HA's extracted by 1% NaF and 0.4% NaOH solutions from soils and peats. Similarly, Smith and Lorimer (1964) report that HA's extracted with dilute $Na_4P_2O_7$ from peat soils resembled in all respects HA's extracted with dilute NaOH solution. Schnitzer and Skinner (1968a) extracted FA from a Spodosol Bh horizon under N_2 with 0.5 N NaOH and with 0.1 N HCl. Following purification, each extract was characterized by chemical and spectrophotometric methods and by gel filtration. The elementary composition of the two materials was very similar and oxygen-containing functional groups were of same order of magnitude. Also, IR spectra of both preparations and their fractionation behaviour on Sephadex gels were practically identical.

The concentration of the NaOH solution affects the yield of the humic material extracted as well as its ash content. Ponomarova and Plotnikova (1968) and Levesque and Schnitzer (1966) found 0.1 N NaOH to be more efficient than higher NaOH concentrations. However, the most suitable extractant for isolating humic materials low in ash was either 0.4 N or 0.5 N NaOH solution (Levesque and Schnitzer, 1966).

Neutral salts of mineral and organic acids have been used for the extraction of humic substances, but yields are usually low. Bremner and Lees (1949) suggested the use of 0.1 M Na—pyrophosphate solution at pH 7 as the most efficient extractant. The action of the neutral salt was thought to depend on the ability of the anion to interact with polyvalent cations bound to humic materials to form either insoluble precipitates or soluble metal complexes, and the formation of a soluble salt of the humic material by reacting with the cation of the extractant as illustrated by the following reaction:

$$R(COO)_4 Ca_2 + Na_4P_2O_7 \rightarrow R(COONa)_4 + Ca_2P_2O_7 \tag{1}$$

According to Alexandrova (1960), $Na_4P_2O_7$ solution extracts not only humic substances but also organo-mineral complexes without destroying non-silicate forms of sesquioxides. The efficiency of extraction can be improved by raising the pH from 7.0 to 9.0 (Kononova, 1966) and increasing the temperature (Livingston and Moe, 1969; Lefleur, 1969). Kononova and Bel'chikova (1961) recommend the use of a combination of 0.1 M $Na_4P_2O_7$ + 0.1 N NaOH (pH $\approx$13). Use of this mixture also avoids decalcification of soils with high pH prior to extraction. Humic materials extracted by the mixture are low in N, (Dormaar, 1972; Vila et al., 1974) and show lower molecular weights and E_4/E_6 ratios, different electrophoretic patterns and behavior on gel filtration than do humic materials extracted from similar soils with 0.1 N NaOH (Vila et al., 1974). Schnitzer et al. (1958) showed that pyrophosphate was difficult to remove from humic materials during purification.

Other approaches that have been employed for the extraction of organic matter from soils involve treatment with chelating resins (Levesque and Schnitzer, 1967; Dormaar, 1972; De Serra and Schnitzer, 1972). Humic materials extracted with the aid of a chelating resin were more polymerized than those extracted by dilute alkali. Another technique that has been used by a number of workers is ultrasonic dispersion (Edwards and Bremner, 1967; Leenheer and Moe, 1969; Watson and Parsons, 1974; Anderson et al., 1974).

Several attempts have been made to extract humic substances with organic solvents. Martin and Reeve (1957a, b) found that acetyl acetone was an effective extractant for organic matter from Spodosol Bh horizons. Porter (1967) used an acetone–water–HCl system, while Parsons and Tinsley (1960) employed anhydrous formic acid + 10% acetyl acetone to extract organic matter from a calcareous meadow soil. Hayes et al. (1975) compared humic materials extracted from an organic soil by thirteen extractants, which included dipolar aprotic solvents, pyridine, ethylenediamine, organic chelating agents, ion exchange resins, $Na_4P_2O_7$ and NaOH. Of the two reagents that were most efficient, ethylenediamine was found by Electron Spin Resonance Spectrometry and elementary analyses to alter the chemical nature and composition of the extract while dilute NaOH solution was regarded as the more reliable extractant. The danger with using organic solvents containing C and N for extracting organic matter is that under these conditions C and N may be added irreversibly to the humic materials and so alter their composition and properties.

A number of workers have extracted humic substances by sequential extraction, using different reagents (Duchaufour and Jacquin, 1963; Smith and Lorimer, 1964; Gascho and Stevenson, 1968; Goh, 1970). Felbeck (1971) suggests the following sequence: (a) benzene-methanol; (b) 0.1 N HCl; (c) 0.1 M $Na_4P_2O_7$; (d) 6 N HCl at 90°C; (e) 5 : 1 chloroform-methanol; and (f) 0.5 N NaOH. By using a sequence of solvents rather than one solvent, a series of fractions can be obtained which may be more homogeneous than the material extracted by one extractant only.

FRACTIONATION AND PURIFICATION

The classical method of fractionation of humic substances is based on differences in solubility in aqueous solutions at widely differing pH levels, in alcohol and in the presence of different electrolyte concentrations (Fig. 1). The major humic fractions are HA, FA and humin. Fractionation of HA into hymatomelanic acid or into gray HA and brown HA is not done very often. One may wonder how useful such separations are.

Additional methods of fractionation of humic substances that have been tried over the years include treatment with tetrahydrofuran, containing increasing percentages of water (Salfeld, 1964; Martin et al., 1963), mixtures of dimethylformamide and water (Otsuki and Hanya, 1966), salting out with

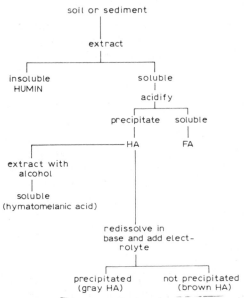

Fig. 1. Fractionation of humic substances.

ammonium sulfate (Theng et al., 1968), varying the ionic strength and pH of pyrophosphate and sodium hydroxide extracting solutions (Lindqvist, 1968), addition of increasing amounts of metal ions such as Pb^{2+}, Ba^{2+} and Cu^{2+} (Sowden and Deuel, 1961) and adding increasing volumes of ethanol to alkaline solutions containing HA's (Kyuma, 1964).

Freezing methods have also been used (Karpenko and Karavayev, 1966; Archegova, 1967) for this purpose.

In recent years, gel filtration has been widely used for the fractionation of soil humic materials. This technique has also been employed for the separation of aquatic humus (Gjessing, 1976). Schnitzer and Skinner (1968b) prepared seven fractions from a FA by carrying out a series of sequential column chromatographic separations using different Sephadex gels. The fractions differed in elementary analysis and functional group content, number-average molecular weights and IR and NMR spectra. Swift and Posner (1971) studied the behavior of HA's on Sephadex and a number of other gels with a variety of eluants. They found that fractionation based solely on molecular weight differences could be achieved by using alkaline buffers containing large amino cations. They warn that in cases where gel-solute interactions could occur, fractionations based on differences in molecular weights would not be possible.

Column chromatography on activated charcoal has been used by Forsyth (1947) for the separation of HA's. Other workers (Dragunov and Murzakov, 1970) have employed Al_2O_3 in addition to charcoal.

Barton and Schnitzer (1963) separated methylated FA over Al_2O_3 with organic solvents of increasing polarities into several fractions, which differed in molecular weights, oxygen-containing functional groups, and spectroscopic properties. At a later date, the author and his coworkers modified and extended this approach. These investigations included solvent extraction of humic materials, followed by exhaustive methylation and separation of benzene-soluble fractions by column-, thin-layer- and gas-chromatography and identification of individual components by mass spectrometry and micro-IR spectrophotometry.

Several workers (Kononova, 1966) have used electrophoretic methods for the separation of humic substances. Continuous zone electrophoresis in free films of buffer has also been employed (Leenheer and McKinley, 1971; Leenheer and Malcolm, 1972).

HA's can be purified efficiently by shaking at room temperature with dilute solutions of HCl-HF (0.5 ml conc. HCl + 0.5 ml of 48% HF + 99 ml of H_2O). After shaking for 24 to 48 h, the acid mixture is removed by filtration and the residue is washed with distilled water until free of Cl^- and then dried. This method has been in use in the author's laboratory for many years, and the ash content of HA's can be reduced in this manner to $<1.0\%$. Another method of purification of HA's that has been used widely is dialysis. While salts and low-molecular weight organic compounds are readily removed, the method cannot separate complexed or strongly adsorbed metals or metal hydroxides from humic materials. FA's are readily purified by passage over Amberlite IR-120 or Dowex-50 exchange resins in H-forms (Schnitzer and Skinner, 1968).

Ultrafiltration has been used for the desalting, concentration and fractionation of humic materials in surface waters (Schindler et al., 1972; Schindler and Alberts, 1974; Ogura, 1974). Gjessing (1970) reports that there is generally more retention of humic materials during ultrafiltration than can be accounted for by the nominal molecular weight cut-off values of the membranes. Alberts et al. (1976) have warned that care should be exercised in any attempt to determine molecular weights of humic materials by ultrafiltration but they found the technique efficient for the preparation and fractionation of humic materials. The solute retention of humic materials on ultrafilter membranes may depend on the charge as well as on the molecular weight and asymmetry of the material to be separated, membrane-solute interaction and solute aggregation in solution.

THE CHARACTERIZATION OF HUMIC MATERIALS

Elementary analysis

Elementary analysis provides information on the distribution of major elements (C, H, N, S and O) in humic substances. Elementary analyses of HA's extracted from soils formed under widely differing geographic and pedologic conditions such as those prevailing in the Arctic, the cool temperate, subtropical and tropical climatic zones are shown in Table I. When more than one set of data is available, the results are shown as ranges. The C content of the HA's ranges from 53.8 to 58.7%, the O content from 32.8 to 38.3%; percentages of H and N vary from 3.2 to 6.2% and 0.8 to 4.3%, respectively. The S content ranges from 0.1 to 1.5%.

Elementary analyses of FA's extracted from the same soils are shown in Table II. Compared to HA's, FA's contain more O and S but less C, H and N than do HA's. The C content of FA's ranges from 40.7 to 50.6%, that of O from 39.7 to 49.8%. Thus, on the average HA's contain 10% more C but 10% less O than do FA's. It is noteworthy that in both HA's and FA's, C and O are the major elements.

Compared to soil humic compounds, humic substances in waters contain less C and N (Gjessing, 1976).

Oxygen-containing functional groups

The major oxygen-containing functional groups in humic substances are carboxyls, hydroxyls and carbonyls. Analytical data for these groups in HA's extracted from widely differing soils are shown in Table III. The total acidity equals the sum of CO_2H + phenolic OH groups. Similar data for FA's are presented in Table IV. The total acidity, and especially the CO_2H content, of FA's are considerably higher than those of HA's. The C=O content varies relatively widely, especially in the case of HA's.

Means of ranges in elementary analyses (Tables I and II) and functional groups (Tables III and IV) are shown in Table V. These data may be considered as approximations of the elementary composition and functional group content of a "model" HA and FA. A more detailed analysis of the data in Table V indicates that: (a) the "model" HA contains approximately 10%

TABLE I

Elementary analysis of HA's extracted from soils from widely differing climates (From Schnitzer, 1977)

Element (%)	Arctic	Cool, temperate		Subtropical	Tropical
		acid soils	neutral soils		
C	56.2	53.8—58.7	55.7—56.7	53.6—55.0	54.4—54.9
H	6.2	3.2— 5.8	4.4— 5.5	4.4— 5.0	4.8— 5.6
N	4.3	0.8— 2.4	4.5— 5.0	3.3— 4.6	4.1— 5.5
S	0.5	0.1— 0.5	0.6— 0.9	0.8— 1.5	0.6— 0.8
O	32.8	35.4—38.3	32.7—34.7	34.8—36.3	34.1—35.2

TABLE II

Elementary analysis of FA's extracted from soils from widely differing climates
(From Schnitzer, 1977)

Element (%)	Arctic	Cool, temperate		Subtropical	Tropical
		acid soils	neutral soils		
C	47.7	47.6—49.9	40.7—42.5	42.2—44.3	42.8—50.6
H	5.4	4.1— 4.7	5.9— 6.3	5.9— 7.0	3.8— 5.3
N	1.1	0.9— 1.3	2.3— 2.8	3.1— 3.2	2.0— 3.3
S	1.6	0.1— 0.5	0.8— 1.7	2.5	1.3— 3.6
O	44.2	43.6—47.0	47.1—49.8	43.1—46.2	39.7—47.8

TABLE III

Functional group analysis and E_4/E_6 ratios of HA's extracted from soils from widely differing climates
(From Schnitzer, 1977)

Functional group (meq./g)	Arctic	Cool, temperate		Subtropical	Tropical
		acid soils	neutral soils		
Total acidity	5.6	5.7—8.9	6.2—6.6	6.3—7.7	6.2—7.5
CO_2H	3.2	1.5—5.7	3.9—4.5	4.2—5.2	3.8—4.5
Phenolic OH	2.4	3.2—5.7	2.1—2.5	2.1—2.5	2.2—3.0
Alcoholic OH	4.9	2.7—3.5	2.4—3.2	2.9	0.2—1.6
Quinonoid C=O	2.3	0.1—1.8	4.5—5.6	0.8—1.5	1.4--2.6
Ketonic C=O	1.7				0.3—1.4
OCH_3	0.4	0.4	0.3	0.3—0.5	0.6—0.8
E_4/E_6	5.3	3.8—5.0	4.0—4.3	3.9—5.1	5.0—5.8

TABLE IV

Functional group analysis and E_4/E_6 ratios of FA's extracted from soils from widely differing climates
(From Schnitzer, 1977)

Functional group (meq./g)	Arctic	Cool, temperate		Subtropical	Tropical
		acid soils	neutral soils		
Total acidity	11.0	8.9—14.2	ND	6.4—12.3	8.2—10.3
CO_2H	8.8	6.1— 8.5	ND	5.2— 9.6	7.2—11.2
Phenolic OH	2.2	2.8— 5.7	ND	1.2— 2.7	0.3— 2.5
Alcoholic OH	3.8	3.4— 4.6	ND	6.9— 9.5	2.6— 5.2
Quinonoid C=O	2.0	1.7— 3.1	ND	1.2— 2.6	0.3— 1.5
Ketonic C=O	2.0		ND		1.6— 2.7
OCH_3	0.6	0.3— 0.4	ND	0.8— 0.9	0.9— 1.2
E_4/E_6	11.5	9.0	ND	8.4— 9.5	7.6—11.2

TABLE V

Analysis of "model" HA and FA (from means of all data)
(From Schnitzer, 1977)

Element (%)	HA	FA
C	56.2	45.7
H	4.7	5.4
N	3.2	2.1
S	0.8	1.9
O	35.5	44.8
	100.4	99.7

Functional groups (meq./g)	HA	FA
Total acidity	6.7	10.3
CO_2H	3.6	8.2
Phenolic OH	3.9	3.0
Alcoholic OH	2.6	6.1
Quinonoid C=O Ketonic C=O	2.9	2.7
OCH_3	0.6	0.8
E_4/E_6	4.8	9.6

more C but 10% less O than does the "model" FA; (b) there is relatively little difference between the two materials in H, N and S content; (c) the total acidity and CO_2H content of the "model" FA are appreciably higher than those of the "model" HA; (d) both materials contain approximately the same concentrations of phenolic OH, total C=O and OCH_3 groups per unit weight, but the FA is richer in alcoholic OH groups; and (e) about 78% of the oxygen in the HA can be accounted for in functional groups, but all of the O in the FA is similarly distributed (see also Tables I and II).

Distribution of N in humic materials

Between 20 and 50% of the N in humic substances appears to consist of amino acid-N and 1—10% as amino sugar-N. (Stevenson, 1960; Bremner, 1965, 1967; Sowden and Schnitzer, 1967; Khan and Sowden, 1971, 1972). Small amounts of purine and pyrimidine bases have also been identified in acid hydrolysates of humic substances (Anderson, 1957, 1958, 1961). Humic materials from widely differing soils do not appear to vary markedly in amino acid composition, but a considerable percentage of the total N in humic materials is neither "protein-like" nor amino sugar nor ammonia. This "unknown" N, much of which is not released by acid and base hydrolysis, needs to be identified, and it should be possible to do this with the sophisticated analytical methods that are now available. For a more complete review of the nature and distribution of N in organic matter, the reader is referred to Bremner (1965, 1967), Flaig et al. (1975) and to Chapter 3 of this book.

THE ANALYSIS OF HUMIC SUBSTANCES — NON-DEGRADATIVE METHODS

Aside from elementary and functional group analyses, the methods most frequently used for the characterization of humic substances can be divided into non-degradative and degradative ones. Non-degradative methods (Table VI) include spectrophotometric, spectrometric, X-ray, electron microscopy, electron diffraction, viscosity, surface tension and molecular weight measurements as well as electrometric titrations (Table VI).

Each of the major methods will be discussed in some detail in the following paragraphs.

Spectrophotometry in the UV and visible region

Generally, humic substances yield uncharacteristic spectra in the UV and visible regions. Absorption spectra of alkaline and neutral aqueous solutions of HA's and FA's and of acidic, aqueous FA solutions are featureless, showing no maxima or minima; the optical density usually decreases as the wavelength increases.

The light absorption of humic substances appears to increase with increase in: (i) the degree of condensation of the aromatic rings that these substances contain (Kononova, 1966); (ii) the ratio of C in aromatic "nuclei" to C in aliphatic side chains (Kasatochkin et al., 1964); (iii) total C content; and (iv) molecular weight.

The ratio of optical densities or absorbances of dilute aqueous HA and FA

TABLE VI

Non-degradative methods used for the characterization of humic substances

---

Spectrophotometry in the UV and visible, spectrophotofluorometry
Infrared (IR) spectrophotometry
Nuclear Magnetic Resonance (NMR) spectrometry
Electron Spin Resonance (ESR) spectrometry, X-ray analysis
Electron microscopy; electron diffraction analysis
Viscosity measurements
Surface tension measurements
Molecular weight measurements
Electrometric titrations

---

solutions at 465 and 665 nm is widely used by soil scientists for the characterization of these materials. This ratio, usually referred to as E_4/E_6, has been
reported to be independent of concentrations of humic materials but to vary
for humic materials extracted from different soil types (Kononova, 1966;
Schnitzer and Khan, 1972). For example, according to Kononova (1966),
E_4/E_6 ratios for HA's extracted from Spodosols are about 5.0, ratios for
HA's extracted from Boralfs are 3.5, those for HA's from Haploborolls are
3.0—3.5, for HA's from Aridic Haploborolls they are 4.0—4.5, and those for
HA's from Inceptisols and Oxisols are about 5.0. For FA's E_4/E_6 ratios range
between 6.0 and 8.5 (Kononova, 1966). E_4/E_6 ratios for HA's and FA's extracted from soils formed under widely differing conditions are shown in
Table III and IV. Kononova (1966) believes that the magnitude of the
E_4/E_6 ratio is related to the degree of condensation of the aromatic C network, with a low ratio indicative of a relatively high degree of condensation
of aromatic humic constituents. Conversely, a high E_4/E_6 ratio reflects a low
degree of aromatic condensation and infers the presence of relatively large
proportions of aliphatic structures. Analytically, the determination of
E_4/E_6 ratios of HA's and FA's is a rapid and convenient procedure that does
not require complex equipment and advanced technical skills but which,
nonetheless, can provide potentially valuable information on these materials.
According to Chen et al. (1977) the E_4/E_6 ratio of HA's and FA's is: (i)
mainly governed by the particle size (or particle or molecular weight); (ii)
affected by pH; (iii) correlated with the free radical concentration, contents
of O, C, CO_2H and total acidity in as far as these parameters are also functions of the particle size or particle or molecular weight; (iv) apparently not
directly related to the relative concentration of aromatic condensed rings; (v)
independent of HA and FA concentrations at least in the 100—500 ppm
range. Chen et al. (1977a) found the following relationship between the
slope of the log optical density (OD) vs log wavelength (λ) curve for FA and
the E_4/E_6 ratio:

$$\text{slope} = \frac{d \log OD}{d \log \lambda} = -6.435 \log E_4/E_6 \qquad (1a)$$

Because the slope is a direct function of particle or molecular size, the E_4/E_6
ratio is also a direct function of the particle or molecular size which, in turn,
is related to particle or molecular weight. Chen et al. (1977) were unable to
find any direct relationship between the E_4/E_6 ratio and the concentration of
condensed aromatic rings in HA's and FA's.

HA's and FA's fluoresce under UV and visible light. Seal et al. (1964)
observed green fluorescence with more or less flat maxima in the 500—540
nm region. Hansen (1969) found that the fluorescence of humic materials
was affected by pH. When dissolved in methanol, FA exhibited a fluorescence
maximum near 507 nm; in 0.01 M CH_3ONa the maximum was lowered to

465 nm. Datta et al. (1971) noted that Na-humates in aqueous solutions produced a fluorescence maximum at 470 nm, when dissolved in ether, pyridine, acetone and dimethylformamide the maximum shifted to 370 nm; in alcohols it was 400 nm. Thus, the fluorescence of HA's and FA's is affected by pH and the polarity of the solvent.

Infrared spectrophotometry

IR spectra of humic substances show bands at the following frequencies: 3,400 cm^{-1} (hydrogen-bonded OH), 2,900 cm^{-1} (aliphatic C—H stretch), 1,725 cm^{-1} (C=O of CO$_2$H, C=O stretch of ketonic C=O), 1,630 cm^{-1} (aromatic C=C (?), hydrogen-bonded C=O of carbonyl or quinone, COO$^-$), 1,450 cm^{-1} (aliphatic C—H), 1,400 cm^{-1} (COO$^-$, aliphatic C—H), 1,200 cm^{-1} (C—O stretch of OH-deformation of CO$_2$H) and 1,050 cm^{-1} (Si-O of silicate impurities). Representative HA and FA spectra are shown in Fig. 2. The bands are broad, likely because of extensive overlapping of individual absorptions. The IR spectra reflect the preponderance of oxygen-containing functional groups, that is, CO$_2$H, OH and C=O in the humic materials. While IR spectra of humic materials provide worthwhile information on the distribution of functional groups, they tell little about the chemical structure of humic "nuclei". Nevertheless, IR spectrophotometry is useful for the gross characterization of humic materials of diverse origins, for the evaluation of the effects of different chemical extractants, chemical modifications such as methylation, acetylation, saponification and the formation of derivatives. It can also be used to detect changes in the chemical structure of humic materials following oxidation, pyrolysis and similar treatments, to ascertain and characterize the formation of metal-humate and clay-humate complexes and to indicate possible interactions of pesticides and herbicides with humic materials (Sullivan and Felbeck, 1968; Stevenson and Goh, 1971; Schnitzer and Khan, 1972).

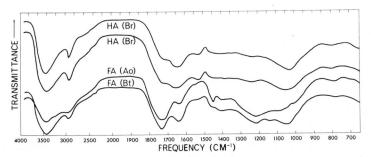

Fig. 2. IR spectra of HA's and FA's extracted from a Boralf (from Schnitzer and Gupta, 1964). Published with permission of the Soil Science Society of America.

Nuclear magnetic resonance spectrometry

Because untreated humic materials are not soluble in organic solvents such as $CHCl_3$ and CCl_4, the use of NMR spectrometry has been confined to methylated humic fractions or to degradation products (Barton and Schnitzer, 1963; Schnitzer and Skinner, 1968b). The most remarkable observations about proton-NMR spectra of methylated humic fractions is the absence of aromatic and olefinic protons. This may be due to the fact that aromatic "nuclei" or "cores" of humic substances are fully substituted by atoms other than hydrogen or that relaxing effects of spins of unpaired electrons (free radicals) interfere with NMR measurements. So far proton-NMR has provided little information on the chemical structure of humic materials. Recently, Neyroud and Schnitzer (1974b) attempted to characterize a methylated high-molecular weight FA fraction by C—13 NMR spectrometry, a method that shows great promise in structural organic chemistry. The resulting C—13 NMR spectrum was practically a straight line with a small but extended peak near 53 ppm (most likely due to C—13 in different types of OCH_3 groups) and a hump near 166 ppm (due to aromatic C—13 bonded to a OCH_3 group). Compared to C—13 NMR spectra of beech and spruce lignins, which consist of large numbers of well-defined peaks, the spectrum for the FA fractions was most disappointing. In a very recent study, Vila et al. (1976) obtained more encouraging results with C—13 NMR spectrometry. C—13 NMR spectra of two HA's and one FA (dissolved in 0.1 N NaOD in D_2O) could be divided into three regions. Between 160 and 200 ppm carbonyl resonance was observed, aromatic C was detected between 100 and 160 ppm, and aliphatic C between 10 and 100 ppm. The resolution of the spectra was restricted by the signal to noise ratio. Vila et al. (1976) suggest that higher magnetic fields or larger diameter probes be used for higher resolution, but from their results it appears that with additional development C—13 NMR spectrometry could eventually provide useful structural information on humic materials.

Electron spin resonance spectrometry

Humic substances are known to be rich in stable free radicals (Rex, 1960; Steelink and Tollin, 1967; Riffaldi and Schnitzer, 1972) which most likely play important roles in polymerization—depolymerization reactions, in reactions with other organic molecules, including pesticides and toxic pollutants, and in the physiological effects that these substances are known to exert (Schnitzer and Khan, 1972). ESR spectra of aqueous HA and FA solutions usually consist of single lines (see Fig. 3) devoid of hyperfine splitting, with g-values ranging from 2.0031 to 2.0045, line widths from 2.0 to 3.6 G and free radical concentrations from $1.4 \cdot 10^{17}$ to $37.4 \cdot 10^{17}$ spins/g (Senesi and Schnitzer, 1977). In a recent investigation Senesi and Schnitzer (1977) deter-

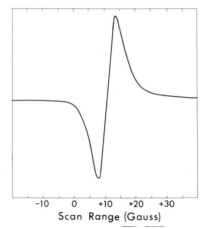

-10   0   +10   +20   +30
Scan Range (Gauss)

Fig. 3. ESR spectrum of HA.

mined ESR parameters of powders and solutions of FA and of a number of molecular weight fractions separated from it. Major objectives of their investigation were to evaluate effects of pH, reaction time, chemical reduction and irradiation on the ESR parameters and to obtain information on the identity (ies) of the free radicals. Two types of free radicals were detected in all FA-preparations: (a) permanent ones, with long life spans; and (b) transient ones, with short lives, which were generated in large concentrations by different treatments in the following order of decreasing efficiencies: chemical reduction > irradiation > raising pH. Spectroscopic splitting factors (*g*-values) of permanent and transient radicals were similar, indicating that the radicals had similar structures. From the magnitude of *g*-values it appeared that the radicals were substituted semiquinones which, in alkaline solutions were stabilized as semiquinone anions as Steelink (1964) had suggested.

Atherton et al. (1967) are so far the only workers who have reported hyperfine splitting of ESR signals of HA's when these were dissolved in 0.1 N NaOH. The HA's were extracted from acidic organic soils and from inorganic soils rich in organic matter and boiled for 24 h with 6 N HCl prior to ESR analyses. ESR spectra of some of the resulting materials showed four lines, indicative of an interaction of the unpaired electron with two nonequivalent protons. In a recent study, Senesi et al. (1977a) found that treatment at room temperature of a 1% aqueous FA solution with dilute H_2O_2 at pH 7 and with Ag_2O at pH 13 produced an unsymmetrical 3-line spectrum with coupling constants of 1.45 G. The three equally spaced lines of the triplet indicated interaction of the unpaired electron with two equivalent protons. After 2 h of oxidation, the signal began to decay. When additional oxidant was added, a well resolved triplet reappeared. Senesi et al. (1977a) rationalized their results in terms of a 2-step oxidation mechanism, with each step consisting of a one-electron oxidation as illustrated by the following scheme:

Hydroquinone (1) is oxidized via semiquinone anion intermediates (2) and (3) to quinone (4). R_1 and R_2 could be CO_2H, alkyl or more complex C-containing groups. The fact that ESR signals of HA's and FA's are difficult to resolve probably means that more than one type of radical is present and that their signals overlap. Thus, mild oxidation of aqueous FA solutions offers an opportunity of revealing the identities of free radicals in FA that contribute to unresolved ESR signals of these materials.

Humic substances can act either as electron donors or electron acceptors via free radical intermediates and so participate in oxidation-reduction reactions with transition metals and biological systems in soils. The oxidation-reductions are reversible, proceed via semiquinones and involve transient free radicals mainly (Senesi et al., 1977b). The latter are usually not sufficiently long-lived under oxidative conditions to be detected but sufficiently stable for this purpose under reducing conditions and after irradiation. The reactions may be summarized under acid, neutral and alkaline conditions by the following scheme:

The findings of Senesi et al. (1977b) predict a number of conditions of interest to soil scientists: (a) water-logged or poorly-drained soils, where reducing conditions prevail, or soils with high pH should contain high concentrations of free radicals; and (b) humic materials near soil surfaces frequently exposed to sunlight should be rich in free radicals. Slawinska et al. (1975)

believe that humic substances can act as photosensitizers for bonded or sorbed substances, so that sorbed herbicides might be detoxified by free radicals whose formation is stimulated by light and oxygen (air). It is likely that important practical applications based on our increasing knowledge of free radical reactions involving humic substances will be forthcoming in the not too distant future.

X-ray analysis

X-ray analysis has been used by several workers (Kasatochkin and Zilberbrand, 1956; Kasatochkin et al., 1964; Tokudome and Kanno, 1965; Kodama and Schnitzer, 1967) for elucidating the structure of soil humic substances. Diffraction patterns of HA's usually show broad bands near 3.5 Å, whereas those for FA's exhibit halos in the 4.1—4.7 Å region.

Kasatochkin et al. (1964) conclude from X-ray studies that humic substances contain flat condensed aromatic networks to which side chains and functional groups are attached. Diffraction patterns of HA's extracted from Haploborolls show distinct 002 bands, indicating that most of their C is in the condensed "nucleus" but little in side chains. By contrast, patterns of FA's extracted from Boralfs exhibit small 002 reflections but distinct γ-bands, which indicates that most of the C is in side chains and little in condensed "nuclei". Diffraction patterns of HA's extracted from Boralfs indicate an intermediate position for these materials.

Naturally occurring humic substances are non-crystalline. Results of diffraction studies for such materials can be expressed as a radial distribution function, which specifies the density of atoms or electrons as a function of the radial distance from any reference atom or electron in the system. The X-ray diffraction pattern of a non-oriented flat powder specimen of FA, examined by Kodama and Schnitzer (1967), exhibited a diffuse band at about 4.1 Å, accompanied by a few minor humps. Radial distribution analysis of the FA showed peaks at 1.6 and 2.9 Å and shoulders at 4.2 and 5.2 Å. The peak maxima were similar to those of carbon black but the electron distribution for the FA peaks was different. This may mean that FA has a considerable random structure, in which in addition to C atoms, O atoms are also major structural components. Kodama and Schnitzer (1967) conclude that the C skeleton of FA consists of a broken network of poorly condensed aromatic rings with appreciable numbers of disordered aliphatic or alicyclic chains around the edges of the aromatic layers.

Small angle X-ray scattering has been used by Wershaw et al. (1967) for measuring the particle size of Na-humate. They conclude that either particles of two or more different sizes exist in solutions or that all of the particles have the same size but consist of a dense core and a less dense outer shell.

Electron microscopy and electron diffraction

Several workers (Flaig and Beutelspacher, 1951, 1954; Visser, 1963, 1964; Wiesemuller, 1965; Dutta et al., 1968; Dudas and Pawluk, 1970; Khan, 1971; Schnitzer and Kodama, 1975; Chen and Schnitzer, 1976a) have used the electron microscope for observing shapes and sizes of HA and FA particles. Flaig and Beutelspacher (1951, 1954) have shown that HA particles are tiny spheres which are capable of joining into chains and of forming racemose aggregates through hydrogen-bonding at low pH. Electron micrographs of HA's published by Khan (1971) show a loose spongy structure with many internal spaces. Diameters of HA particles have been estimated to be of the order of 100—160 Å (Flaig and Beutelspacher, 1951; Visser, 1964; Wiesemuller, 1965).

According to Schnitzer and Kodama (1975) who examined FA under the electron microscope, the crystallinity, shapes, dimensions and extent of aggregation of the particles depend on the pH. At pH 2.5, three types of particles can be observed: small spheroids (15—20 Å in diameter), aggregates of spheroids (200—300 Å in diameter) and amorphous material of low contrast, perforated by voids (500—1,100 Å in diameter). The spheroidal aggregates tend to form elongated, irregularly shaped structures, 20,000—30,000 Å long. At pH 3.5, which is the natural pH of a dilute FA solution, electron micrographs show a sponge-like structure of variable thickness (100—300 Å) punctured by voids, 200—1,000 Å in diameter. At pH 4.5 and higher, electron micrographs show flat sheet-like lamellae of very low contrast, perforated by voids, 200—2,000 Å in diameter.

Chen and Schnitzer (1976a) used a scanning electron microscope to explore the effect of pH on the shape, size and degree of aggregation of HA and FA particles. Compared to the conventional transmission electron microscope (TEM), the scanning electron microscope (SEM) offers the following advantages: (a) it yields three-dimensional pictures of samples; (b) surfaces can be directly observed; and (c) the orientation of particles with respect to each other and to other sample features can be observed.

As viewed under the scanning electron microscope (Fig. 4), FA at pH 2—3 occurs mainly as elongated fibers and bundles of fibers, forming a relatively open structure. With increase in pH, the fibers tend to mesh into a finely-woven network to yield a sponge-like structure. Above pH 7, a distinct change in the structural arrangement and an improved orientation can be observed. At pH 8, the FA forms sheets which tend to thicken at pH 9. At pH 10, fine, homogeneous grains are visible. The effect of pH on the HA structure (Fig. 5) is similar to that observed on FA, except that because of low solubility in water, the pH range had to be narrowed to between 6 and 10, and the pH at which the major transitions occur is higher. Thus, there is a gradual transition from a fibrous structure at low pH to a more sheet-like one at higher pH. Simultaneously, the particles become smaller as the pH increases. The

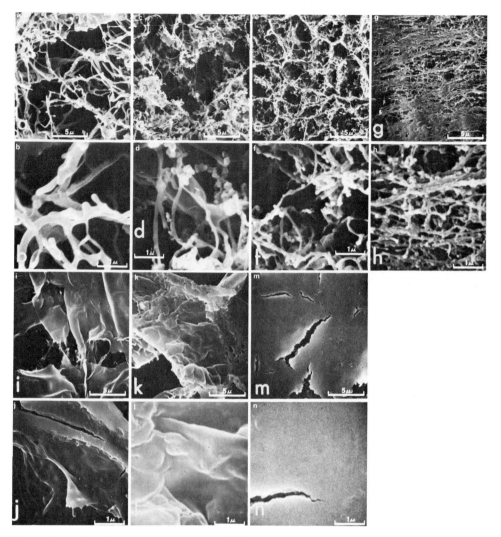

Fig. 4. Scanning electron micrographs of FA at various pH's: a, b. pH 2; c, d. pH 4; e, f. pH 6; g, h. pH 7; i, j. pH 8; k, l. pH 9; m, n. pH 10 (from Chen and Schnitzer, 1976a). Published with permission of the Soil Science Society of America.

aggregation of FA particles at low pH can be explained as being due to hydrogen-bonding, Van der Waal's interactions and interactions between π-electron systems of adjacent molecules. As the pH increases, these forces become weaker, and because of increasing ionization of CO_2H and phenolic OH groups, particles separate and begin to repel each other electrostatically, so that the molecular arrangements become smaller and smaller but better

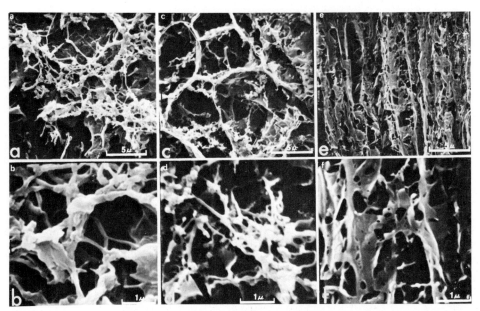

Fig. 5. Scanning electron micrographs of HA at various pH's: a, b. pH 6; c, d. pH 8, e, f. pH 10 (from Chen and Schnitzer, 1976a). Published with permission of the Soil Science Society of America.

oriented. Similar aggregation—dispersion phenomena can be observed for HA, although over a narrower range.

Table VII lists electron diffraction data for FA at pH 2.5 (Schnitzer and Kodama, 1975). These data show the presence of crystalline materials plus other components that produce diffuse patterns. Most of the spacings in Ta-

TABLE VII

Electron diffraction data for FA aggregates prepared at pH 2.5
(From Schnitzer and Kodama, 1975)

No. of rings	Feature of ring	d (Å)	I
1	diffuse	2.5	weak
2	spotty	2.12	strong
3	spotty	1.82	medium strong
4	diffuse	1.5	very weak
5	spotty	1.28	strong
6	diffuse	1.2	weak
7	spotty	1.10	medium strong
8	spotty	0.84	medium weak
9	spotty	0.76	weak

ble VII resemble those of disordered C (Frondel and Marvin, 1967) except for the basal spacing. Since crystallinity could be detected at pH 2.5 only, it seems that low pH favours the formation of crystalline structures from at least parts or certain components of FA molecules or aggregates.

Viscosity measurements

Viscosity measurements can provide important information on particle shapes and sizes, particle weights and polyelectrolytic behavior of macromolecules in aqueous solutions. The method has been applied to HA's by a number of workers whose contributions have recently been summarized by Flaig et al. (1975). There is, however, considerable disagreement in the literature on particle shapes of HA's derived from viscometric measurements. According to Flaig and Beutelspacher (1954), the particles are globular. Visser (1964) and Orlov and Gorshkova (1965) report that the particles are spherical, while Piret et al. (1960) maintain that they are elongated ellipses. Khan (1971) states that HA's consist of mixtures of spherical and linear particles. On the other hand, Kumada and Kawamura (1968) conclude that viscometric measurements cannot tell whether HA particles are spherical or linear; the only information that these measurements provide is that the particles are fairly flexible. From the above it becomes apparent that the literature contains much contradictory information on HA characteristics that can be measured by viscometry. One major reason for this unsatisfactory situation is that different workers have used widely differing methods for the extraction, separation and purification of humic materials. What are really FA's are often referred to in the literature as HA's, or mixtures of HA's and FA's are designated as HA's. Chen and Schnitzer (1976b) examined the effect of pH on particle shapes and dimensions, particle weights and polyelectrolytic behavior of HA's and FA's. The pH of the HA solutions ranged from 7.0 to 10.5, that of the FA solution from 1.0 to 10.5. HA at pH 7.0 and FA at pH 1.0 and 1.5 behaved like uncharged polymers. At higher pH levels, both HA's and FA's exhibited strong polyelectrolytic characteristics. The viscosity data fitted an equation developed for linear, flexible polyelectrolytes.

In the case of FA, it is possible to assess the effect of pH over a wide range of pH values (Table VIII). At very low pH, FA has the highest particle weight and particle volume. With increase in pH, the two parameters decrease to a minimum at pH 3, and then increase moderately. An analysis of the data for particle shapes and dimensions shows that the most likely particle configuration is rods (Chen and Schnitzer, 1976b). Axial ratios for FA's range from 8.8 to 11.9 and those for HA's are close to 15.0 (Table VIII). The minimum intrinsic viscosity and parameters derived from it at pH 3.0 signalize a minimum particle size for FA at that pH. As the pH is lowered, aggregation or association of particles occurs, so that the viscosity increases; the FA particles are practically uncharged under these conditions and electrostatic repul-

TABLE VIII

Parameters computed from viscosity measurements
(From Chen and Schnitzer, 1976b)

Humic substance	pH	[η] 100 ml/g	M	V (Å^3)	Particle dimensions (Å)		
					a/b	a	b
FA	1.0	0.107	2,790	2,879	11.9	80.4	6.8
FA	1.5	0.080	1,795	1,852	9.7	60.5	6.2
FA	2.0	0.072	1,550	1,599	9.0	54.8	6.1
FA	3.0	0.070	1,471	1,517	8.8	53.1	6.0
FA	4.0	0.072	1,550	1,599	9.0	54.8	6.1
FA	6.0	0.077	1,712	1,766	9.5	58.8	6.2
FA	8.0	0.079	1,754	1,810	9.6	59.7	6.2
FA	10.0	0.086	1,997	2,060	10.2	64.9	6.4
HA	7.0	0.158	2,745	3,145	14.4	94.0	6.5
HA	8.5	0.181	3,361	3,851	15.8	107.0	6.8
HA	10.5	0.158	2,745	3,145	14.4	94.0	6.5

[η] = intrinsic viscosity
M   = molecular or particle weight
V   = molecular or particle volume
a   = diameter or major axis of molecular particle
b   = thickness of molecule or particle

sion is not a factor. As the pH is raised above 3.0, increased dissociation of oxygen-containing functional groups takes place, which leads to increased repulsion of FA particles. This is accompanied by water molecules clustering around ionized functional groups, so that the net result is a gradual increase in viscosity. Chen and Schnitzer (1976b) conclude that humic substances behave in aqueous solutions like flexible, linear polyelectrolytes, so that one does not deal here with structures exclusively composed of condensed rings, but that there must be numerous linkages about which free rotation can occur.

While many problems still remain to be resolved, viscometry offers an almost unique opportunity of studying a number of important characteristics of humic substances such as particle weights, particle volumes and dimensions and polyelectrolytic behavior in aqueous solutions over a wide pH range. Drying, heating and exposure to high vacuum which may result in molecular modifications can so be avoided.

Surface tension measurements

Visser (1964) and Tschapek and Wasowski (1976) have shown that HA's are surface-active. HA's and FA's are predominantly hydrophilic but, as will be shown later in this chapter, they also contain substantial concentrations of aromatic rings, fatty acid esters, aliphatic hydrocarbons and other hydrophobic substances which, together with the hydrophilic groups account for the surface activity of these materials. In a recent investigation, Chen and Schnitzer (1977) measured the surface tension (γ) of HA's and FA's at various pH's and concentrations. Both HA's and FA's were found to lower the γ of water as the pH and concentrations of humic materials increased. The lowest γ values measured were 44.2 and 43.2 dynes/cm for HA (at pH 12.7, 2% w/v) and FA (pH = 12.0, 3% w/v), respectively; γ for water is 72.0 dynes/ cm. Chen and Schnitzer (1977) calculated the maximum number of molecules per 100 Å^2 of liquid-air interfacial surface with the aid of the Gibbs adsorption equation. These values were 2.47 and 1.04 for a HA and FA, respectively. Hydrophilic oxygen-containing functional groups (CO_2H, OH, C=O) in the humic materials were thought to play significant roles in lowering the γ of water and in so increasing soil wettability. FA's should be especially active in this respect because they are soluble in water at any pH normally found in soils in contrast to HA's, which are water-soluble at pH $>$ 6.5 only. Chen and Schnitzer (1977) suggest that water repellency, which is found in some soils, may be due to a lack of sufficient FA in the soil solution or on surfaces of soil particles, so that hydrophobic HA surface sites play a greater role than they do normally.

Molecular weight measurements

A wide variety of methods have been used for measuring molecular weights of humic substances (Schnitzer and Khan, 1972). These can be grouped into three classes: (a) those measuring number-average ($\overline{M}n$) molecular weights (osmotic pressure, cryoscopic, diffusion, isothermal distillation); (b) those determining weight-average ($\overline{M}w$) molecular weights (viscosity, gel filtration); and (c) those measuring z-average ($\overline{M}z$) molecular weights (sedimentation). Molecular weights reported range from a few hundred for FA's to several millions for HA's. There is considerable disagreement in the literature between methods measuring the same type of molecular weight. There are wide discrepancies between results obtained by dialysis, diffusion and osmometry which all measure $\overline{M}n$ nor is there any better agreement between methods measuring $\overline{M}w$ such as gel filtration and viscosity. This is not too surprising when differences in origin, extractants, degree of purification, etc. are taken into consideration. Also, as has been pointed out in previous paragraphs, molecular weights of humic materials change with pH, and also with ionic strength, so that unless experimental conditions are described in great detail, even the same types of molecular weights measured in different laboratories will not agree with each other. We shall now briefly discuss a number of methods that are often used.

Vapor pressure osmometry

This method is very suitable for measuring number-average ($\overline{M}n$) molecular weights of water-soluble humic substances, especially FA's. Rapid molecular weight measurements can be done in water but it is necessary to correct for the dissociation of acidic functional groups. Hansen and Schnitzer (1969a) developed a correction system which overcomes these difficulties. Table IX shows experimental and corrected $\overline{M}$n values for unfractionated FA and for fractions derived from it by gel filtration. All $\overline{M}$n data were determined by vapor pressure osmometry.

The ultracentrifuge

Flaig and Beutelspacher (1968) have used the ultracentrifuge to measure sedimentation and diffusion constants, molecular weights, radius and friction coefficient of a HA at different pH levels in the absence and presence of NaCl. When the pH of an aqueous HA solution increased from 4.5 to 6.0, the molecular weight decreased from 4,850 to 2,050. Addition of NaCl to give 0.2 M solutions increased the molecular weight to 60,400 at pH 4.5 and to 77,000 at pH 6.0. These data demonstrate the profound effects that pH, ionic strength or salt concentration exert on the magnitude of $\overline{M}$w's of HA's.

Gel filtration

This is experimentally the simplest and most convenient method, and for these reasons it has been used widely. Weight-average molecular weights of the order of 300—>200,000 have been reported for soil HA's (Posner, 1963; Dubach et al., 1964; Bailly and Margulis, 1968). For HA's and FA's extracted from marine sediments $\overline{M}$w's ranging from 700 to >2,000,000 (Rashid and King, 1969, 1971) and for humic substances extracted from natural waters $\overline{M}$w's ranging from <700 to >50,000 have been proposed (Gjessing, 1976). Some of these molecular weights appear to be excessively high and it

TABLE IX

Experimental and corrected molecular weights for FA
(From Hansen and Schnitzer, 1969a)

Compound	Molecular weight ($\overline{M}$n)		Weight fraction (fx)
	exp.	corr.	
Fulvic acid	460	951	1.0000
Fraction I	910	2,110	0.0668
II	675	1,815	0.1088
III	570	1,181	0.5670
IV	449	883	0.1079
IV-1	207	311	0.0939
IV-2	257	275	0.0149

is likely that gel-solute interactions interfere with separations on the basis of molecular weights. Swift and Posner (1971) have suggested that this difficulty could be overcome by using an alkaline buffer containing a large amino cation.

Other methods

Cameron and Posner (1974) determined the molecular weight distribution of four HA's by density gradient ultracentrifugation. $\overline{M}$n's of the HA's ranged from 5,900 to 13,500; $\overline{M}$w's from 49,000 to >216,000 and $\overline{M}$z's from >300,000 to >1,100,000.

Wershaw et al. (1967) have used small-angle X-ray scattering for measuring molecular weights of aqueous HA solutions. They found two types of particles. The larger ones were ellipsoidal with a molecular weight of 1,000,000 while the smaller particles were nearly spheroidal with a molecular weight of 210,000. Wershaw et al. (1967) caution that they may have measured molecular weights of micelles of hydrated molecules rather than of true molecular species. R.L. Wershaw (personal communication, 1976) is continuing research on this method and believes that it can provide useful information on molecular weights, association and dissociation behavior and shapes of HA and FA molecules.

Electrometric titrations

Potentiometric (Van Dijk, 1960; Pommer and Breger, 1960; Wright and Schnitzer, 1960; Schnitzer and Desjardins, 1962; Posner, 1964; Gamble, 1970) and conductimetric (Van Dijk, 1960; Datta and Mukherjee, 1968; Gamble, 1970) titrations have been employed for the determination of acidic functional groups in humic materials. Potentiometric titration curves are usually sigmoidal, suggesting an apparent monobasic character. This is due to the difficulty in distinguishing by titration between the two major types of functional groups, that is, CO_2H and phenolic OH, because the dissocation of protons from the two groups overlaps. To overcome these difficulties, non-aqueous titrations (Van Dijk, 1960; Wright and Schnitzer, 1960), high-frequency titrations (Van Dijk, 1960) and discontinuous titrations (Pommer and Breger, 1960; Schnitzer and Desjardins, 1962) have been used for this purpose. While some of these methods have led to some slight improvements in separating CO_2H from phenolic OH groups, the major problem still remains to be resolved.

Schnitzer and Desjardins (1962) noted that the ratio of molecular to equivalent weight of FA approximated the sum of CO_2H + phenolic OH groups. Posner (1964) observed variations in titration curves of HA's with ionic strength and concluded that HA's were not typical polyelectrolytes in which ionization was influenced by the charge on the molecule as it was neutralized. Titration curves in the presence of LiCl, KCl and NaCl were identical, indicating the absence of specific complex formation.

Gamble (1970) assumes that FA has two types of CO_2H groups: type I, which is ortho to a phenolic OH group, and type II which is not adjacent to a phenolic OH group. Gamble (1970) calculated mass action quotients, K_1 and K_2 for each of the two types of CO_2H groups. The mass action quotients correspond roughly to the dissociation constants which Katchalsky and Spitnik (1947) have used for CO_2H groups of the vth state of ionization of polymethylacrylic acid. Both K_1 and K_2 decrease with increasing degree of ionization.

Gamble (1970) concludes that FA shows the potentiometric behavior of a low-molecular weight polyelectrolyte. In another study, Gamble (1973) measured the electrostatic binding of Na^+ and K^+ by FA in aqueous solution with cation electrodes. Binding equilibria were followed during the course of acid-base titrations. Distinct binding regions were observed in the titration curves. In one of these regions Na^+ was more strongly bound to FA than was K^+. Standard free energies of binding were $11.7 \cdot 10^3$ to $20.5 \cdot 10^3$ J/mol $\pm$ 2% for Na^+, and $11.5 \cdot 10^3$ to $17.8 \cdot 10^3$ J/mol $\pm$ 1% for K^+. Gamble (1973) explains his results by postulating that FA in aqueous solution contains cages lined by solvated COO^- groups. If the incoming cation is able to share solvating molecules with the COO^- group, then the smaller bare-ion diameter of Na^+ could be favoured over that of K^+.

DEGRADATIVE METHODS

Degradative methods used for the characterization of humic materials include oxidations in alkaline and acidic media, reduction, hydrolysis, thermal, radiochemical and biological degradations (Table X). With complex materials such as HA's and FA's, degradation is often a useful approach to obtaining information on their chemical structures. The expectation here is to produce simpler compounds that can be identified and whose chemical structures can be related to those of the starting materials. The method of choice should not be too drastic or lead to the formation of unwanted byproducts and/or artifacts. Recent advances in the development of efficient gas chromato-

TABLE X

Degradative methods used for the characterization of humic substances

Oxidation
Reduction
Hydrolysis
Thermal degradation
Radiochemical degradation
Biological degradation

graphic-mass spectrometric-computer systems, that make possible the separation and the qualitative and quantitative identification of micro-amounts of organic compounds in complex mixtures, have greatly enhanced the efficacy of chemical and possibly also of biological degradation as structural tools. Applications to humic substances of the methods listed in Table X will now be described.

Oxidative degradation

Oxidation of unmethylated and methylated humic substances with alkaline $KMnO_4$ solution has been extensively employed by the author and his collaborators (Wright and Schnitzer, 1960; Schnitzer and Desjardins, 1964; Hansen and Schnitzer, 1966; Schnitzer and Desjardins, 1970; Khan and Schnitzer, 1971a, 1972a, b; Matsuda and Schnitzer, 1972; De Serra and Schnitzer, 1972, 1973a,b; Schnitzer and Riffaldi, 1973; Neyroud and Schnitzer, 1974b; Schnitzer and Vendette, 1975; Griffith and Schnitzer, 1975b, Schnitzer, 1976, 1977). The somewhat milder degradation with alkaline CuO has been used on humic materials by Greene and Steelink (1962), Schnitzer and De Serra (1973a), Schnitzer (1974) and Neyroud and Schnitzer (1974a). Sequential oxidation with CuO-NaOH + $KMnO_4$ and with CuO-NaOH + $KMnO_4$ + H_2O_2 has also been employed (Schnitzer and De Serra, 1973a; Neyroud and Schnitzer, 1974a). Several workers have degraded humic materials under acidic conditions, using peracetic acid (Meneghel et al., 1972; Schnitzer and Skinner, 1974a,b) and nitric acid (Schnitzer and Wright, 1960a,b; Hayashi and Nagai, 1961; Hansen and Schnitzer, 1967). Other oxidants used include alkaline nitrobenzene (Morrison, 1958; Wilding et al., 1970), H_2O_2 (Savage and Stevenson, 1961; Mehta et al., 1963; Schnitzer and De Serra, 1973a; Griffith and Schnitzer, 1977), and sodium hypochlorite solution (Chakrabartty et al., 1974).

The general procedure used by the author and his associates for the degradation of humic substances and identification of degradation products is shown in Fig. 6. Following degradation, the products are extracted into organic solvents, methylated and then separated by column and t.l.c. chromatography. Portions of the fractions are then further separated by preparative gas chromatography into well-defined compounds which are identified by comparing their mass- and micro- I.R.-spectra with those of authentic specimens. Other portions of each fraction are injected directly into a GLC-MS-computer system. Preliminary identification of gas chromatographic peaks is first made by recording "mass chromatograms" for fragments characteristic of specific compounds or groups of compounds expected to occur in the mixture by searching through the stored spectral data for specific m/e ratios (Skinner and Schnitzer, 1975). The identity of the compound in each peak is then confirmed by: (a) running its mass spectrum; (b) eluting it from the gas chromatograph and recording its micro-I.R.-spectrum; (c) matching its

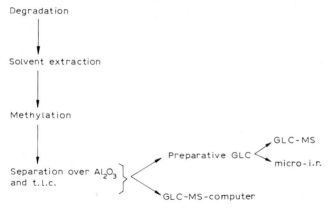

Fig. 6. Procedure used for the chemical degradation of HA's and FA's and for the identification of oxidation products.

mass and micro-I.R.-spectrum with the known compound to which it corresponds; and (d) co-chromatography (on the gas chromatograph) of known and unknown.

Major degradation products

Major compounds produced by the oxidation of methylated and unmethylated humic substances from widely differing environments in alkaline as well as in acidic solutions are aliphatic carboxylic, benzenecarboxylic and phenolic acids. In addition, smaller amounts of n-alkanes, substituted furans and dialkyl phthalates are also isolated. The most abundant aliphatic degradation products consist of n-fatty acids, especially the n-C_{16} and n-C_{18} acids, and di- and tri-carboxylic acids (Fig. 7). Prominent benzenepolycarboxylic acids are the tri-, and tetra-, penta- and hexa-forms (Fig. 8). Major phenolic acids isolated include those with between 1 and 3 OH groups and between 1

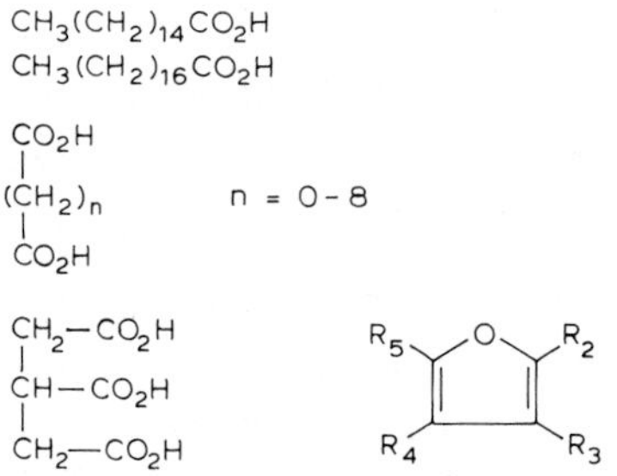

$CH_3(CH_2)_{14}CO_2H$
$CH_3(CH_2)_{16}CO_2H$

CO_2H
|
$(CH_2)_n$ n = 0-8
|
CO_2H

$CH_2—CO_2H$
|
$CH—CO_2H$
|
$CH_2—CO_2H$

Fig. 7. Major aliphatic compounds produced by the oxidative degradation of humic substances.

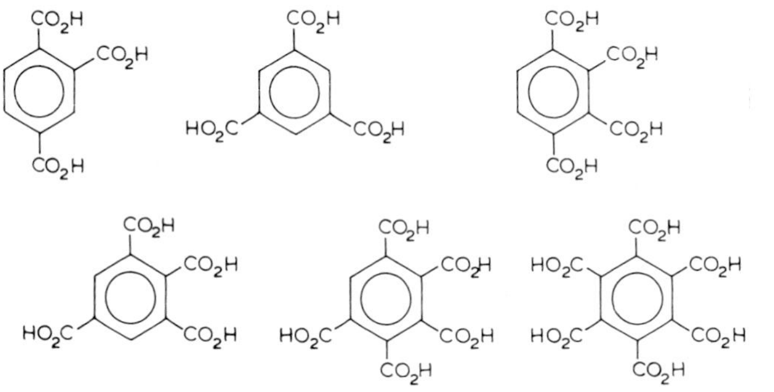

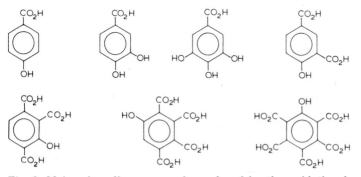

Fig. 8. Major benzenecarboxylic compounds produced by the oxidative degradation of humic substances.

and 5 CO$_2$H groups on the aromatic ring (Fig. 9). The main chemical structures in six tropical volcanic HA's and FA's, as revealed by the KMnO$_4$ oxidation of unmethylated as well as methylated materials (Griffith and Schnitzer, 1975b) are shown in Fig. 10. In this investigation 52 degradation products were identified. In structures *31, 38, 39, 48,* and *51* the aromatic ring is substituted by C-atoms only. Structures *36, 43* and *50* are produced by the oxidation of both unmethylated and methylated HA's and FA's . Because these structures resist destruction by electrophilic alkaline KMnO$_4$, it is very probable that the OCH$_3$-groups on the rings of these structures occur in that form in the initial humic materials, rather than as OH groups. By contrast, the OH-groups in *28, 35, 37* and *46* must have occurred as OH or, partly as OCH$_3$ and as OH, in the initial HA's and FA's, since these structures were found among oxidation products from methylated humic materials only, and were apparently destroyed in unmethylated HA's and FA's by KMnO$_4$. The structures shown in Figs. 7—10 may be considered to constitute

Fig. 9. Major phenolic compounds produced by the oxidative degradation of humic substances.

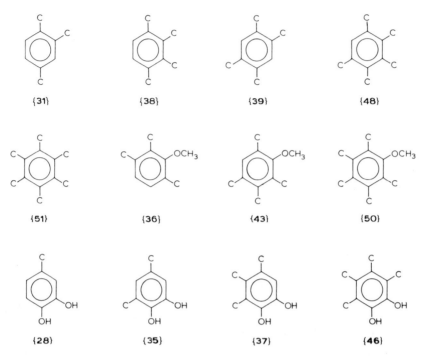

Fig. 10. Major aromatic compounds produced by the KMnO$_4$ oxidation of methylated and unmethylated HA's and FA's extracted from tropical volcanic soils. C stands for a CO$_2$H group or a group forming CO$_2$H on oxidation (from Griffith and Schnitzer, 1975b). Published with permission of the Agricultural Institute of Canada.

the major humic "building blocks". These are similar in HA's, FA's and humins, regardless of whether the humic materials are extracted from surface or subsurface soils.

Major types of products resulting from the oxidation of HA's and FA's extracted from soils formed under widely differing climatic environments

Table XI lists the major types of products that resulted from the KMnO$_4$ oxidation of 1.0 g of methylated HA's. The Arctic HA produced considerably higher yields of aliphatic compounds than did any of the other HA's. Weight ratios of benzenecarboxylic to phenolic compounds, which may be considered to reflect the interrelationship between major humic "building blocks", are, however, of the same magnitude, except that the ratio for HA's from subtropical soils is higher than the remaining ratios. Aliphatic compounds account for only a small portion of the total products resulting from the degradation of FA's (Table XII). Weight ratios of benzenecarboxylic to phenolic compounds are similar for all FA's and are of the same magnitude

TABLE XI

Major types of products (mg) resulting from the $KMnO_4$ oxidation of 1.0 g of methylated HA
extracted from soils from various climates
(From Schnitzer, 1977)

Type of product	Arctic	Cool, temperate		Subtropical	Tropical
		acid soils	neutral soils		
Aliphatic	129.2	9.8— 51.6	7.0— 15.8	0.6— 1.0	8.3—116.7
Phenolic	36.8	70.4— 79.0	68.1— 96.5	16.2— 31.2	32.3—111.7
Benzenecarboxylic	50.8	80.2—122.4	147.5—173.5	99.2—164.1	49.6—183.3
Total identified	224.2	194.4—282.4	248.6—281.3	118.1—198.5	100.2—350.4
Weight ratio: benzenecarboxylic / phenolic	1.4	1.1— 1.6	1.8— 2.2	5.3— 6.1	1.2— 2.4

as those for most HA's. With one exception, there is no significant effect of
climate on the weight ratios of major aromatic structures for the different
HA's and FA's. The only distinct effect of climate is that the Arctic HA appears to contain considerably greater amounts of aliphatic structures than do
the other HA's.

Products resulting from the alkaline CuO oxidation of HA's and FA's

Major types of products resulting from the alkaline CuO oxidation of 1.0 g
of HA and FA are shown in Tables XIII and XIV. Again, the Arctic HA pro-

TABLE XII

Major types of products (mg) resulting from the $KMnO_4$ oxidation of 1.0 g of methylated
FA extracted from soils from various climates
(From Schnitzer, 1977)

Type of product	Cool, temperate		Subtropical	Tropical
	acid soils	neutral soils		
Aliphatic	9.2	6.8— 8.9	0 — 1.6	5.7— 27.1
Phenolic	30.5	43.1— 46.5	7.4— 60.0	15.9— 26.3
Benzenecarboxylic	46.9	77.6— 95.4	20.6— 86.9	37.9— 68.4
Total identified	86.6	129.6—147.8	28.0—148.1	64.6—108.0
Weight ratio: benzenecarboxylic / phenolic	1.5	1.8— 2.1	1.2— 2.8	1.4— 3.5

TABLE XIII

Major types of products (mg) resulting from the CuO—NaOH oxidation of 1.0 g of HA
extracted from soils from various climates
(From Schnitzer, 1977)

Type of product	Arctic	Cool, temperate		Tropical
		acid soils	neutral soils	
Aliphatic	104.2	26.2	46.2	19.6— 55.3
Phenolic	63.6	51.0	68.1	62.7— 98.0
Benzenecarboxylic	24.8	28.4	20.2	20.9— 34.9
Total identified	192.6	105.6	135.0	103.2—188.2
Weight ratio: benzenecarboxylic / phenolic	0.4	0.6	0.3	0.3— 0.5

duces larger amounts of aliphatic compounds than do any of the other HA's.
Aside from the Arctic HA, the most abundant compounds formed by the
alkaline CuO oxidation of HA's and FA's are phenolics. This confirms earlier findings (Schnitzer and De Serra, 1973a) that alkaline cupric oxide oxidation is an especially efficient method for releasing phenolic structures.
Relatively low yields of benzenecarboxylic acids result from the alkaline
CuO oxidation of the HA's and FA's . This is also indicated by the low benzenecarboxylic to phenolic weight ratios (Tables XIII and XIV). Again
climate does not appear to affect the yields of phenolic and benzenecarboxylic oxidation products.

TABLE XIV

Major types of products (mg) resulting from the CuO-NaOH oxidation of 1.0 g of FA
extracted from soils from various climates
(From Schnitzer, 1977)

Type of product	Cool, temperate acid soils	Tropical
Aliphatic	51.6	15.6— 44.2
Phenolic	45.2	22.7— 55.4
Benzenecarboxylic	24.0	19.5— 43.8
Total identified	120.8	57.8—143.4
Weight ratio: benzenecarboxylic / phenolic	0.5	0.8— 0.9

Maximum yields of principal products

To uncover which method of oxidation would produce the highest yield of major products, a HA extracted from the Ah horizon of a Haploboroll and a FA extracted from the Bh horizon of a Spodosol were degraded by the following methods or combination of methods: (a) alkaline CuO oxidation; (b) alkaline CuO + KMnO$_4$ oxidation; (c) alkaline CuO + KMnO$_4$ + H$_2$O$_2$ oxidation; (d) KMnO$_4$ oxidation; and (e) peracetic acid oxidation.

The two humic fractions were selected for this purpose because they were among the best developed naturally occurring HA's and FA's that the author was able to find.

All methods were used on methylated and unmethylated HA and FA. Alkaline CuO oxidation of unmethylated humic materials was found to be especially efficient for releasing phenolic structures but was relatively inefficient for degrading aromatic structures bonded by C—C bonds and poor in oxygen (Neyroud and Schnitzer, 1974a). The latter types of structures were more readily degraded by the more drastic peracetic acid oxidation (Schnitzer and Skinner, 1974a) and by oxidation of methylated HA and FA with alkaline KMnO$_4$ (Khan and Schnitzer, 1971a). Maximum yields of major oxidation products resulting from the degradation of the HA and the FA referred to above and the methods that were most efficient for this purpose are listed in Table XV. The data in the table were multiplied by 2 because it was assumed that losses during the lengthy oxidation, separation and isolation procedures were ~50%. Thus, the data in Table XV show that the "model" HA contains about equal proportions of aliphatic and phenolic structures but a greater percentage of benzenecarboxylic structures or structures producing benzenecarboxylic acids on oxidation. By contrast, the "model" FA contains more phenolic than benzenecarboxylic and aliphatic structures. It is interesting that both the HA and FA contain approximately equal pro-

TABLE XV

Major chemical structures in "model" HA and FA
(From Schnitzer, 1977)

Major products		HA (%)	FA (%)	Method
Aliphatic		24.0	22.2	CuO-NaOH + KMnO$_4$
Phenolic		20.3	30.2	CuO-NaOH + KMnO$_4$
Benzenecarboxylic		32.0	23.0	CH$_3$CO$_2$OH, KMnO$_4$
Total		76.3	75.4	
Ratio	$\dfrac{\text{benzenecarboxylic}}{\text{phenolic}}$	1.6	0.8	
Aromaticity		69	71	

portions of aliphatic structures. Aromaticity values were approximated by expressing yields of phenolic and benzenecarboxylic acids as percentages of total yields. The aromaticity calculated in this manner is about 70% for both materials. Thus, as far as chemical structure is concerned, the "model" HA and FA are quite similar, except that the FA is richer in phenolic but poorer in benzenecarboxylic structures than the HA. The main difference between the two materials is that the "model" HA contains more C and N, fewer CO_2H groups per unit weight but has a higher particle or molecular weight than the "model" FA, and is thus less soluble in aqueous solutions at pH >7.0.

Hypohalite oxidation

Chakrabartty et al. (1974) treated an HA extracted from a Haploboroll with sodium hypochlorite solution. Over 50% of the total C in the HA was converted to CO_2. Among the non-volatile oxidation products was a high proportion of aliphatic mono-, di-, and tri-carboxylic acids. Judging from the known chemistry of hypohalite oxidation, Chakrabartty et al. (1974) concluded that soil HA's represented a special group of mixed alkyl-aryl cyclo-alkyl compounds in which Sp^3 C constituted over 80% of the total C.

Reductive degradation

Zn-dust distillation and fusion

A number of workers (Cheshire et al., 1967; Hansen and Schnitzer, 1969b, c) have used Zn-dust distillation to obtain information on the basic skeletons of HA's and FA's. Cheshire et al. (1967) subjected an acid-boiled Histo-sol HA to Zn dust distillation at 500°C under a stream of H_2. A yellow oil was obtained from which anthracene and 2,3-benzofluorene were isolated in crystalline forms. The presence in the oil of the following polycyclics was indicated by UV, gas chromatographic and mass spectrometric analyses: naphthalene, α-methylnaphthalene, β-methylnaphthalene and higher homologues of naphthalene. Other polycyclics identified were pyrene, perylene, 1,2-benzopyrene, 3,4-benzopyrene, triphenylene, chrysene, 1, 12-benzoperylene and coronene. Hansen and Schnitzer (1969b,c) did Zn-dust distillation and fusion on a HA and FA. The major reaction products included 1,2,7-tri-methylnaphthalene, other polysubstituted naphthalenes, 1-, and 9-methyl-anthracene, 2-methylphenanthrene, 1-, 4-methylpyrene, pyrene, perylene and substituted perylene. In addition, small amounts of the following poly-cyclic hydrocarbons were also identified: fluoranthene, 1,2-benzanthrene, 1,2- and 2,3-benzofluorene, 1,2- and 3,4-benzopyrene and naphtho(2′,3′:1,2)-pyrene. Products of the Zn-dust distillation and fusion accounted for 0.66 and 0.62% of the initial HA, respectively. The Zn-dust distillation and fusion of FA produced the same types of products as that of HA, but at somewhat lower yields. Calculated on a functional group-free basis, yields of products resulting from the Zn-dust fusion of the HA and FA were almost identical, accounting for close to 1% of the initial materials. It is noteworthy that the Zn-dust distillation of alkaloids often produces yields of the order of 1% or less; polysubstituted quinones may yield degradation products accounting for 10—20% of the starting materials (Hansen and Schnitzer, 1969b). Thus, assuming a yield of 10% of theory, the reduction products identified from the HA and FA could account for about 12% of the HA "nucleus" and for 25% of the FA "nucleus". This may mean that "nuclei" of soil humic substances either contain significant amounts of polycyclic aromatics or that the polycyclic aromatics which were isolated were formed from less condensed structures during the drastic experimental conditions of Zn-dust distillation and fusion.

Cheshire et al. (1967) believe that the results of Zn-dust distillation indicate the existence of a polynuclear aromatic core in the initial HA. Hydrolyzable substances such as carbohydrates, polypeptides, phenolic acids and metals are attached to the core. On the other hand, Hansen and Schnitzer (1969b) emphasize the drastic reaction conditions associated with Zn-dust distillation which are known to cause excessive bond breaking and molecular rearrangements so that caution should be exercised in interpreting the results of Zn-dust distillation.

Na-amalgam reduction

A number of workers (Zetsche and Reinhart, 1939; Burges et al., 1963, 1964; Hurst and Burges, 1967; Mendez and Stevenson, 1966; Martin et al., 1967; Stevenson and Mendez, 1967; Dormaar, 1969; Martin and Haider, 1969) have used Na-amalgam reduction to degrade humic substances into relatively simple organic compounds, hoping that these would provide information on the chemical structure of the parent materials. The earliest workers to use this procedure on humic substances were Zetsche and Reinhart (1939) who reduced synthetic and natural humic acids with Na-amalgam in aqueous solutions at about 100°C. The solutions underwent a series of color changes which ranged from dark brown through green, reddish, orange to light yellow. The reduction products were found to be extremely unstable and were readily reoxidized by air-oxygen. Thus, it was necessary to completely exclude oxygen. The reduced materials could be stabilized by methylation, followed immediately by reduction with Zn-dust in glacial acetic acid and additional methylation. Zetsche and Reinhart (1939) concluded that the reductive action of Na-amalgam was limited mainly to the removal of oxygen; no relatively simple organic compounds were identified. Hurst and Burges (1967) report that the reductive degradation of humic acids from different soils releases as many as 40 different degradation products, with many of these retaining side chains and functional groups. The degradation products include possible lignin residues in addition to flavonoid units. Reductive degradation of humic acids also yields acid-labile p-hydroxybenzyl alcohols and 3-methoxy-4-hydroxyphenylpropanols (Hurst and Burges, 1967). The latter workers admit that the number of compounds identified in the reduction mixtures is far from complete and that they may have exaggerated the relative importance of the more stable and easily identifiable methoxylated derivatives. It is well known that compared with lignin, the methoxyl content of humic and fulvic acids is low.

Mendez and Stevenson (1966) and Stevenson and Mendez (1967) were unable to confirm the findings of Burges et al. (1963, 1964) and of Hurst and Burges (1967). They noted the occurrence of substantial amounts of aliphatic substances in the reduction products from humic acids. No phenols were detected; only vanillic and syringic acid were positively identified and strong evidence was obtained for the presence of vanillin and syringaldehyde among the reduction products. These compounds were estimated to account for no more than 1% of the original humic acid. Mendez and Stevenson (1966) also showed that known phenols and phenolic acids were degraded by Na-amalgam and that substances rich in aliphatic structures were formed. This suggests that phenolic constituents released from humic acids during reductive cleavage could be chemically modified, so that the chemical structures of the compounds isolated may have little or no relation to those occurring in the initial humic acids. Dormaar (1969) used Na-amalgam to reduce humic acids extracted from the Ah and Bm horizons of a Haploboroll and identified a number of phenolic acids. One C_6C_3-type structure was detected but no flavonoid derived units. Dormaar (1969) concludes that reductive cleavage cannot be used to differentiate between HA's from morphologically different soils as had been suggested by Burges et al. (1964). Similarly, Schnitzer and De Serra (1973b) were unable to confirm claims by Burges et al. (1964) that Na-amalgam reduction differentiates between HA's from different origins and that it provides a "fingerprint" for the characterization of humic substances. Schnitzer and De Serra (1973b) conclude that compared to alkaline $KMnO_4$ oxidation of methylated HA's and FA's, Na-amalgam reduction is a relatively inefficient method that tells little about the chemical structure of humic materials. This appears to be related to the fact that humic substances, which contain relatively large amounts of oxygen, are difficult to reduce but are more amenable to oxidation. Also, Na-amalgam reduction is a complicated method. It appears to work well on some HA's and in the hands of some workers, but does not do so on other HA's and in other laboratories.

Martin and Haider (1969) found Na-amalgam reduction to work well on fungal "HA's" synthesized in the laboratory. They separated 14 phenols from an ether extract of reduced "HA" synthesized by *E. nigrum* and up to 25 phenols from reduction products of "HA's" synthesized by *S. atra* and *S. chartarum*. These phenolic compounds were estimated to account for between 2 and 6% of the initial materials, but how similar fungal "HA's" are to soil HA's is still a matter of conjecture.

Hydrogenation and hydrogenolysis

Gottlieb and Hendricks (1946) treated organic soil extracts with H_2 in the presence of a Cu-chromite catalyst. Dioxane extracted a colorless oil from which they were unable to isolate any identifiable compounds. Kukharenko and Savelev (1951, 1953) hydrogenated HA's suspended in dioxane with Ni as catalyst. They isolated a number of carboxylic acids, phenols and neutral compounds. Murphy and Moore (1960) used Raney Ni as catalyst and produced oils from which they were unable to isolate any identifiable compounds. Felbeck (1965) hydrogenated non-hydrolyzable fractions of an organic soil at 350°C with kaolin as catalyst. A second hydrogenation of a benzene extract of the first reaction products yielded a series of oils, 65% of which could be distilled; n-C_{25} and n-C_{26} hydrocarbons were isolated from the distillable products. Felbeck (1965) believes that the central structure of HA is the source of the aliphatic hydrocarbons, and that it is a linear polymer consisting primarily of 4-pyrone units connected by methylene bridges at 2,6-positions.

In general, hydrogenation has not provided much structural information on humic materials. The experimental conditions are drastic; temperatures up to 350°C and pressures up to 5,000 psi are used, which may lead to molecular rearrangements and condensation reactions.

Other degradation methods

Jackson et al. (1972) have attempted to depolymerize HA with phenol in the presence of p-toluenesulphonic acid or boron trifluoride as catalysts. This method has been developed by coal chemists for the depolymerization of coal through replacement of aromatic units by phenols, yielding new compounds from which the original inter-aromatic linkages can be identified. A number of bridging units were postulated for HA's which include:

$$-CH_2-CH_2-, \quad -\overset{\displaystyle CH_3}{\underset{\displaystyle CH}{|}}, \quad CH_2-CH_2-CH_2-,$$

$$CH_2-\overset{\displaystyle }{\underset{\displaystyle \overset{\|}{O}}{C}}-, \quad -\overset{\displaystyle }{\underset{\displaystyle \overset{\|}{O}}{C}}-, \quad -\overset{\displaystyle }{\underset{\displaystyle |}{C}}H, \quad \overset{\displaystyle }{>}CH-CH_2-, \quad -\overset{\displaystyle |}{\underset{\displaystyle |}{C}}-$$

Hydrolysis with water

Boiling HA's with water releases polysaccharides (Duff, 1952), small amounts of phenolic acids and aldehydes (Jakab et al., 1962), polypeptides

(Haworth, 1971), alkanes and fatty acids (Schnitzer and Neyroud, 1975; Neyroud and Schnitzer, 1975)

Hydrolysis with acid

Between 1/3 and 1/2 of the total organic matter in most mineral and organic soils is dissolved by refluxing with hot acids (Cheshire et al., 1967). Among the soluble products are proteins, peptides, amino acids, sugars, uronic acids and pigmented substances as well as phenolic acids and aldehydes (Jakab et al., 1962; Shivrina et al., 1968).

Hydrolysis with base

A number of workers (Coffin and De Long, 1960; Steelink et al., 1960) subjected humic materials to KOH fusion. Ether extracts of the fusion products contained phenolic acids and phenols. According to Cheshire et al. (1968) phenolic acids can be produced during the fusion procedure from non-aromatic compounds, so that the products of KOH fusion are not diagnostic of the structure of the initial humic material. The net result is that little structural information is provided by this method.

Jakab et al. (1963) degraded a number of HA's and FA's by hydrolysis for 3 h at 170°C and 250°C with 2N and 5N NaOH. The following compounds were identified among the degradation products: 1,2- and 1,3-dihydroxybenzene, 3- and 4-hydroxybenzoic acid, vanillic acid and vanillin.

Neyroud and Schnitzer (1975) subjected a HA and a FA to four successive hydrolyses with 2N NaOH at 170°C for 3 h in an autoclave. Following each hydrolysis, the degradation products were extracted into ethyl acetate, methylated, separated by chromatographic techniques and identified by mass spectrometry and micro-IR spectrophotometry.

One g of HA yielded 113.5 mg of aliphatic compounds (mainly n-C_{16} and n-C_{18} fatty acids), 72.1 mg of phenolics (principally guaiacyl and syringyl derivatives), 17.1 mg of benzenecarboxylic acids and 30.4 mg of N-containing compounds. Repeated attacks by the alkali on 1.0 g of FA released 103.8 mg of aliphatics, 133.8 mg of phenolics, 27.0 mg of benzenecarboxylic but 181.8 mg of N-containing compounds. Most of the latter appeared to have been formed during the second, third and fourth hydrolysis which produced reactive compounds that could abstract N from diazomethane which was used as methylating reagent. While 25% of the initial HA resisted repeated attacks by the alkali, practically all of the FA was converted into ethyl acetate-soluble products. The alkali-resistant HA residue produced high yields of benzenecarboxylic acids on subsequent alkaline permanganate oxidation.

Alkaline hydrolysis, which is known to cleave C—O bonds, was found by Neyroud and Schnitzer (1975) to be relatively specific, although not very efficient, for the degradation or liberation of structural phenolic HA- and

FA- components and for concomitant freeing of bonded and/or adsorbed aliphatics and N-containing compounds. The method was found to be ineffective for degrading aromatic structures linked by C—C bonds.

Hayes et al. (1972) studied the degradation of a HA with 10% Na_2S solution under autoclave conditions for 2 h at 250°C. About 60% of the degradation products were soluble in ether and ether-ethanol. Several neutral volatile degradation products were observed by gas-liquid chromatography. Ethyl esters of aliphatic dibasic acids and possibly similar esters of phenyl propionic acid derivatives were isolated from ether-soluble acidic degradation products. Hayes et al. (1972) consider the technique promising and worthy of further application.

Thermal degradation

TG (thermogravimetry), DTG (differential thermogravimetry), DTA (differential thermal analysis) and isothermal heating have been used for investigating the mechanism of thermal decomposition of humic materials. Schnitzer and Hoffman (1964) heated a HA and a FA at a constant rate under air from room temperature to 540°C. Samples were withdrawn at regular intervals and analyzed by chemical and IR methods. The C content of the chars increased with temperature, accompanied by a simultaneous decrease in O. Chars of both HA and FA heated to 540°C contained identical percentages of C and H but no O. Some of the N and S in the initial preparations was so stable that it was recovered in the chars heated to the highest temperatures. Phenolic OH groups were more stable than CO_2H groups but both were degraded between 250 and 400°C. The two types of functional groups were more heat-resistant in the FA than in the HA. DTG curves for HA and FA are shown in Fig. 11. The low-temperature peaks result from the elimination of functional groups, whereas the high-temperature maxima are due to the decomposition of the "nuclei" (Schnitzer and Hoffman, 1964).

Van Krevelen's graphical-statistical method for studying coal decomposition was applied by Schnitzer and Hoffman (1964) to the pyrolysis of HA and FA. The main reactions governing the pyrolysis of HA are: (a) dehydrogenation (up to 200°C); (b) a combination of decarboxylation and dehydration (between 200° and 250°C); and (c) further dehydration (up to the highest temperature). The main reaction governing the pyrolysis of FA is dehydration.

Pyrolysis—gas chromatography

This method is useful for the characterization of polymers and for structural elucidations of non-volatile organic materials. Nagar (1963) was the first to use this technique for the characterization of HA's from widely differ-

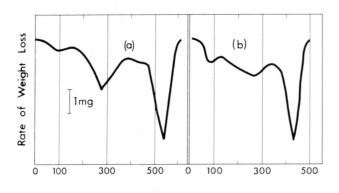

Fig. 11. Differential thermogravimetric (DTG) curves of (a) HA and (b) FA (from Schnitzer and Hoffman, 1964). Published with the permission of the Soil Science Society of America.

ing environments. Numerous peaks could be observed on gas chromatograms but techniques to identify these peaks were not yet well developed at that time. Wershaw and Bohner (1969) extended the scope of the method by identifying the gas chromatographic peaks by mass spectrometry. They propose a pattern of peaks that can "fingerprint" different HA's and FA's.

Kimber and Searle (1970a, b) report that HA's with highly condensed structures can be distinguished readily from those with less condensed structures by pyrolysis-gas chromatography. Also, yields of benzene and toluene are directly related to the acid-hydrolyzable amino acid-N content and inversely related to absorbances at 260 and 450 nm. Martin (1976) used pyrolysis-gas chromatography to evaluate effects of different extractant on the chemical characteristics of FA's. Bracewell and Robertson (1973) and Bracewell et al. (1976) used the technique to distinguish between mull and more humus types and to study the effect of climate on the nature of the organic matter in A horizons of seven New Zealand soils.

Recent advances in the pyrolysis technique such as the development of Curie point pyrolysis in conjuction with low voltage electron impact mass spectrometry have enhanced the usefulness of this technique. Nagar et al. (1975) report that HA's extracted from three Indian soils have similar structures and resemble a fungal HA rather than lignin. According to Haider et al. (1977), HA's from different soil types give complex pyrolysis spectra. Although the identity of individual compounds cannot be assigned with certainty, the spectra show a marked qualitative and quantitative correspondence through the presence of a similar series of homologous ions. These are probably derived from protein-like materials, aromatic and phenolic compounds and polysaccharides. Spectra for FA's differ from those of

HA's and appear to be more closely related to polysaccharides. With some exceptions spectra of fungal HA's resemble those of soil HA's. Additional investigations are under way and it still is premature to assess the usefulness of the method for providing structural information on humic materials.

High-energy irradiation of humic substances

Senesi et al. (1977c) irradiated a HA and a FA as solids with electrons and γ-rays, employing dosages of up to 93 Mrads. In addition, they also irradiated the humic materials in aqueous solutions at different pH levels with γ-rays. The effects of irradiation were evaluated by chemical, spectrophotometric, spectrometric and thermal analyses. On the whole, HA and FA structures were found to be stable toward irradiation. E_4/E_6 ratios and ESR measurements were found to be the most sensitive probes for following the minute changes caused by irradiation. These were limited in the main to decreases in particle size with increasing irradiation when HA and FA were exposed as solids. When exposed in solution, irradiation at neutral and acid pH tended to bring about aggregation (formation of larger particles), whereas irradiation at high pH favoured dispersion (formation of smaller particles). Senesi et al. (1977c) concluded that the major effects of irradiating HA and FA solutions were pH-dependent and involved mainly dispersion (depolymerization) and aggregation (polymerization). At the levels employed (up to 93 Mrads), irradiation did not appear to cause major changes in the chemical structures of the HA and FA.

Radiocarbon dating

Humic substances extracted from Ap horizons of a Haploboroll and a Boralf were characterized by radiocarbon dating and chemical analysis by Campbell et al. (1967). The data showed that: (1) humin was more stable than "mobile" HA, and that the latter was more resistant to decomposition than was FA; (2) humic fractions extracted from a Boralf were less stable than were similar fractions extracted from a Haploboroll; (3) fractions derived from Haploboroll HA's ranged from very stable Ca-humates (1,400 years) to labile HA-hydrolysates (25 years), which consisted of amino acids, peptides, carbohydrates, etc. "Mobile" HA (extracted with 0.5 N NaOH without acid pretreatment) was intermediate (780 years) and FA was still younger (550 years). Campbell et al. (1967) found an inverse relationship between absorbance ratios at 465 and 650 nm and mean residence times, which appeared to indicate that the more condensed and aromatic humic components were more stable than the aliphatic and alicyclic ones. Paul et al. (1964) report a mean residence time of 1,000 years for organic matter from two Udic Borolls. The organic matter of one of these soils was fractionated into HA, humin and FA, which had mean residence times of $1,308 \pm 64$, $1,240 \pm 60$ and 630 ± 60 years, respectively. A soil sample from a Boralf had a mean residence time that was only one third of that of a Haploboroll sample. Paul et al. (1964) recommend the use of carbon dating for studying the dynamics of soil organic matter. It is, however, advisable to consider effects of rejuvenating components such as cellulose and tissue remnants, which are often insoluble in dilute alkali, on the mean residence times of unfractionated soil organic matter.

Biological degradation

Burgess and Latter (1960) examined large numbers of fungal strains for their ability to decolorize HA's and to reduce the CO_2H group in *m*-hydroxybenzoic acid. They found a positive correlation between the two functions and conclude that the fungal degradation of HA's involves reduction of CO_2H groups but that the reducing power of the organisms is produced by aerobic growth on substrates other than HA's. Mishustin and Nikitin (1961) report that the degradation of HA by a pseudomonad is due to the peroxidase activity of the organism. Mathur and Paul (1967a, b) examined the role of *Penicillium frequentans* in utilizing HA as the sole C source. After seven weeks under restricted aeration, salicylaldehyde and salicyl alcohol were identified in the growth medium. They also found that the total OH content of the degraded material was higher than that of the initial HA. Mathur (1969) tested the ability of a number of fungi to decompose FA. All organisms were inhibited by 1% FA. Within 48 h, *Poria subacida* used 66% of a 0.05% FA solution present as sole C source in a static replacement culture, and up to 45% in 24 days in a static culture. Phenol oxidase did not appear to play a role in the degradation of the FA but darkened the color of a well aerated medium through oxidative polymerization of FA products.

There are a number of serious limitations to the use of biological degradation in structural studies on humic materials. One is never quite sure what type of compounds are produced by the organisms, which compounds originate from the humic materials, and how the organisms modify these compounds prior to their isolation and identification. Thus, while biological degradation is certainly a potentially most valuable and desirable approach, it will be first necessary to overcome the many difficulties and complexities currently associated with it before it can provide useful information on the chemical structure of humic materials.

THE ISOLATION OF ALKANES AND FATTY ACIDS FROM HUMIC SUBSTANCES

Earlier work in the author's laboratory (Ogner and Schnitzer, 1970; Schnitzer and Ogner, 1970b) had shown that HA's and FA's contained small amounts of *n*-alkanes and *n*-fatty acids. To obtain more definite information on the qualitative and quantitative distribution of these compounds in humic substances, Schnitzer and Neyroud (1975) assessed effects of the following pretreatments on the extractability by organic solvents of alkanes and fatty acids from a HA and FA: (a) no pretreatment; (b) ultrasonic dispersion; (c) hydrolysis with H_2O at $170°C$ in an autoclave; (d) a combination of ultrasonic dispersion and H_2O-hydrolysis; (e) methylation; and (f) saponification with $2 N$ NaOH at $170°C$.

As shown in Table XVI, H_2O-hydrolysis at $170°C$ and a combination of ultrasonic dispersion + H_2O-hydrolysis are the most efficient pretreatments for the extraction of n-alkanes from the HA and FA. Maximum amounts of n-alkanes that can be extracted are 3.7 mg/g of HA and 17.0 mg/g of FA. The alkanes extracted from the two humic preparations vary from n-C_{12} to n-C_{38}, the majority being in the n-C_{18} to n-C_{38} range, and with an odd to even C ratio of 1.0. The distribution of n-alkanes and the odd to even C ratio are similar to those of microbial hydrocarbons (Jones, 1969).

The most effective pretreatment for the release of fatty acids from the HA and FA is saponification, which makes it possible to extract between 90 and 100 mg of fatty acids per gram of each humic material. The other pretreatments are considerably less effective in this respect. The fatty acids range

TABLE XVI

Extraction of alkanes and fatty acids from HA and FA after different pretreatments (From Schnitzer and Neyroud, 1975)

Type of pre-treatment *	HA		FA	
	alkanes (mg/g)	fatty acids (mg/g)	alkanes (mg/g)	fatty acids (mg/g)
(a)	0.4	1.4	0.4	1.2
(b)	1.3	4.5	3.0	7.6
(c)	3.7	4.3	2.6	7.4
(d)	2.7	4.6	17.0	24.4
(e)	3.4	4.7	1.6	1.0
(f)	0.1	98.2	0.1	91.7

* (a) no pretreatment; (b) 500 mg of HA or FA, suspended or dissolved in 50 ml of distilled water in a 100 ml beaker, was dispersed for 2 h with the aid of a Blackstone BP-2 ultrasonic probe operated at 20 kHz and equipped with a cooling system; (c) 500 mg of HA or FA in 50 ml of distilled water was heated in a 300 ml stainless steel autoclave, equipped with a magnetic stirrer, at $170°C$ for 3 h; (d) 500 mg of HA or FA in 50 ml of distilled water was first dispersed for 2 h by ultrasonics as described under (b), then hydrolysed with water at $170°C$ as in pretreatment (c), and finally further dispersed by ultrasonics for $\frac{1}{2}$ h; (e) 500 mg of HA or FA was suspended in 10 ml of methanol and methylated with diazomethane; (f) 500 mg of HA or FA was hydrolysed in the autoclave with 50 ml of $2 N$ NaOH solution at $170°C$ for 3 h. Following solvent extraction, the residues were hydrolysed for the second time.

from n-C_{12} to n-C_{38}, with most being in the C_{14}—C_{22} range. Fatty acids extracted from the HA and FA have even to odd C ratios of 9.6 and 2.6, respectively; n-C_{16} and n-C_{18} acids constitute between 60 and 80% of the fatty acids extracted from the two humic preparations, which suggests a microbiological origin (Breger, 1960).

The data in Table XVI can be interpreted in the following manner: the relatively mild pretreatments (ultrasonic dispersion, H_2O-hydrolysis, methylation) facilitate the extraction of "loosely-held" fatty acids, whereas the harsher saponification liberates total fatty acids. Thus, in the case of the HA, about 5% of the fatty acids is "loosely held", whereas 95% is "tightly bound". In the case of the FA, about 25% of the fatty acids isolated is "loosely held", whereas 75% is "tightly bound". Two points are of special interest: (a) the two humic materials investigated come from widely differing pedological and geochemical environments, geographically 3,000 miles apart; yet they contain almost identical amounts of fatty acids per unit weight; and (b) yields of fatty acids reported here are considerably higher than any reported so far for humic substances.

TABLE XVII

Phenolic and straight-chain fatty acids released by two successive hydrolyses of HA and FA with 2 N sodium hydroxide at 170°C for 3 h (mmol/g of humic material) (From Schnitzer and Neyroud, 1975)

Compound	HA	FA
Phenolic acids:		
4-OH-benzenecarboxylic acid	0.031	0.084
3,4-diOH-acetophenone	0.018	0.010
3,5-diOH-benzenecarboxylic acid	0.037	0.206
3,4-diOH-benzenecarboxylic acid	0.054	0.164
3,4,5-triOH-benzenecarboxylic acid	0.100	0.073
3,4,5-triOH-acetophenone	0.035	0.024
4-OH-benzene-1,2-dicarboxylic acid	0.003	0.025
2-OH-benzene-1,3,5-tricarboxylic acid	0	0.025
Total	0.278	0.611
Straight-chain fatty acids:		
C_{14}	0.013	0.004
C_{15}	0.016	0.011
C_{16}	0.233	0.144
C_{17}	0.016	0.076
C_{18}	0.058	0.052
C_{19}	0.003	0.006
C_{20}	0.008	0.029
C_{22}	0.017	0.005
Total	0.364	0.327

Along with fatty acids, saponification also releases a number of phenolic acids from the HA and FA (Table XVII). The data show a phenolic to fatty acid molar ratio of 0.76 for HA and of 1.90 for FA. This suggests that in the humic substances fatty acids react with phenolic OH groups to form esters of the following type:

$$R_1 - \text{(ring with } R_2, R_3 \text{ top and } R_6, R_5 \text{ bottom)} - O - \overset{O}{\underset{\|}{C}} - (CH_2)_n - CH_3$$

where R_1 (Table XVII) = CO_2H or $COCH_3$ or OH; R_2 = H or OH or CO_2H; R_3 = H or OH or OCH_3 or CO_2H; R_4 = OH esterified to fatty acid; R_5 = H or OH or OCH_3; R_6 = H or CO_2CH_3 and n equals mainly 14 and 16 for the HA and 14, 15, 16 and 18 for the FA. Esters could form via any free OH group on the aromatic ring, that is, at R_1, R_2, R_3 and R_5 in addition to R_4, so that in humic substances the occurrence of a considerable number of different phenol-fatty acid esters is a distinct possibility. The isolation of a p-hydroxy-propiophenone ester of stearic acid from a peat by Morita (1975) provides additional evidence for the occurrence of phenol esters in humic materials.

Thus, while small proportions of the fatty acids are "loosely-held", possibly physically adsorbed, on the large humic surfaces and in internal voids, most of the fatty acids form esters with phenolic humic "building blocks". By this type of reaction mechanism humic substances may fix, stabilize and preserve over long periods of time relatively large amounts of hydrophobic organic compounds, provided that these contain at least one CO_2H group. These observations suggest that the role of humic substances in the transport, dispersion and sedimentation in soils and waters of hydrophobic organic compounds, including toxic pollutants and petroleum source materials, may have been underestimated.

THE CHEMICAL STRUCTURE OF HUMIC SUBSTANCES

From the data presented in the preceding paragraphs it appears that up to 50% of the aliphatic structures in HA's and FA's consist of n-fatty acids esterified to phenolic OH groups. The remaining aliphatics are made up of more "loosely" held fatty acids and alkanes that seem to be physically adsorbed on the humic materials and which are not structural humic components, and possibly of aliphatic chains joining aromatic rings. As shown by a wide variety of chemical degradation experiments on HA's and FA's extracted from soils differing widely in locations and pedological histories, the major HA and FA degradation products are phenolic and benzenecarboxylic acids.

These could have originated from more complex aromatic structures or could have occurred in the initial humic materials in essentially the same forms in which they were isolated but held together by relatively weak bonding. If the latter hypothesis is correct, then phenolic and benzenecarboxylic acids would be the "building blocks" of humic materials and future research should be directed toward finding out how the "building blocks" fit together and what type of structural arrangement is produced. If, on the other hand, the degradation products originate from more complex chemical structures, further research should be concerned with developing mild degradative methods that would permit the isolation and identification of larger fragments of the total structure.

Some clues as to what type of structural arrangement may be the most probable one are provided by physico-chemical methods. Scanning electron microscopy and viscosity measurements show that both the shape and size of HA and FA particles are strongly affected by pH. At low pH, HA and FA tend to aggregate, forming elongated fibers and bundles of fibers. The aggregation appears to be brought about by hydrogen bonding, Van der Waal's forces and interactions between π-electrons of adjacent aromatic rings, as well as by homolytic reactions between free radicals. As the pH increases, these forces become weaker, and because of increasing ionization of CO_2H and phenolic OH groups, particles separate and begin to repel each other electrostatically, so that the molecular arrangements become smaller and smaller but better oriented. Thus, one witnesses aggregation at low pH and dispersion at high pH, which is also the basis for separating humic substances into HA's and FA's. HA's and FA's behave like flexible, linear, synthetic polyelectrolytes, so that one deals here not with chemical structures largely or exclusively composed of condensed rings, but there must be present numerous linkages about which relatively free rotation can occur.

It becomes thus more and more apparent that humic substances are not single molecules but rather associations of molecules of microbiological, polyphenolic, lignin and condensed lignin origins. These are the benzene-carboxylic and phenolic acids that are referred to above as "building blocks". There appear to be some differences in the strengths of bonding between HA's and humins and the lower-molecular weight FA's. In the higher-molecular weight fractions, the "building blocks", although made up of the same types of benzenecarboxylic and phenolic acids, appear to be more complex and more stable than in FA's. The increased stability of each "building block" may arise from either more energetic linkages of the types discussed above or from additional chemical bonding via C—O and C—C bonds. Thus, whereas the "building blocks" in the higher-molecular weight fractions may be chemically more complex than those in lower-molecular weight ones, the author believes that the molecular forces holding the "building blocks" together are similar, consisting mainly of hydrogen bonds, Van der Waal's forces and π-bonding.

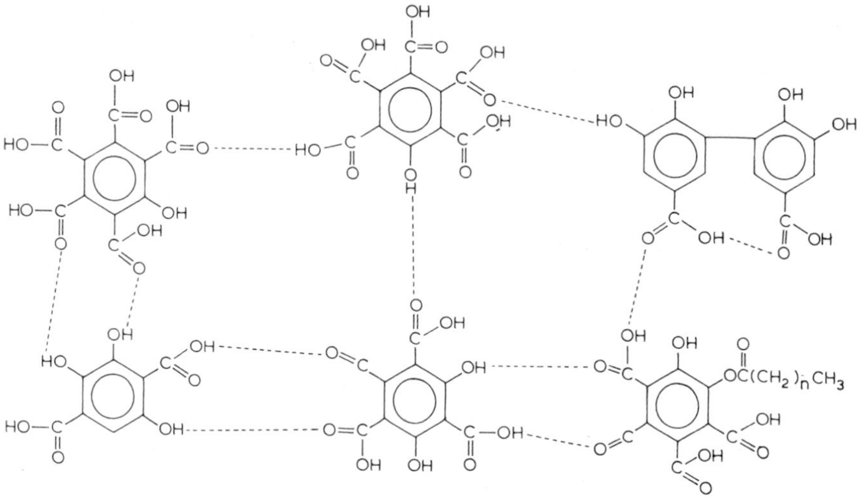

Fig. 12. A partial chemical structure for FA.

It is noteworthy that X-ray analysis, electron microscopy and viscosity measurements of FA point to a relatively "open", flexible structure perforated by voids of varying dimensions that can trap or fix organic and inorganic compounds that fit into the voids, provided that the charges are complementary. A chemical structure that is in harmony with many of the requirements discussed in this chapter has been proposed by the author (Fig. 12). This molecular arrangement may account for a significant part of the FA structure. Each of the compounds that make up the structure has been isolated from FA without and after chemical degradation. Bonding between "building blocks" is by hydrogen-bonds, which makes the structure flexible, permits the "building blocks" to aggregate and disperse reversibly, depending on pH, ionic strength etc., and also allows the FA to react with inorganic and organic soil constituents either via oxygen-containing functional groups on the large external and internal surfaces, or by trapping them in internal voids.

Although the advent of the gas-chromatograph mass-spectrometer computer system has provided soil chemists with a powerful tool for the qualitative and quantitative analysis of very complex HA and FA degradation mixtures, further research is needed to provide more specific and detailed information on the structural arrangement(s) of HA's and FA's.

REACTIONS OF HA'S AND FA'S WITH METALS AND MINERALS

Humic substances are capable of interacting with metal ions, metal oxides, metal hydroxides and more complex minerals to form metal-organic associations of widely differing chemical and biological stabilities and characteristics. As has been mentioned in previous paragraphs, humic materials contain, per unit weight, relatively large numbers of oxygen-containing functional groups (CO_2H, phenolic OH, C=O), through which they can attack and degrade soil minerals by complexing and dissolving metals and transporting these within soils and waters.

Interactions between humic substances and metal ions have been described as ion-exchange, surface adsorption, chelation, coagulation and peptization reactions (Mortensen, 1963). The importance of organic matter in reactions with Zn^{2+} has been demonstrated by Himes and Barber (1957) who showed that oxidation of the organic matter with H_2O_2 destroyed the ability of the soil to complex Zn^{2+}, but that the removal of hydrous silicates had little effect. A more adequate knowledge of metal-HA and metal-FA interactions is desirable from the standpoint of soil genesis, soil structure formation, nutrient availability and especially the chemistry of micro- and toxic elements and their mobilization, transport and immobilization in terrestrial and aquatic environments.

Potentiometric titrations of humic substances in the absence and presence of various metal ions have been done by several workers. Beckwith (1959) reports that metals of the first transition series of the periodic table form complexes with humic substances, and that the order of stabilities of the different metal complexes follows that of the Irving-Williams series (1948) ($Pb^{2+} > Cu^{2+} > Ni^{2+} > Co^{2+} > Zn^{2+} > Cd^{2+} > Fe^{2+} > Mn^{2+} > Mg^{2+}$). Similar results were obtained by Khan (1969, 1970) and Khanna and Stevenson (1962) for metal-HA complexes. By contrast, De Borger (1967) and Van Dijk (1971) note that the order of pH drop when HA's and FA's are titrated in the presence of metals does not follow the Irving-Williams series of stabilities. According to Van Dijk (1971) there is no large difference in bond strength at pH 5 for metal-HA complexes involving Ba^{2+}, Ca^{2+}, Mg^{2+}, Mn^{2+}, Co^{2+}, Ni^{2+}, Fe^{2+} and Zn^{2+} (increasing only slightly in this order). Pb^{2+}, Cu^{2+} and Fe^{3+} (in that order) are more firmly bound by HA's, while Al forms the hydroxide. Fe^{3+} and Al^{3+} are the most abundant metallic elements in soils and their interactions with humic materials are of special concern to soil scientists. When Fe^{3+} and Al^{3+} are added to HA at low pH and the solutions are titrated with dilute base, precipitates are formed which, however, begin to dissolve as the titration proceeds and the pH rises above 7.

Van Dijk (1971) rationalizes the titration of HA + Fe^{3+} with base in the following manner:

$$\text{(5)}$$

dihydroxo-humato-ferrate

According to this mechanism, a proton is displaced from an acidic OH group of the HA at low pH. At higher pH, a proton dissociates from water covalently bonded to the metal ion and a hydroxo complex is formed which becomes soluble as more base is added.

Similarly, Van Dijk (1971) considers the reaction between Cu^{2+} and HA to proceed in the following manner:

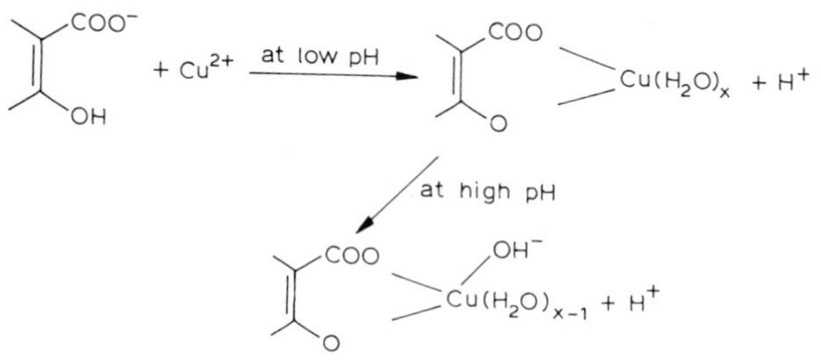

(6)

Thus, at low pH a proton is displaced from an acidic OH group of the HA, whereas at higher pH protons dissociate from water molecules covalently bound to Cu^{2+}, so that a hydroxo complex is formed.

Van Dijk (1971) also notes that the capacity of HA's to bind metal ions equals approximately the number of titratable H^+-ions divided by the valency of the interacting metal ions. Schnitzer and Skinner (1965) showed that either one acidic CO_2H and one phenolic OH groups or two acidic CO_2H groups, in both instances in close proximity, reacted simultaneously with metal ions and hydrous oxides. The involvement of CO_2H and phenolic OH groups of humic substances in metal complexing has been confirmed by Van Dijk (1971) and Stevenson (1976a,b). Both authors mention the formation of mixed complexes. This possibility had already been mentioned earlier by Schnitzer (1969) who considered the reaction at pH 5.0 between Cu^{2+} and FA to proceed in the following manner:

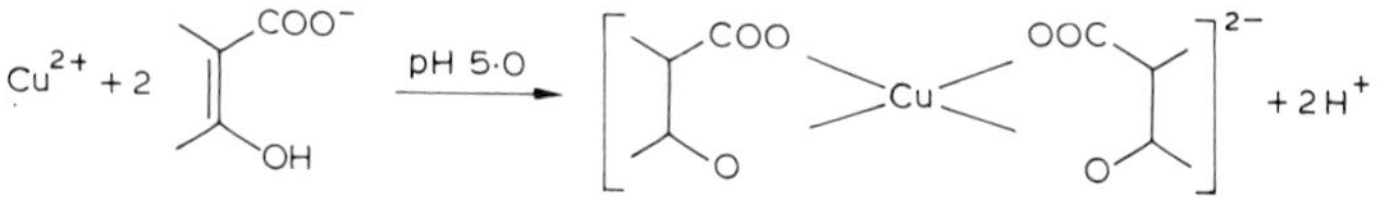

(7)

Metal-HA and metal-FA stability constants have been determined by a number of workers (Schnitzer and Skinner, 1966, 1967; Courpron, 1967; Schnitzer and Hansen, 1970; Van Dijk, 1971; Stevenson and Ardakani, 1972; Gamble and Schnitzer, 1973; Stevenson, 1976a,b; Adhikari and Ray, 1976) but the results vary widely. There is still considerable uncertainty about how to define the problem, which method to use and how useful such constants are once they have been determined. The main reason for the current state of uncertainty in this field is that we do not know enough about the chemical structure of humic materials, and this is a serious obstacle to an intelligent understanding of the reactions of these materials. Stability constants, expressed as log K, for metal-FA complexes were determined by Schnitzer and Hansen (1970) by the method of continuous variations and by the ion exchange equilibrium method. These data are shown in Table XVIII. Stability constants measured by the two methods were in good agreement with each other, increased with increase in pH but decreased with increase in ionic strength. Of all metals investigated, Fe^{3+} formed the most stable complex with FA. The order of stabilities at pH 3.0 was: $Fe^{3+} > Al^{3+} > Cu^{2+} > Ni^{2+} > Co^{2+} > Pb^{2+} > Ca^{2+} > Zn^{2+} > Mn^{2+} > Mg^{2+}$. At pH 5.0, stability constants of Ni—FA and Co—FA complexes were slightly higher than that of the Cu—FA complex. The stability constants shown in Table XVIII are considerably lower than those for complexes formed between the same metal ions and synthetic complexing agents such as EDTA. This may mean that metals complexed by FA should be more readily available to plant roots, microbes and small animals than when sequestered by EDTA or similar reagents.

Langford and Khan (1975) determined the rate of binding of Fe^{3+} by FA in aqueous solutions. The rate of complex formation was similar to that for sulfosalicylic acid. The dissociation was slow ($t_{1/2} > 10$ s). The binding of Fe^{3+} by FA in acid solution (pH 1.0—2.5) was investigated by kinetic analysis in which the reaction of free Fe^{3+} with sulfosalicylic acid was followed by stopped flow spectrophotometry and compared to the release of Fe^{3+} by FA. Conditional equilibrium constants were $1.5 \pm 0.3 \cdot 10^4$ at pH 1.5 and 2.5, and $2.8 \pm 0.3 \cdot 10^3$ at pH 1.0 at $25°C$ and at an ionic strength of 0.1.

TABLE XVIII

Stability constants expressed as log K of metal-FA complexes (CV = method of continuous variations; IE = ion-exchange equilibrium method)
(From Schnitzer and Hansen, 1970)

Metal	pH 3.0		pH 5.0	
	CV	IE	CV	IE
Cu^{2+}	3.3	3.3	4.0	4.0
Ni^{2+}	3.1	3.2	4.2	4.2
Co^{2+}	2.9	2.8	4.2	4.1
Pb^{2+}	2.6	2.7	4.1	4.0
Ca^{2+}	2.6	2.7	3.4	3.3
Zn^{2+}	2.4	2.2	3.7	3.6
Mn^{2+}	2.1	2.2	3.7	3.7
Mg^{2+}	1.9	1.9	2.2	2.1
Fe^{3+}	6.1 *	—	—	—
Al^{3+}	3.7 **	3.7 **	—	—

* Determined at pH 1.70.
** Determined at pH 2.35.

Gamble et al. (1976) report on the binding of Mn^{2+} to FA based on the competition with K^+ in an ion exchange equilibration experiment. The free energy of binding of $\frac{1}{2} Mn^{2+}$ to FA was only 1—2 kJ/equivalent more favourable than that for K^+. This suggests an outer sphere electrostatic structure for the complex. The suggestion was confirmed by observation of minimal change, in the presence of FA, of the effect of paramagnetic Mn^{2+} on the NMR spectra of water. The interpretation of the NMR spectra was supported by comparison with NMR measurements of Mn^{2+} complexes with simple ligands and contrasts with NMR measurements on Fe^{3+}-FA complexes, which were confirmed as inner sphere complexes.

Recently, Gamble et al. (1977) used Electron Spin Resonance (ESR) spectrometry to determine weighted-average equilibrium functions ($\bar{K}_c$) of water-soluble complexes formed between FA and divalent Mn.

The reaction between FA and Mn^{2+} was assumed to proceed according to eq. 8. The complexing equilibrium was calculated according to eq. 9:

$$\bar{K}_c = \frac{m_C m_H}{m_{TH} m_M} = \frac{1}{X_{TH}} \int_0^{X_{TH}} K_c \, dX_{TH} \tag{9}$$

As described elsewhere in detail (Gamble, 1970; Gamble and Schnitzer, 1973; Gamble et al., 1976), $\bar{K}_c$ and K_c are the weighted average and differential equilibrium functions, respectively; m_C, m_H, m_{TH} and m_M are molalities of complexed metal, H-ions, total protonated carboxyl groups, and free metal, respectively, while X_{TH} is the mole-fraction of total carboxyl groups still protonated.

The differential equilibrium function K_c can be defined as follows:

$$K_c = \frac{m_H \delta m_C}{m_M \delta m_{TH}} \tag{10}$$

where m_C and m_{TH} are infinitesimal concentration increments of complexed $\frac{1}{2} Mn^{2+}$ and carboxyl groups which are still protonated.

K_c is related to ΔG_c^o in the following manner:

$$\Delta G_c^b = -RT \ln K_c \tag{11}$$

ΔG_c^b differs from ΔG_c^o only to the extent of a small activity coefficient term. R and T are the gas constant and the absolute temperature, respectively.

Gamble et al. (1977) found $\bar{K}_c$ values for Mn^{2+}—FA complexes that ranged from $0.2 \cdot 10^{-2}$ (at pH 6.15) to $0.5 \cdot 10^{-3}$ (at pH 6.45). These values were in excellent agreement with $\bar{K}_c$ values determined previously for Mn—FA complexes by an ion-exchange method, but the ESR method was found to be more sensitive than the ion exchange method. Gamble et al. (1977) concluded from their experiments that Mn^{2+} appeared to be bound simultaneously electrostatically and by hydrogen-bonding as $Mn(OH_2)_6{}^{2+}$ to FA donor groups in outer sphere complexing sites in an unsymmetrical complex in the following manner:

$$
\begin{array}{c}
\text{H}\text{O}^- \\
\vdots| \\
Mn^+ \ \ \text{O}-\text{H}\text{------}\text{O}=\text{C} \\
| \\
\text{R}
\end{array}
\qquad (12)
$$

where Mn^+ stands for Mn^{2+} and $RCOO^-$ for partly ionized FA. There are six H_2O molecules in the primary hydration shell of Mn^{2+}. Mn complexed in this way may be more readily available to plant roots and microbes than when complexed as chelate. Other transition metals may react with FA by a similar mechanism to form water-soluble complexes.

Dissolution of minerals

In view of their ability to complex mono-, di-, tri- and tetra-valent metal ions, FA's and at pH >6.5 also HA's can attack and degrade minerals to form water-soluble and water-insoluble metal complexes. If metal/HA or metal/FA ratios are low, the complexes are water-soluble, but if the ratios are high, the metal-HA or -FA complexes become water-insoluble (Schnitzer and Khan, 1972). Strong solvent activity has been reported for HA's for hematite, pyrolusite, feldspar, biotite, enstatite, actinolite and epidote (Baker, 1973). Mn^{3+}- and Mn^{4+}-oxides and Mn^{2+}-hydroxides (Rosell and Babcock, 1968) and basalt (Singer and Navrot, 1976) are also gradually dissolved. Similarly, FA's have been shown to be efficient for dissolving metals from goethite, gibbsite and soils (Schnitzer and Skinner, 1963), chlorites (Kodama and Schnitzer, 1973) and micas (Schnitzer and Kodama, 1976).

A point of considerable interest is that minerals rich in iron are most susceptible to attack by humic materials (Schnitzer and Kodama, 1976). The high affinity of HA or FA for iron and the low crystal field stabilization energies of Fe^{3+} and Fe^{2+} provide possible explanations for these findings. The complexing of iron by humic materials appears to have an adverse effect on the structural stability of Fe-rich minerals as, along with Fe, other major constituent elements such as Mg, Al, K and Si are also more readily released and brought into solution. Recently, Senesi et al. (1977d) have combined ESR and Mössbauer spectroscopy with chemical treatments to obtain information on oxidation states and site symmetries of iron bound by HA and FA. At least two, possibly three, different binding sites for iron were found to occur in humic materials: (a) Fe^{3+} was strongly bound and protected by tetrahedral and/or octahedral coordination; this form of iron exhibited considerable resistance to complexation by known sequestering agents and reduction; and (b) Fe^{3+} was adsorbed on external surfaces of humic materials, weakly bound octahedrally, and easily complexed and reduced. These observations were made on laboratory-prepared Fe—HA and Fe—FA complexes and on a Fe-pan taken from a soil rich in organic matter.

Adsorption on external mineral surfaces

The extent of adsorption of humic materials on mineral surfaces depends on the physical and chemical characteristics of the surface, the pH of the system and its water content. One can visualize the formation of a wide range of mineral-humic associations, involving chemical bonding with widely differing strengths. Another mechanism of considerable importance is hydrogen-bonding, the occurrence of which is clearly indicated in IR spectra of HA- and FA-mineral complexes (Schnitzer and Khan, 1972). These reactions are likely to involve H and O of CO_2H and OH groups in HA and FA and O and H on mineral surfaces and edges. It is probable, as has been mentioned above, that cations with high solvation energies on mineral surfaces react via water-bridges with HA- and FA- functional groups. Van der Waal's forces may also contribute to the adsorption of humic substances on mineral surfaces.

The importance of surface geometry and surface chemistry has been pointed out by a number of workers. Inoue and Wada (1971) studied reactions between humified clover and imogolite and postulated the following two mechanisms: (1) incorporation of CO_2H groups into the coordination shell of Al atoms, and (2) relatively weak bonding by H-linkages and Van der Waal's forces. The two adsorption mechanisms were thought to assist each other and the number of bonds formed was considered to depend on the orientation of the humic molecules on the clay. Kodama and Schnitzer (1974) report high adsorption of FA on sepiolite surfaces. Sepiolite has a channel-like structure formed by the joining of edges of long and slender talc-like structures. In untreated sepiolite the channels are occupied by bound and/or zeolitic water which apparently can be displaced by undissociated FA. IR spectra of FA-sepiolite interaction products suggest the formation in the channels of COO^- groups which are linked to Mg^{2+} at edges that have been exposed by the displacement of water by FA.

Adsorption in clay interlayers

The evidence currently available shows that the interlayer adsorption by expanding clay minerals of humic materials is pH-dependent, being greatest at low pH, and no longer occurring at pH >5.0 (Schnitzer and Kodama,

1966). Adsorbed FA cannot be displaced from clay interlayers by leaching with 1 N NaCl, and an inflection occurs in the adsorption-pH curve near the pH corresponding to the pK of the acid species of FA (Schnitzer and Kodama, 1966). On the basis of these criteria, the adsorption could be classified as a "ligand-exchange" reaction (Greenland, 1971). In this type of adsorption the anion is thought to penetrate the coordination shell of the dominant cation in the clay interlayer and displace water coordinated to it. The ease with which water can be displaced will depend on the affinity for water of the dominant cation with which the clay is saturated and also on the degree of dissociation of the FA. Since the latter is very low at low pH, interlayer adsorption of FA is greatest at low pH levels as Schnitzer and Kodama (1966) have observed. Concurrently the FA can dissolve a proportion of the dominant cation in the clay by forming a soluble complex and replacing the removed cation by H^+. If this process continues over long periods of time, the FA will eventually degrade the clay structure.

REACTIONS OF HUMIC SUBSTANCES WITH ORGANIC COMPOUNDS

The solubilization in water by humic materials of organic compounds which are otherwise water-insoluble is a matter of considerable interest to soil scientists. Wershaw et al. (1969) have shown that the solubility of DDT in 0.5% aqueous sodium humate solution is at least twenty times greater than that in water. Solubilizing effects of humic materials on pesticides are described in greater detail in Chapter 4.

There has been considerable evidence (Ogner and Schnitzer, 1970a; Khan and Schnitzer, 1971a,b) that suggests that humic substances can "complex" dialkyl phthalates and similar compounds and so modify their behavior and activity. These interactions have been investigated in some detail by Matsuda and Schnitzer (1971). They found that amounts of dialkyl phthalates solubilized by aqueous FA solutions at pH 2.35 depended on the type of phthalate. For example, 950 g of FA could "complex" up to 1,560 g of bis(2-ethylhexyl)phthalate but only 495 g of dicyclohexyl and 278 g of dibutyl phthalate. Analysis of IR spectra showed that there was no chemical interaction between the FA and the phthalate but that the latter was firmly adsorbed, possibly by hydrogen-bonding, on FA surfaces. It is noteworthy that FA can "fix" high-molecular weight water-insoluble organic compounds and make them water-soluble. FA may so act as a vehicle for the mobilization, transport and immobilization of such substances in terrestrial and aquatic environments.

Other reactions of special interest to soil scientists are those between humic substances and N-containing compounds, including fertilizers. Stepanov (1969) has shown that ammonia can be adsorbed by reactive humic surfaces. Others (Lindbeck and Young, 1965; Flaig, 1950) have suggested that HA polymerizes as it adsorbs ammonia and that heterocyclic rings are formed. Urea can also be adsorbed by humic materials (Pal and Banerjee, 1966), resulting in decreased exchange capacity and increase in pH.

Stepanov (1969) reports that amino acids and glycocol are only weakly adsorbed on HA surfaces. Similar conclusions were reached by Schnitzer et al. (1974) who studied interactions between five HA's and seventeen amino acids commonly occurring in proteins in aqueous solutions at pH 3.0 and 6.5. They found that the reactions were affected by: (a) the nature of the HA; (b) the pH of the system; and (c) length of contact between the reactants. The principal reaction appeared to be the microbiological oxidative degradation of the amino acids, leading to the formation of substantial amounts of ammonia. While there was no evidence that HA's per se interacted with the amino acids, the humic materials did not appear to interfere with the microbial degradation of the amino acids.

There is still considerable controversy in the literature on reactions of "protein-like" materials with humic substances. It has been suggested that peptides or proteins react via NH_2-groups with phenolic lignin degradation products or with phenols resulting from metabolic reactions of microorganisms. During the course of the reactions the phenolic compounds form quinones to which the polypeptides become attached, and this combination resists acid hydrolysis (Haider et al., 1965). Thus, Flaig (1964) believes that N is an integral component of humic substances. On the other hand, Haworth (1971) proposes that polypeptides are linked to humic substances by hydrogen-bonding because the peptides can be removed by boiling water. Similar observations have been made by other workers (Roulet et al., 1963; Sowden and Schnitzer, 1967) who report that the N-content of humic materials can be lowered significantly by passage over a cation exchange resin in the H^+-form. This suggests that at least a substantial portion of the N-components is either bonded relatively loosely to the humic substances or not bonded at all.

PHYSIOLOGICAL EFFECTS OF HUMIC SUBSTANCES

Humic substances exert indirect and direct effects on plant growth. Indirect effects relate to humic materials acting as suppliers and regulators of plant nutrients similar to synthetic ion exchangers, whereas direct effects involve the uptake by plant roots of humic substances per se. Kononova (1966) reports that small concentrations (up to 60 ppm) of HA's enhance root development and plant growth. According to Khristeva and Luk'yanenko (1962) HA's enter the plant during early stages of growth and act as respiratory catalysts. Flaig and Otto (1951), Guminski (1957) and Flaig (1958) suggest that HA's enter plants, mediate in respiration and act as hydrogen acceptors, thus affecting oxidation-reduction reactions. Flaig and Saalbach (1959) note that HA's alter the carbohydrate metabolism of plants and may promote the accumulation of reducible sugars. Guminski et al. (1965) attribute the stimulation of growth of tomato seedlings by HA under anaerobic conditions not to enhanced respiration but to its acting as a chelating agent which makes iron more available to roots. Other workers report stimulating effects of HA's on peroxidase activity (Khristeva and Luk'yanenko, 1962), invertase development in beet storage disks under aseptic conditions (Vaughan, 1969), cell elongation in excised pea roots at low HA concentrations up to 50 mg/l (Vaughan, 1974) and on the growth of both roots and shoots of wheat plants (Vaughan and Linehan, 1976). Vaughan and Linehan (1976) believe that HA has a direct effect on growth processes. They observed that C^{14}-labelled HA was taken up by wheat roots but that virtually no transport of the labelled HA to the shoot occurred. Only some 30—40% of the incorporated radioactivity was associated with root cell walls but more than 60% was in the cytoplasm and may have influenced biochemical processes regulating plant growth.

Vaughan and MacDonald (1976) report that the absorbing capacity of beet disks aged in aqueous solution is modified by the presence of HA (100—200 mg/l at pH 5.6). The capacity for absorbing Na and Ba increases relative to the control but that for absorbing Ca and Zn decreases. On the other hand, the rate of cation absorption is affected differently. HA inhibits the rate of Na and Zn uptake, but has no effect on Ba and Ca uptake. The effect on Zn is ascribed to it being complexed by HA. Vaughan and MacDonald (1976) also point out that HA has selective effects on changes in protein synthesis. They believe that only low-molecular weight humic materials are biologically active.

Schnitzer and Poapst (1967) examined the effect of a low-molecular weight humic material, FA, on root formation in bean stem segments. They note that root formation is increased by over 300% when between 3,000 and 6,000 ppm of FA is administered. They believe that the stimulating effect of FA is due to its ability to form stable complexes with metal ions and that FA may aid in the movement of metal ions, such as Fe^{3+}, which in the absence of FA are transported within the plant with difficulty only. It is also possible that stable free radicals in FA are active in stimulating root formation.

In another investigation, Poapst et al. (1970) observed that FA appeared to block the uptake of gibberellic acid by peas when the two substances were applied simultaneously to the leaves, but when the two substances were applied separately, the FA had no effect on gibberillic acid-stimulated growth. Interactions between FA and natural growth regulators need further investigation because the results reported so far (Poapst and Schnitzer, 1971) are inconclusive. From the incomplete summary presented in these paragraphs it becomes apparent that relatively little is known about physiological effects exerted by HA's and FA's, but it appears that these effects play important roles in influencing soil fertility and productivity.

SUMMARY

Humic substances constitute the bulk of the organic matter in soils and sediments and also account for high proportions of the soluble organic matter in fresh and sea water. The chemical structure and reactions of humic materials have been the subject of numerous researches for over 200 years, yet much remains to be learned about these materials. While the recent availability of advanced and sophisticated instruments has assisted in generating important information on the "building blocks" that make up humic materials, little is known on how the "building blocks" align and what structural arrangement they produce. The main task that confronts researchers in this field today is to develop a valid concept of the chemical structure of humic materials. Once this has been achieved, it will be possible to understand the behaviour of these materials in soils and waters, and this will allow us to foresee what will happen under certain conditions. For example, it will then be possible to determine clearly defined metal-HA or metal-FA stability constants from which we shall be able to predict whether or not a metal will be available to plant roots or microbes or, conversely, what should be done to increase metal availability in case of deficiencies. Similarly a more adequate knowledge of the chemical structure of humic materials will assist us in better understanding the physiological effects that these materials are known to exert. How do humic substances affect cell division, cell elongation, the action of growth regulators, etc.? Thus, what is needed at this time is more fundamental research in order to solve practical problems of great significance, problems that affect food production and food quality.

The application of organic chemicals to soils and plants is increasing at an alarming rate. It is known that humic substances react with these chemicals by providing large external surfaces for adsorption in addition to firmer retention in internal spaces. By means of these mechanisms, humic materials can accumulate large amounts of these chemicals which are often toxic. Two questions arise: (a) what will happen when the adsorption or retention capacities of humic materials are exhausted, and (b) what will occur when by some unknown reactions the relatively large amounts of "complexed" chemicals are suddenly released? These two questions should certainly concern every person interested in the preservation of the environment as we know it today and in maintaining or increasing our current agricultural productivity.

As has been shown throughout this chapter, many of the methods used for the characterization of humic materials provide relatively little clear-cut and precise information on humic materials in contrast to the wealth of information that these same methods yield on pure compounds or better defined materials. This state of affairs does not facilitate the task of the chemist, biochemist or biologist active in this field. Admittedly, working with humic materials is often frustrating, always laborious and seldom rewarding. Yet, these materials, so widely distributed on the earth's surface, control directly or indirectly many reactions that affect man's survival on this planet and they continue to challenge the curiosity and ingenuity of scientists from many disciplines. Judging from the increasing activities in this field of research in recent years, the challenge does not and will not go unanswered.

REFERENCES

Adhikari, M. and Ray, J.N., 1976. J. Indian Chem. Soc., 53: 238—241.
Alberts, J.J., Schindler, J.E., Nutter, D.E. and Davis, E., 1976. Geochim. Cosmochim. Acta, 40: 369—372.
Alexandrova, L.N., 1960. Sov. Soil Sci., 190—197.
Anderson, D.W., Paul, E.A. and St. Arnaud, R.J., 1974. Can. J. Soil Sci., 54: 317—323.
Anderson, G., 1957. Nature, 180: 287—288.
Anderson, G., 1958. Soil Sci., 86: 169—174.
Anderson, G., 1961. Soil Sci., 91: 156—161.
Archegova, L.B., 1967. Sov. Soil Sci., 757—763.
Atherton, N.M., Cranwell, P.A., Floyd, A.J. and Haworth, R.D. 1967. Tetrahedron, 23: 1653—1667.
Bailly, J. and Margulis, H., 1968. Plant Soil, 29: 343—361.
Baker, W., 1973. Geochim. Cosmochim. Acta, 37: 269—281.
Barton, D.H.R. and Schnitzer, M., 1963. Nature, 198: 217—218.
Beckwith, R.S., 1959. Nature, 184: 745.
Bohn, H., 1976. Soil Sci. Soc. Am. J., 40: 468—469.
Borger, R. de, 1967. Rev. Agric., 20: 555—565.
Bracewell, J.M. and Robertson, G.W., 1973. J. Soil Sci., 24: 421—428.
Bracewell, J.M., Robertson, G.W. and Tate, K.R., 1976. Geoderma, 15: 209—215.
Breger, I.A., 1960. J. Am. Oil Chemist Soc., 43: 197—202.
Bremner, J.M., 1965. In: W.V. Bartholomew and F.E. Clark (Editors), Soil Nitrogen. Am. Soc. Agron., Madison, Wisc., pp. 93—149.
Bremner, J.M., 1967. In: A.D. McLaren and G.H. Peterson (Editors), Soil Biochemistry, Dekker, New York, N.Y., pp. 19—90.
Bremner, J.M. and Lees, H., 1949. J. Agric. Sci., 39: 274—279.
Burges, N.A., Hurst, H.M., Walkden, S.B., Dean, F.M. and Hirst, M., 1963. Nature, 199: 696—697.
Burges, N.A., Hurst, H.M. and Walkden, S.B., 1964. Geochim. Cosmochim. Acta, 28: 1547—1554.
Burges, N.A. and Latter, A., 1960. Nature, 186: 404.
Cameron, R.S. and Posner, A.M., 1974. Trans. Int. Congr. Soil Sci., 10th, Moscow, II, 325—331.
Campbell, C.A., Paul, E.A., Rennie, D.A. and McCallum, K.J., 1967. Soil Sci., 104: 217—224.
Chakrabartty, S.K., Frotschmer, H.O. and Cherwonka, S., 1974. Soil Sci., 117: 318—322.
Chen, Y. and Schnitzer, M., 1976a. Soil Sci. Soc. Am. J., 40: 682—686.
Chen, Y. and Schnitzer, M., 1976b. Soil Sci. Soc. Am. J., 40: 866—872.
Chen, Y. and Schnitzer, M., 1977. Soil Sci. In press.
Chen, Y., Senesi, N. and Schnitzer, M., 1977. Soil Sci. Soc. Am. J., 41: 352—358.
Cheshire, M.V., Cranwell, P.A., Falshaw, C.P., Floyd, A.J. and Haworth, R.D., 1967. Tetrahedron, 23: 1669—1681.
Cheshire, M.V., Cranwell, P.A. and Haworth, R.D., 1968. Tetrahedron, 24: 5155—5160.
Coffin, D.E. and De Long, W.A., 1960. Trans. Int. Congr. Soil Sci., 7th, Madison, Wisc., 2: 91—97.

Courpron, C., 1967. Am. Agron., 18: 623—638.

Datta, C. and Mukherjee, S.K., 1968. Indian J. Chem., 45: 555—562.

Datta, C. and Mukherjee, S.K., 1970. J. Indian Chem. Soc., 47: 979—985.

Datta, C., Ghosh, K. and Mukherjee, S.K., 1971. J. Indian Chem. Soc., 48: 279—287.

De Serra, M.I. and Schnitzer, M., 1972. Can. J. Soil Sci., 52: 365—374.

De Serra, M.I. and Schnitzer, M., 1973a. Soil Biol. Biochem., 5: 287—296.

De Serra, M.I. and Schnitzer, M., 1973b. Can. J. Chem., 51: 1554—1566.

Dormaar, J.F., 1964. Can. J. Soil Sci., 44: 232—236.

Dormaar, J.F. 1969. Plant Soil, 31: 182—184.

Dormaar, J.F. 1972. Can. J. Soil Sci., 52: 67—77.

Dragunov, S.S. and Murzakov, B. 1970. Sov. Soil Sci., 200—225.

Dubach, P., Mehta, N.C., Jakab, T., Martin, F. and Roulet, N., 1964. Geochim. Cosmochim. Acta, 28: 1567—1578.

Duchaufour, P. and Jacquin, F., 1963. Ann. Agron., 14: 885—918.

Dudas, M.J. and Pawluk, S., 1970. Geoderma, 3: 19—36.

Duff, R.B., 1952. J. Sci. Food Agric., 3: 140—144.

Dutta, S., Mukherjee, S. and Roy, H., 1968. Technology, 5: 10—15.

Edwards, A.P. and Bremner, J.M., 1967. J. Soil Sci., 18: 47—63.

Felbeck, G.T., 1965. Soil Sci. Soc. Am. Proc. 29: 48—55.

Felbeck, G.T., 1971. In: A.D. McLaren and J. Skujins (Editors), Soil Biochemistry, Vol. 2. Dekker, New York, N.Y., pp. 36—59.

Flaig, W., 1950. Z. Pflanzenernähr. Düng. Bodenk., 193—212.

Flaig, W., 1958. Trans. Comm. Intern. Soc. Soil Sci., 2nd and 4th, Hamburg, II, 11—45.

Flaig, W., 1964. Z. Chemie, 4: 253—265.

Flaig, W. and Beutelspacher, H., 1951. Z. Pflanzenernähr. Düng. Bodenk., 52: 1—21.

Flaig, W. and Beutelspacher, H., 1954. Landbouwk. Tijdschr., 66: 306—336.

Flaig, W. and Beutelspacher, H., 1968. In: Isotopes and Radiation in Soil Organic Matter Studies. International Atomic Energy Agency, Vienna, pp. 23—30.

Flaig, W. and Otto, H. 1951., Landwirtsch. Forsch., 3: 66—89.

Flaig, W. and Saalbach, E., 1955. Z. Pflanzenernähr. Düng. Bodenk., 71: 215—225.

Flaig, W., Beutelspacher, H. and Rietz, E., 1975. In: J. Gieseking (Editor), Soil Components. Springer-Verlag, New York, N.Y., pp. 1—211.

Forsyth, W.G.C., 1947. Biochem. J., 41: 176—181.

Frondel, C. and Marvin, U.B., 1967. Nature, 214: 587—589.

Gamble, D.S., 1970. Can. J. Chem., 48: 2662—2668.

Gamble, D.S., 1973. Can. J. Chem., 51: 3217—3222.

Gamble, D.S. and Schnitzer, M. 1973. In: P.C. Singer (Editor), Ann Arbor Science. Ann Arbor, Mich., pp. 265—302.

Gamble, D.S., Langford, C.H. and Tong, J.P.K., 1976. Can. J. Chem., 54: 1239—1245.

Gamble, D.S., Schnitzer, M. and Skinner, D.S., 1977. Can. J. Soil Sci., 57: 47—53.

Gascho, G.J. and Stevenson, F.J., 1968. Soil Sci. Soc. Am. Proc., 32: 117—119.

Gjessing, E.T., 1970. Environ. Sci. Tech., 4: 437—438.

Gjessing, E.T., 1976. Physical and Chemical Characteristics of Aquatic Humus. Ann Arbor Science. Ann Arbor, Mich., 120 pp.

Goh, K.M., 1970. New Zealand, J. Sci., 13: 669—686.

Gottlieb, S. and Hendricks, S.B., 1946. Soil Sci. Soc. Am. Proc., 10: 117—125.

Greene, G. and Steelink, G., 1962. J. Org. Chem., 27: 170—174.

Greenland, D.J., 1971. Soil Sci., 111: 34—41.

Griffith, S.M. and Schnitzer, M., 1975a. Soil Sci. Soc. Am. Proc., 39: 861—867.

Griffith, S.M. and Schnitzer, M., 1975b. Can. J. Soil Sci., 55: 251—267.

Griffith, S.M. and Schnitzer, M., 1977. Can. J. Soil Sci., 57: 223—231.

Guminski, S., 1957. Pochvovedenie, 12: 72—78.

Guminski, S., Guminska, A. and Sulej, J., 1965. J. Exp. Bot., 16: 151—162.

Haider, K., Frederick, L.R. and Flaig, W., 1965. Plant. Soil, 22: 49—64.
Haider, K., Nagar, B.R., Saiz, C., Meuzelaar, H.L.C. and Martin, J.P., 1977. Proc. Int.
    Symp. Soil Organic Matter, 3rd, Braunschweig, International Atomic Energy Agency,
    Vienna, pp. 213—220.
Hansen, E.H. and Schnitzer, M., 1966. Soil Sci. Soc. Am. Proc., 30: 745—748.
Hansen, E.H. and Schnitzer, M., 1967. Soil Sci. Soc. Am. Proc., 31: 79—85.
Hansen, E.H. and Schnitzer, M., 1969a. Anal. Chim. Acta, 46: 247—254.
Hansen, E.H. and Schnitzer, M., 1969b. Soil Sci. Soc. Am. Proc., 33: 29—36.
Hansen, E.H. and Schnitzer, M., 1969c. Fuel, 48: 41—46.
Haworth, R.D., 1971. Soil Sci., 111: 71—79.
Hayashi, T. and Nagai, T., 1961. Soil Plant Food (Jap.), 6: 170—175.
Hayes, M.H.B., Stacey, M. and Swift, R.S., 1972. Fuel, 51: 211—213.
Hayes, M.H.B., Swift, R.S., Wardle, R.E. and Brown, J.K., 1975. Geoderma, 13: 231—
    245.
Himes, F.L. and Barber, S.A., 1957. Soil Sci. Soc. Am. Proc., 21: 368—373.
Hurst, H.M. and Burges, N.A., 1967. In: A.D. McLaren and G.H. Peterson (Editors),
    Soil Biochemistry, Vol. 1. Dekker, New York, N.Y., pp. 276—286.
Inoue, T. and Wada, K., 1971. Clay Sci., 4: 61—70.
Irving, H. and Williams, R.P.J., 1948. Nature, 162: 746.
Jackson, M.P., Swift, K.S., Posner, A.M. and Knox, J.R., 1972. Soil Sci., 114: 75—78.
Jakab, T., Dubach, P., Mehta, N.C. and Deuel, H., 1962. Z. Pflanzenernähr. Düng.
    Bodenk., 96: 213—217.
Jakab, T., Dubach, P., Mehta, N.C. and Deuel, H., 1963. Z. Pflanzenernähr. Düng.
    Bodenk., 102: 8—16.
Jones, J.G., 1969. J. Gen. Microbiol., 59: 145—152.
Karpenko, N.P. and Karavayev, M.M., 1966. Sov. Soil Sci., 1154—1156.
Kasatochkin, V.I., Kononova, M.M., Larina, N.K. and Egorova, O.I., 1964. Trans. Int.
    Congr. Soil Sci., 8th, Bucharest, III, pp. 81—86.
Kasatochkin, V.I. and Zilberbrand, O.I., 1956. Pochvovdenie, 80.
Katchalsky, A. and Spitnik, P., 1947. J. Polym. Sci., 2: 432—446.
Khan, S.U., 1969. Soil Sci. Soc. Am. Proc., 33: 851—854.
Khan, S.U., 1970. Z. Pflanzenernähr. Düng. Bodenk., 127: 121—126.
Khan, S.U., 1971. Soil Sci., 112: 401—409.
Khan, S.U. and Schnitzer, M., 1971a. Can. J. Chem., 49: 2302—2309.
Khan, S.U. and Schnitzer, M., 1971b. Soil Sci., 112: 231—238.
Khan, S.U. and Schnitzer, M., 1972a. Can. J. Soil Sci., 52: 43—51.
Khan, S.U. and Schnitzer, M., 1972b. Geoderma, 7: 113—120.
Khan, S.U. and Sowden, F.J., 1971. Can. J. Soil Sci., 51: 185—193.
Khan, S.U. and Sowden, F.J., 1972. Can. J. Soil Sci., 52: 116—118.
Khanna, S.S. and Stevenson, F.J., 1962. Soil Sci., 93: 298—305.
Khristeva, L.A. and Luk'yanenko, N.V., 1962. Sov. Soil Sci., 1137—1141.
Kimber, R.W.L. and Searle, P.L., 1970a. Geoderma, 4: 57—71.
Kimber, R.W.L. and Searle, P.L., 1970b. Geoderma, 4: 47—55.
Kodama, H. and Schnitzer, M., 1967. Fuel, 46: 87—94.
Kodama, H. and Schnitzer, M., 1970. Soil Sci., 109: 265—271.
Kodama, H. and Schnitzer, M., 1973. Can. J. Soil Sci., 53: 240—243
Kodama, H. and Schnitzer, M., 1974. Inter. Congr. Soil Sci., 10th, Moscow, II, pp. 51—
    56.
Kononova, M.M., 1966. Soil Organic Matter. Pergamon Press, Oxford, 544 pp.
Kononova, M.M. and Bel'chikova, N.P., 1961. Sov. Soil Sci., 1112—1121.
Kukharenko, T.A. and Savelev, A.S., 1951. Dokl. Akad. Nauk SSSR, 76: 77; read in:
    Chem. Abstr., 45: 845L, 1951.
Kukharenko, T.A. and Savelev, A.S., 1953. Dokl. Akad. Nauk USSR, 86: 729. read in:
    Chem. Abstr., 47: 8037L.

Schnitzer, M. and Wright, J.R., 1959. Can. J. Soil Sci., 39: 44—53.
Schnitzer, M. and Wright, J.R., 1960a. Soil Sci. Soc. Am. Proc., 24: 273—276.
Schnitzer, M. and Wright, J.R., 1960b. Int. Congr. Soil Sci., 7th, Madison, Wisc., 2: 112—
    119.
Schnitzer, M., Wright, J.R. and Desjardins, J.G., 1958. Can. J. Soil Sci., 38: 49—53.
Schnitzer, M., Sowden, F.J. and Ivarson, K.C., 1974. Soil Biol. Biochem., 6: 401—407.
Seal, B.K., Roy, K.B. and Mukherjee, S.K., 1964. J. Indian Chem. Soc., 41: 212—214.
Senesi, N. and Schnitzer, M., 1977. Soil Sci., 123: 224—234.
Senesi, N., Chen, Y. and Schnitzer, M., 1977a. Soil Biol. Biochem. In press.
Senesi, N., Chen, Y. and Schnitzer, M., 1977b. Soil Biol. Biochem. In press.
Senesi, N., Chen, Y. and Schnitzer, M., 1977c. Fuel., 56: 171—176.
Senesi, N., Griffith, S.M., Schnitzer, M. and Townsend, M.G., 1977d. Geochim. Cosmo-
    chim. Acta, 41: 969—976.
Shivrina, A.N., Rydalevskaya, M.D. and Tereshenkova, I.A., 1968. Sov. Soil Sci., 62—67.
Singer, A. and Navrot, J., 1976. Nature, 252: 479—480.
Skinner, S.I.M., and Schnitzer, M., 1975. Anal. Chim. Acta, 75: 207—211.
Slawinska, D., Slawinski, J. and Sarna, T., 1975. J. Soil Sci., 26: 93—99.
Smith, D.G. and Lorimer, J.W., 1964. Can. J. Soil Sci., 44: 76—87.
Sowden, F.J. and Deuel, H., 1961. Soil Sci., 91: 44—48.
Sowden, F.J. and Schnitzer, M., 1967. Can. J. Soil Sci., 47: 111—116.
Steelink, C., 1964. Geochim. Cosmochim. Acta, 28: 1615—1622.
Steelink, C., Berry, J.W., Ho, A. and Nordby, H.E., 1960. Sci. Proc. R. Dublin. Soc., Ser.
    A, pp. 59—67.
Steelink, C. and Tollin, G., 1967. In: A.D. McLaren and G.H. Peterson (Editors), Soil
    Biochemistry. Dekker, New York, N.Y., Vol. I, pp. 147—169.
Stepanov, V.V., 1969. Sov. Soil Sci., pp. 167—173.
Stevenson, F.J., 1960. Soil Sci. Soc. Am. Proc., 24: 472—477.
Stevenson, F.J., 1976a. In: J.O. Nriagu (Editor), Environmental Biogeochemistry. Ann
    Arbor Science, Ann Arbor, Michigan pp. 519—540.
Stevenson, F.J., 1976b. Soil Sci. Soc. Am. J., 40: 664—672.
Stevenson, F.J. and Ardakani, M.S., 1972. In: J.J. Mortvedt, P.M. Giardano and W.L.
    Lindsay (Editors). Soil Science Society of America, Madison, Wisconsin, pp. 79—114.
Stevenson, F.J. and Goh, M.K., 1971. Geochim. Cosmochim. Acta, 35: 471—483.
Stevenson, F.J. and Mendez, J., 1967. Soil Sci., 103: 383—388.
Sullivan, J.D. and Felbeck, G.T., 1968. Soil Sci., 106: 42—52.
Swift, R.S. and Posner, M.A., 1971. J. Soil Sci., 22: 237—249.
Theng, B.K.G., Wake, J.H.R. and Posner, A.M., 1968. Plant Soil, 29: 305—316.
Tokudome, S. and Kanno, I., 1965. Soil Sci. Plant Nutr. (Jap.), 11: 193—199.
Tschapek, M. and Wasowski, C., 1976. Geochim. Cosmochim. Acta, 40: 1343—1345.
Van Dijk, H., 1960. Sci. Proc. R. Dublin Soc., Ser. A, pp. 163—176.
Van Dijk, H., 1971. Geoderma, 5: 53—67.
Vaughan, D., 1969. Soil Biol. Biochem., 1: 15—28.
Vaughan, D., 1974. Soil Biol. Biochem., 6: 241—247.
Vaughan, D. and Linehan, D.J., 1976. Plant Soil, 44: 445—449.
Vaughan, D. and MacDonald, I.R., 1976. Soil Biol. Biochem., 8: 415—421.
Vila, F.J.G., Saiz-Jimenez, C. and Martin, F., 1974. Agrochimica, 18: 164—172.
Vila, F.G.J., Lentz, H. and Ludemann, H.D., 1976. Biochem. Biophys. Res. Comm., 72:
    1063—1070.
Visser, S.A., 1963. Soil Sci., 96: 353—356.
Visser, S.A., 1964. J. Soil Sci., 15: 202—219.
Visser, S.A., 1964. Nature, 204: 581.
Watson, J.R. and Parsons, J.W., 1974. J. Soil Sci., 25: 9—15.
Welte, E., 1952. Z. Pflanzenernähr. Düng. Bodenk., 56: 105—139.

Wershaw, R.L. and Bohner, G.E., 1969. Geochim. Cosmochim. Acta, 33: 757—762.
Wershaw, R.L., Burcar, P.J. and Goldberg, M.C., 1969. Environ. Sci. Tech., 3: 271—273.
Wershaw, R.L., Burcar, P.J., Sutula, C.L. and Wiginton, B.J., 1967. Science, 157: 1429—
    1430.
Wiesemuller, W., 1965. Albrecht Thaer Arch., 9: 419—436.
Wildung, R.E., Chesters, G. and Behmer, D.E., 1970. Plant Soil, 32: 221—237.
Wright, J.R. and Schnitzer, M. 1960. Int. Congr. Soil Sci., 7th, Madison, Wisc., 2: 120—
    127.
Zetsche, F. and Reinhart, H., 1939. Brennstoff Chem., 20: 84—87.

Chapter 2

CARBOHYDRATES IN SOIL

L.E. LOWE

INTRODUCTION

Carbohydrates are a significant component of the organic matter of all soils, commonly accounting for 5—20% of soil organic matter. Carbohydrates comprise 50—70% of the dry weight of most plant tissues, and hence are the most abundant materials added to soil in the form of plant residues. Furthermore, carbohydrates are important constituents of soil micro-organisms, being present both as structural constituents, and as exocellular and intracellular components.

While much is known of the structure and function of carbohydrates in living organisms, soil carbohydrates are less well understood. Interest in the latter initially arose largely from their suggested role in development of soil structure, following the demonstration that bacterial polysaccharides could cause aggregation of sand—clay mixtures (Martin, 1945, 1946). However, while research on soil carbohydrates received considerable stimulation from interest in problems of soil structure, knowledge of these materials is likely to be important in many other areas; for example, carbohydrates are a primary energy source for many soil micro-organisms, and can also be involved in metal binding reactions. Thus, a knowledge of their properties and behaviour will be important in problems of soil fertility and soil genesis.

GENERAL NATURE OF CARBOHYDRATES; NOMENCLATURE

Carbohydrates are universal constituents of living matter, being involved in both structural and metabolic roles. They exhibit considerable diversity in composition and in properties. In view of this diversity a brief section on classification and nomenclature is included, for the assistance of those unfamiliar with the subject.

Carbohydrates are polyhydroxy compounds which can be represented by the general formula $C_n(H_2O)_m$, hence their name. All carbohydrates are based on a common group of simple monomers (sugars and their derivatives) and are termed monosaccharides, disaccharides, oligosaccharides or polysaccharides, depending respectively on whether one, two, a few or many monomer units are included in the structure. The term homopolysaccharide is used for polymers composed of one type of monomer, whereas hetero-polysaccharides contain two or more monomeric types. While monosaccharides containing varied lengths of carbon chains are known, most contain 5-carbon or 6-carbon chains, termed pentoses and hexoses respectively. The monosaccharides generally exist as ring structures involving an oxygen atom in the ring. Those with 5-membered rings are said to have a furanose struc-

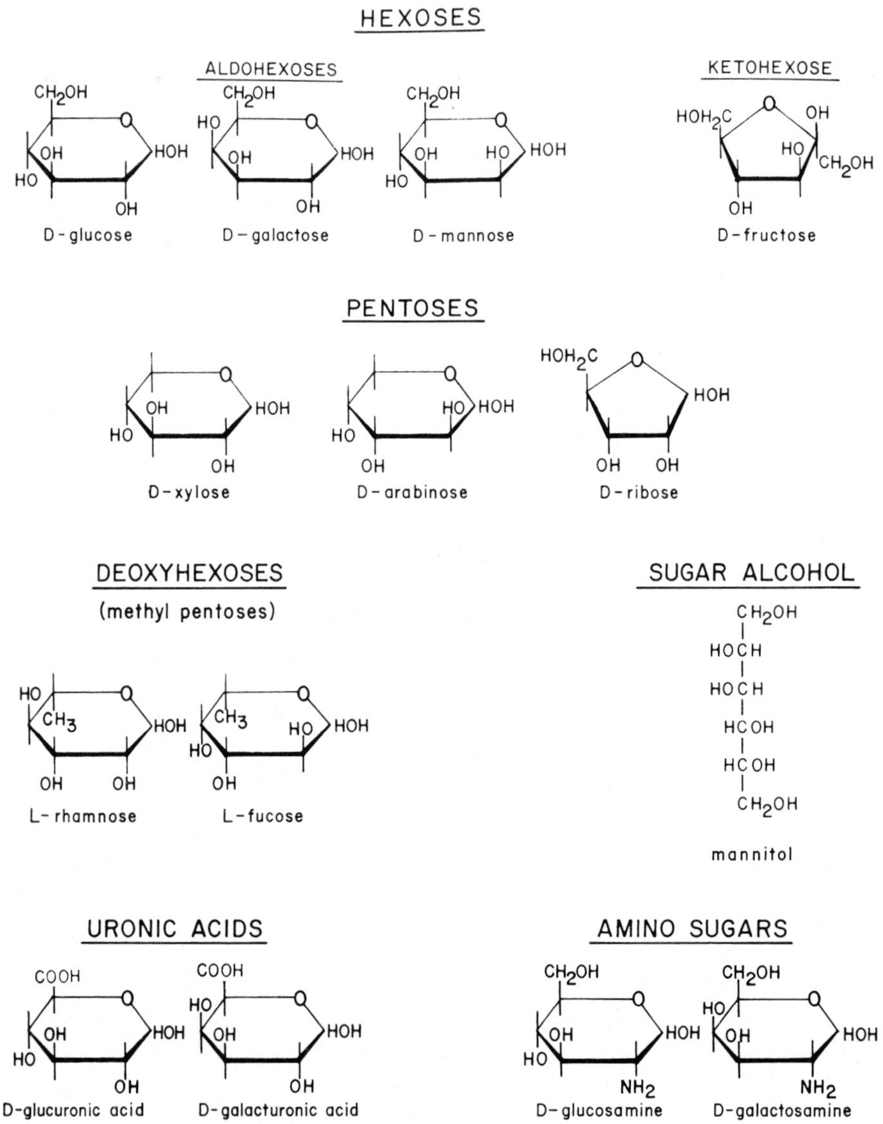

Fig. 1. Structures of common monosaccharides.

ture, while the term pyranose is applied to those with 6-membered rings. These ring structures are commonly represented by Haworth formulae (see Fig. 1). Alternative structural formulae (chair type) are also used to stress conformational features and to indicate the non-planar nature of the molecule. Standard textbooks on carbohydrate chemistry can be consulted for details (Pigman and Horton, 1972). An important feature of monosaccharide structures is the presence of asymmetric carbon atoms, giving rise to optical isomerism. Studies of soil carbohydrates have generally paid little attention to stereoisomers, and this aspect will not be discussed further here. Two other terms are encountered in descriptions of monosaccharide structure, namely aldose and ketose. Of the sugars reported in soils, only fructose is a ketose, the remainder being aldoses. Aldoses in their straight chain form have a terminal aldehyde group, whereas ketoses lack the aldehyde group but exhibit a keto group.

Important derivatives of the simple hexoses and pentoses include aminosugars, uronic acids, sugar alcohols, methylated sugars and deoxy-sugars (where oxygen has been eliminated from a hydroxyl group). The deoxyhexoses (rhamnose and fucose) are also referred to by some as methyl pentoses. The various classes of monosaccharides reported to occur in soils are illustrated in Fig. 1. The monosaccharides can also occur in nature in combination with other non-carbohydrate components. The general term glycoside is applied to these substances. The types of carbohydrates that have been reported to occur in soils can thus be classified as follows:

Monosaccharides and derivatives:
    Hexoses
    Pentoses
    Deoxyhexoses
    O-methylated sugars
    Hexosamines
    Uronic acids
    Sugar alcohols
Disaccharides
Oligosaccharides
Polysaccharides
    Homopolysaccharides (cellulose)
    Heteropolysaccharides (hemicellulose, microbial gums).

CARBOHYDRATE DISTRIBUTION IN SOIL

Total carbohydrate content of soil

There is as yet no entirely satisfactory method for measuring total carbohydrates in soil. The best procedures involve separation of individual monomers, following acid hydroylsis, and separate measurement of the individual constituents. However, this approach is time consuming and estimates of total carbohydrate content are commonly based on colorimetric procedures (such as the anthrone procedure of Brink et al. 1960) applied directly to soil hydrolysates. Despite their limitations, such methods permit comparisons of different soils and give a reasonable picture of the distribution of total carbohydrates and of the proportions of soil organic matter present in carbohydrate form. The limitations of these methods have been discussed in some detail by Greenland and Oades (1975). Table I presents examples of data obtained by such methods.

In general, total soil carbohydrate levels parallel organic matter contents, being highest in organic-rich surface horizons and lowest in subsoil horizons low in organic matter. The carbohydrate fraction generally accounts for about one tenth of the total organic carbon (6—14%). However, there are some notable exceptions, namely samples relatively rich in undecomposed plant debris (e.g. peat, and forest humus L and F horizons) where the proportion of carbohydrates is higher, and podzol B horizons, where the organic matter contains relatively less carbohydrate. Results on organic horizons under deciduous and coniferous forest (Table I) indicate that with increasing humification, the carbohydrate content of organic matter progressively decreases. A similar trend has been observed for a sphagnum peat site. A similar conclusion can also be drawn from the data on reducing sugar content cited by Greenland and Oades (1975).

It is less easy to draw conclusions with respect to changes in the carbohydrate content of organic matter with increasing depth in the mineral profile, partly because of the limited amount of data reported. The results of Graveland and Lynch (1961) on some Alberta soils indicate higher proportions of carbohydrates in subsurface horizons than in surface horizons. However, in nine soils in the same region examined subsequently, the carbohydrate fraction accounted for 9.3% of the organic carbon in Ah horizons (Lowe, 1969) as compared with 9.2% for B horizons (Lowe, unpublished data).

Factors affecting carbohydrate content of soil

As already indicated soil carbohydrate levels appear to depend largely on total organic matter content, hence those factors which influence inputs of organic matter and rates of decomposition, will also be those controlling total carbohydrate levels in soil. Thus, carbohydrate levels in virgin soils will be determined by the natural soil forming factors at any particular location. Under cultivation, however, organic matter content can be drastically modified by the cropping system imposed by man. For example, Oades (1967) has shown that a soil under a four-course rotation including pasture, had nearly twice as high a carbohydrate content as the same soil under a continuous wheat-fallow system. However, changes in carbohydrate content due to cropping systems were not accompanied by significant changes in the pro-

TABLE I

Examples of carbohydrate content of soils determined by anthrone colorimetric methods

Soil		Total organic carbon (%)	Soil carbohydrates (%)	Carbohydrate-C as % organic-C	Reference
Orthic black	Ah	6.3	1.01	9.4	Graveland and Lynch (1961)
Black solod	Ah	4.0	0.73	10.6	Graveland and Lynch (1961)
Black solodized solonetz	Ah	5.7	0.94	9.6	Graveland and Lynch (1961)
Brown forest	mull	14.9	1.24	8.3	Ivarson and Sowden (1962)
Podzol	mor	18.6	1.47	7.9	Ivarson and Sowden (1962)
Dark brown	Ah 0—6"	8.6	1.09	12.7	Ivarson and Sowden (1962)
Black	0—6" cult.	10.5	0.76	7.2	Ivarson and Sowden (1962)
Grey luvisol	Ah	8.9	2.70	12.2	Lowe (1969)
Orthic black	Ah	4.2	0.85	8.1	Lowe (1969)
Orthic brown	Ah	2.8	0.45	6.6	Lowe (1969)
Grey solodized solonetz	H/Ah	21.3	5.70	10.7	Lowe (1969)
Black solonetz	Ah	4.5	1.27	11.4	Lowe (1969)
Brown solonetz	Ah	3.8	0.75	8.0	Lowe (1969)
Krasnozem	0—22.5 cm	6.1	1.60	10.7	Oades (1967)
Red-brown earth	Ap 0—6 cm	2.5	0.30	6.9	Oades (1967)
Podzol	B	2.2	0.10	1.8	Oades (1967)
Peat	0—15 cm	47.5	10.8	9.1	Oades (1967)
Lodgepole pine	L	50.4	29.0	23.0	Lowe (1974)
	F	41.8	17.2	16.5	Lowe (1974)
	H	29.9	8.3	11.1	Lowe (1974)
Aspen poplar	L	48.8	17.9	14.7	Lowe (1974)
	F	34.3	9.0	10.5	Lowe (1974)
	H	15.1	3.0	7.9	Lowe (1974)
	Ah	4.5	0.8	7.2	Lowe (1974)
Sphagnum peat	L (Of)	50.9	41.0	32.2	Lowe (1974)
	F (Om)	50.9	34.4	27.0	Lowe (1974)
	H (Oh)	41.4	10.0	9.7	Lowe (1974)

portion of organic matter present as carbohydrate. This relative constancy of carbohydrate levels in the organic matter was even more marked if undecomposed plant material was removed from the sample prior to analysis.

Thus in general, available data indicates that soil carbohydrate levels lie in the vicinity of one tenth of the organic matter content, except where a large proportion of undecomposed or partially decomposed plant debris is present, and that levels will be influenced by any factors affecting organic matter content or its degree of decomposition.

Carbohydrate fractions in soil

In terms of the behaviour of a soil system, the total carbohydrate content is not readily interpreted, because of the diversity of materials included in this broad fraction. In order to discuss further the properties and significance of soil carbohydrates, it is necessary to distinguish a number of types of carbohydrate material. The classification of soil carbohydrate fractions can be approached in a variety of ways, none of which by themselves is entirely satisfactory.

The total carbohydrate in a soil sample can be partitioned into polysaccharides and free monosaccharides, of which the latter is quantitatively rather insignificant. Further characterization can be achieved by separate determination of hexoses, pentoses, hexosamines and uronic acids, following hydrolysis of the polysaccharides. However, this yields no information on the association of monomeric units with each other or with non-carbohydrate material. A more useful picture of the distribution of carbohydrate fractions is obtained by separating a variety of organic fractions from the soil prior to analysis for the different classes of sugars. Ideally such an approach would be most useful if based on separation of fractions which can be equated with their function in the soil system, or interpreted in terms of origin or role in soil formation. Unfortunately, our understanding of soil organic matter is at present far too incomplete to permit achieving such ideals. In practice methods selected for separating organic fractions in order to characterize their carbohydrate constituent, are to a large extent arbitrary, and influenced in some degree by the amount of relatively undecomposed plant material present in the sample. In this section results obtained by two approaches will be presented, first by fractionation based on proximate analysis procedures designed originally for the study of plant material and secondly by separation of the classical humus fractions.

Proximate analysis procedures (e.g. that described by Stevenson, 1965) generally provide estimates of cellulose, hemicellulose and, in some cases, water soluble polysaccharides. The estimates are based on determination of "reducing sugars" or total sugars following acid hydrolysis. Early work by such methods has been reviewed by Waksman (1938). For samples relatively rich in undecomposed and partially decomposed plant material, such as peat

or forest humus, proximate analysis gives a reasonable estimate of the two
major types of polysaccharide present. However, despite its occasional use
for the purpose, proximate analysis is of very doubtful validity, when applied
to mineral soils where the bulk of the organic matter is well humified. Selected data are presented in Table II. Relative proportions of cellulose and hemicellulose vary, depending on the material. However, with increasing decomposition, the cellulose is generally depleted to a greater extent than the hemicellulose fraction. Furthermore, levels of cellulose based on proximate analysis may well be overestimated in many cases, since Gupta and Sowden
(1964) have shown that direct extraction of cellulose with Schweitzer's reagent yields lower results than those based on hydrolytic methods as used in
proximate analysis. These workers reported cellulose values (direct extraction) in the range 0.3—1.9% of total organic matter.

The major soil organic fractions of mineral soils for which carbohydrate
content has been investigated, are the humic acid (HA), fulvic (FA) and
humin fractions. The latter is sometime referred to as "extraction residue".
In addition some studies have examined the "light fraction", which is the

TABLE II

Polysaccharide fractions in forest humus layers and in peat, based on proximate analysis
(fractions expressed as % of organic matter)

	Hemi-cellulose	Cellulose	Water soluble poly-saccharide	Reference
Spruce L	12.6	25.7		Remezov and Pogrebnyak
F	7.3	18.4		(1965)
H	6.8	16.3		
Birch L	14.5	26.1		Remezov and Pogrebnyak
F	9.7	19.0		(1965)
H	6.3	16.6		
Hardwood spruce F	16.6	10.3		Waksman et al. (1928)
H	13.8	2.8		
Mixed conifers F	17.8	8.4		Waksman et al. (1928)
H	20.3	4.4		
Moss-hemlock (mor) (av. 4 sites)	16.6	10.0	5.1	Klinka and Lowe (1975)
Western red cedar-polystichum (mull) (av. 4 sites)	8.8	3.2	3.8	Klinka and Lowe (1975)
Sphagnum peat F	12.8	14.2		Lowe (1974)
H	5.8	3.9		

relatively undecomposed material separated by flotation on a heavy liquid, prior to humus fractionation. If not so removed, much of this material is normally included in the extraction residue. (See Table III).

The same trend (Residue > FA > HA) has also been noted for forest humus layers (Lowe, 1974), and for three prairie soils (Acton et al., 1963).

A number of studies have examined the distribution of carbohydrates within the FA fraction. Forsyth (1947), using charcoal adsorption, separated four fulvic fractions, one of which contained relatively pure polysaccharide material, and another thought to contain phenolic glycosides. More recent studies (Sequi et al., 1975; Lowe, 1975) have shown that most of the highly coloured polyphenolic components of FA can be removed from the main polysaccharide constitutents by adsorption on polyamide or polyvinylpyrrolidone. Polysaccharide-rich materials have been separated from FA by a number of workers using acetone precipitation. These fractions have commonly been referred to as "microbial gum" or "soluble polysaccharide" fraction. However, in a number of studies it has not been reported what proportion of the total or FA carbohydrates were accounted for. Dormaar (1967) has reported that the precipitated gum accounted for 6% of total carbohydrates and about 14% of the carbohydrates in the FA fraction of Ah and Bm horizons of chernozemic soils. Acton et al. (1963) reported that the acetone precipitated fraction accounted for less than 10% of the soil carbohydrates of four Saskatchewan soils. Thus, it cannot at present be assumed that fulvic polysaccharide fractions obtained by either the Forsyth (1947) procedure, or by acetone precipitation from FA fractions, constitute a major proportion of carbohydrates in FA fractions, although relative freedom from contamination has made these preparations an attractive starting point for studying the properties of a soil polysaccharide.

Carbohydrates, in whatever form, can however account for a substantial proportion of the FA-carbon. Oades (1967) has reported a value of 19.6% for a red-brown earth under continuous pasture. For a range of forest humus samples (mull and mor types), the carbohydrate components (estimated by the phenol-sulphuric acid method) were found to account for 16—40% of the FA-carbon (Lowe, unpublished data).

With regard to the HA fraction, it is clear that carbohydrate components

TABLE III

Carbohydrate distribution in humus fractions * (Dormaar, 1967)

Horizon	FA	HA	Residue
Ah	43	7	50
Bm	42	3	55

* Average for 24 profiles, data expressed as % of total carbohydrate content.

are a minor constituent, which tends to decrease as the HA preparation is subjected to further purification processes. Values of 3% and below have been reported by Oades (1967) and Lowe (1969), the latter for purified HA fractions from a range of Alberta soils. For some tropical volcanic soils, Griffith and Schnitzer (1975) reported carbohydrate contents of HA in the range of 6—10%. Corresponding values for three Australian soils reported by Butler and Ladd (1971) were between 0.8% and 4.4%, with the lowest value associated with a podzol B horizon. These workers also noted that for one soil, the carbohydrate content of high molecular size fractions of both HA and FA was somewhat greater than for low molecular size fractions obtained by Diaflo ultrafiltration.

While little work has been done on the ethanol-soluble portion of HA (hymatomelanic acid), Clark and Tan (1969) concluded from infrared absorption data that this component is a humic acid type material bound to polysaccharides by ester linkage, whereas the ethanol-insoluble HA is essentially free from carbohydrate.

The carbohydrate content of the residue following alkali extraction, depends on whether the main undecomposed material is separated as the light fraction prior to the alkali treatment, and on the abundance of undecomposed material. Few studies have examined the composition of the light fraction, but carbohydrate levels between 13.9% and 28.2% have been reported (Oades, 1967; Whitehead et al., 1975), with the light fraction accounting for 29—43% of the soil carbohydrates (Oades and Swincer, 1968) in some soils of relatively low organic matter content. Where the light fraction has not been separated, the extraction residue will contain both residual plant constituents (including cellulose and hemicellulose) and more resistant humification products that are not extractable by alkali. Little work has been undertaken on distinguishing the carbohydrate components of this material, although a separation has been attempted of the more stable polysaccharides of forest humus layer extraction residues (presumed to be mainly cellulose) from the remaining polysaccharides, on the basis of resistance to hydrolysis by acid of different concentrations (Lowe, 1974). For such materials the stable polysaccharide component of the extraction residue was relatively much more abundant in F layers than in the more decomposed H layers.

Carbohydrate distribution in fractions obtained with other extractants, e.g. anhydrous formic acid (Parsons and Tinsley, 1961), has also been studied, but these fractions are not readily equated with those obtained by the classical humus fractionation scheme.

In summary, it can be stated that carbohydrates are present in soil in a variety of forms as undecomposed plant debris, and in all the classical humus fractions. An understanding of soil carbohydrates requires information on all the various fractions, and on the nature of association between carbohydrate fractions and other soil constituents. The following sections will review the composition and properties of soil carbohydrates in relation to the fractions outlined above. It should be emphasized that the relative abundance of the different carbohydrate fractions shows considerable variation among soil types and horizons.

COMPOSITION OF SOIL CARBOHYDRATES

The monosaccharides identified in soil hydrolysates are essentially the same for all soils studied. Furthermore, only minor variations have been recorded in the proportions of monomers present. On the other hand considerable variation has been noted in the monomer distribution between individual soil fractions. In whole-soil hydrolysates, glucose is consistently the most abundant of the neutral sugars, and galactose, mannose, arabinose, xylose and rhamnose are always present in significant amounts. Smaller amounts of fucose and ribose are also commonly found as well as a number of minor constituents which show high Rf values in paper chromatography. Some of the latter have been identified as 2-O-methylrhamnose, 4-O-methylgalactose (Duff, 1961), 2-O-methylxylose and 3-O-methylxylose (Bouhours and Cheshire, 1969). In addition, hexosamines and uronic acids appear to be of universal occurrence in soils. While much of the earlier work on soil sugars was carried out using paper chromatography, the most reliable quantitative data is probably that obtained by gas liquid chromatography (GLC) or ion exchange chromatography. Selected results on neutral sugar distribution in soil carbohydrates as a whole, are presented in Table IV. The overall variation in composition is remarkably small, considering the range of soils represented. Total hexose levels (glucose, galactose, mannose) are generally 2—4 times higher than pentoses (arabinose, xylose, ribose), while ratios of deoxyhexoses (rhamnose, fucose) to pentoses generally range from 0.2 to 0.7 (Folsom et al., 1974).

Reliable quantitative information on uronic acid levels in soils has proved hard to obtain. This is due partly to serious losses during hydrolysis (Ivarson and Sowden, 1962) and partly to the shortcomings of analytical methods. These have been discussed by Ivarson and Sowden (1962), Lowe (1968), and by Greenland and Oades (1975).

Various workers have estimated that uronic acids account for 1—5% of the organic matter (Greenland and Oades, 1975), but such estimates cannot be regarded as entirely satisfactory. Nevertheless uronic acids are probably an important component of soil carbohydrates, both quantitatively and because of the influence of their acidic groups on polysaccharide properties.

Hexosamines will not be discussed at length in this chapter, since they are dealt with in Chapter 3. In practice hexosamines are most easily studied at the same time as amino-acids, since ion exchange chromatography of hydrolysates permits simultaneous analysis of both classes of compound. Partly for this reason, only rarely has data been published on both hexosamine and

TABLE IV

Neutral sugar distribution in soil hydrolysates (expressed in percent of total neutral sugars recovered)

Soil		Glucose	Galactose	Mannose	Arabinose	Xylose	Rhamnose	Fucose	Ribose	Reference (method)
Longterm cultivated		41.3	14.4	10.8	12.7	12.7	5.1	1.1		Whitehead et al., 1975 (GLC)
Longterm pasture		39.2	14.3	10.4	11.5	16.9	5.8	2.0		
Marshall Ap – developed under prairie		27.6	17.5	17.5	15.0	12.9	5.4	2.3	1.8	Folsom et al., 1974 (GLC)
Grundy Ap – developed under prairie		29.3	16.7	16.8	15.5	10.7	6.4	2.6	2.0	Folsom et al., 1974
Mexico A_1 – developed under prairie		33.1	15.6	19.5	11.2	9.5	6.4	3.4	1.3	Folsom et al., 1974
Menfro Ap – developed under forest		30.7	18.7	16.1	12.4	11.4	6.4	3.2	1.1	Folsom et al., 1974
Weldon Ap – developed under forest		30.1	17.4	13.5	15.1	12.7	6.9	3.3	1.0	Folsom et al., 1974
Marion Ap – developed under forest		28.5	20.0	16.8	11.2	11.3	6.5	4.1	1.6	Folsom et al., 1974
Urrbrae pasture (0–6 cm)		39.3	12.2	19.4	11.7	7.5	4.4	3.6	1.9	Swincer et al., 1968b (GLC)
Urrbrae fallow-wheat (0–6 cm)		43.1	13.0	18.5	9.8	8.7	3.5	2.6	0.8	Swincer et al., 1968b
Orthic grey luvisol (grey wooded)	LH	44	15	12	12	9	5	4*		Gupta et al., 1963 (paper and cellulose column chromatography)
	AH	48	14	13	10	8	5	2*		
	Bt	33	19	14	13	8	9	4*		
Orthic podzol	LH	54	15	15	5	4	7	5*		Gupta et al., 1963
	B	35	16	16	9	9	9	6*		Gupta et al., 1963
Dark brown solodized solonetz	Ah	28	16	18	17	9	8	4*		Gupta et al., 1963
	Bnt	24	14	18	18	10	8	8*		Gupta et al., 1963
Orthic black	Ah	36	14	16	15	8	8	3*		Gupta et al., 1963
	Bm	34	15	14	12	8	12	5*		Gupta et al., 1963

* Includes ribose.

total carbohydrate content for the same samples. Accordingly, it is difficult to assess hexosamine levels in relation to other carbohydrate constituents. Hexosamines have been reported (Parsons and Tinsley, 1975) to account for 2—13% of total soil nitrogen, and based on the results of Sowden and Ivarson (1962) for four soils, hexosamines may well be present in mineral soils at levels in the order of 30—40% or more of the total neutral sugar content. Glucosamine and galactosamine account for virtually all the hexosamines recovered in soil hydrolysates (Sowden, 1959; Stevenson and Braids, 1968). Glucosamine : galactosamine ratios were reported by Sowden (1959) to vary from 1.6 to 4.1.

Free sugars

Several workers have examined the composition of free sugars extracted with water or ethanol (Alvsaker and Michelsen, 1957; Nagar, 1962; Grov, 1963; Gupta and Sowden, 1963). Grov (1963) reported levels representing 0.24% and 0.22% of soil organic matter for H and A horizons of a pine forest soil, with glucose accounting for about 80% of the free sugars. Gupta and Sowden (1963) reported rather lower levels of free sugars in a variety of Canadian soils, but also noted the predominance of glucose. Galactose, mannose, arabinose, xylose, fucose, ribose, rhamnose and fructose have also been reported to be present. Based on the limited data available, free sugars appear to be somewhat more abundant in cool climatic conditions where decomposition rates are low.

Polysaccharides

Polysaccharide composition has been studied both on water or acid soluble isolates relatively free from non-carbohydrate material, and on the major soil organic matter fractions. The major conclusion from such investigations is that soils contain a great variety of polysaccharide constituents, varying not only in monomer composition, but also in the degree of association with non-carbohydrate materials and in the nature of associated materials.

Selected data on monomer composition of polysaccharide fractions are presented in Table V. Marked differences in monomer composition of polysaccharides have been reported (Oades, 1972; Whitehead et al., 1975) between the light fraction and heavy fraction of soils. This can be attributed to the predominance of plant debris in the light fraction, whereas in the heavy fraction, any plant-derived polysaccharides are accompanied by major amounts of materials synthesized within the soil system. The light fraction is generally richer in glucose and xylose, and lower in rhamnose and mannose than the heavy fraction. Differences in relative sugar levels are clearly shown by the ratios glucose/mannose, arabinose/xylose and xylose/rhamnose shown in Table V.

TABLE V

Monomer composition of soil polysaccharide fractions (expressed as percent of total neutral sugars recovered)

	Glu-cose	Galac-tose	Man-nose	Arabi-nose	Xylose	Fucose	Rham-nose	Ribose	Glu./Man.	Arab./Xyl.	Xyl./Rham.
Cultivated soil (Whitehead et al., 1975):											
Light fraction	49.5	7.6	5.0	9.2	26.8	0.6	1.3	—	9.9	0.3	20.6
Heavy fraction	39.9	16.2	12.3	12.7	12.2	1.4	5.3	—	3.2	1.0	2.3
Pasture red-brown earth (Swincer et al., 1968b):											
Extracted 1 N HCl	29.2	19.2	19.9	7.5	5.5	7.0	11.3	0.5	1.5	1.3	0.5
Extracted 0.5 N NaOH	27.9	12.3	24.5	10.9	11.3	4.1	5.2	3.7	1.1	1.0	2.2
Extracted with acetylation	57.5	8.6	8.5	12.7	7.8	2.1	2.4	0.4	6.8	1.6	3.3
Residue	43.4	11.7	21.4	14.6	3.1	2.3	1.8	1.7	2.0	4.7	1.7
Pasture, red-brown earth; molecular size fractions (Swincer et al., 1968b):											
1 N HCl extract:											
4,000—100,000 MW	28.1	21.0	19.4	7.7	5.1	7.8	10.5	0.4	1.4	1.5	0.5
>100,000 MW	31.5	15.5	20.8	7.1	6.5	5.5	12.5	0.5	1.5	1.1	0.5
0.5 N NaOH extract:											
>4,000 MW	27.4	11.1	26.8	12.2	11.0	7.5	9.1	4.7	1.0	1.1	1.2
4,000—100,000 MW	26.8	13.4	19.3	10.7	15.9	3.8	5.4	2.8	1.4	0.7	2.9
>100,000 MW	32.9	15.3	17.2	6.5	11.0	7.5	9.1	0.4	1.9	0.6	1.2

Similarly, Swincer et al. (1968b) demonstrated considerable variation in composition of polysaccharides in four fractions of red brown earths, from which the light fraction had been removed. The four fractions (obtained sequentially) were: (1) extracted with 1 N HCl; (2) extracted with 0.5 N NaOH; (3) extracted with acetylating reagents; and (4) residue. The latter two fractions are expected to contain any undecomposed or slightly decomposed plant materials not removed in the light fraction. These two fractions have the highest glucose and lowest rhamnose and fucose levels. The same investigators also examined variations in polysaccharide composition in molecular size fractions of the 1 N HCl and 0.5 N NaOH extracts obtained by gel filtration (Table V). While differences associated with molecular size were less marked than those between fractions, nevertheless variations in sugar ratios; e.g. glucose/mannose, especially for the NaOH-extractable polysaccharides, indicated that the fractions are far from homogeneous. This conclusion has been confirmed by other methods, as will be discussed in the following section on polysaccharide properties.

In a further extension of the Australian studies outlined above Oades (1972) examined a broader range of soil types, and found very great similarities between the soils with respect to polysaccharide composition for each of the soil fractions investigated. In the acid-extractable fraction, glucose was the dominant sugar in the lower molecular size components, whereas mannose tended to be dominant in the larger molecules. The alkali-extractable polysaccharide was similar for all the soils, with glucose accounting for about one third of the neutral sugars. The residue and acetylation extract were particularly rich in glucose.

Others have examined acid-soluble polysaccharide fractions, either extracted directly or separated from fulvic acid fractions by charcoal adsorption, following the method of Forsyth (1947) (Whistler and Kirby, 1956; Finch et al., 1966, 1968; Thomas et al., 1967; Lowe, 1968; Cheshire et al., 1974). In general these have been found to have rather similar composition (considering the variety of methods involved), but such polysaccharide preparations can be resolved by fractional precipitation (Whistler and Kirby) and by column chromatography on DEAE-cellulose (Thomas et al., 1967) or DEAE-Sephadex (Finch et al., 1966) into several fractions varying somewhat in composition. Finch et al. (1968) for example, separated a range of fractions on DEAE Sephadex with glucose contents varying between 21% and 60% of the neutral sugars, and rhamnose ranging from a trace up to 18%. Clearly many of the preparations are in reality complex mixtures of polysaccharides.

One other fraction whose composition has been examined, is the cellulose fraction extracted directly with Schweitzer's reagent (Gupta and Sowden, 1964). Glucose was found to account for 65—82% of the fraction, with no other sugar present in more than minor amounts.

Variation in the composition of soil polysaccharide fractions has also been demonstrated by analysis of hydrolysates for uronic acids, amino-sugars and amino acids. Table VI illustrates this type of data for light and heavy fractions and for molecular size fractions of fulvic acid polysaccharides. For the latter especially, marked changes in relative uronic acid and amino acid levels were noted, with uronic acids increasing and amino acids decreasing with increasing molecular size. The presence of amino acids, or indeed of hexosamines, in hydrolysates of soils or of their light and heavy fractions, is not evidence that these components are directly associated with polysaccharides in complex molecules. However, in the case of fulvic polysaccharides studied by Swincer et al. (1968a, b), from which the major coloured polyphenolic components had been removed by adsorption on polyvinyl-pyrrolidone, and which were separated into molecular size fractions, it appears likely that nitrogenous components are closely associated with the polysaccharides.

In summary, knowledge of soil polysaccharide composition has made considerable progress, especially since the advent of GLC techniques and developments in fractionation methods. However, present knowledge still remains fragmentary, and difficulties in isolating homogeneous polysaccharide fractions have discouraged any attempt at detailed structural studies. Nevertheless, it is now clear that a great variety of polysaccharide materials are present in soil, exhibiting variations in hexose, pentose, deoxyhexose, uronic acid and hexosamine content, and also in their association with nitrogenous and polyphenolic soil constituents. In general, soil polysaccharide fractions thus far examined, are characterized by the presence of six or more monomeric constituents, in contrast to polysaccharides of higher plants and animals, which are commonly composed essentially of only one to three monomeric constituents.

Finally, it must be emphasized that analytical results on the composition and distribution of soil polysaccharides, as with most biochemicals, are very

TABLE VI

Relative proportions of neutral sugars and associated materials in hydrolysates of soil fractions (neutral sugars = 100)

	Neutral sugars	Uronic acids	Amino sugars	Amino acids
Average for 3 soils (Whitehead et al., 1975)				
Light fraction	100	6	9	34
Heavy fraction	100	8	35	84
Average for 14 soils (Oades and Swincer, 1968)				
FA molecular size fractions:				
<4,000	100	26	13	21
4,000—100,000	100	44	14	60
>100,000	100	2	11	88

dependent on the methods employed, especially with regard to extraction, fractionation and hydrolysis methods. For more detailed discussion of these topics, the reader is referred to the reviews of Swincer et al. (1969) and Greenland and Oades (1975).

PROPERTIES OF SOIL POLYSACCHARIDES

Present knowledge of the physical or physico-chemical properties of soil polysaccharides is based essentially on studies of preparations extracted with water, buffers, dilute acid or alkali, and subsequently purified by removal of much of the accompanying non-carbohydrate materials. It should be emphasized that such isolates normally constitute only a small proportion of the total soil polysaccharides. It should also be stressed that isolated soil polysaccharide fractions are unlikely to consist of a single polysaccharide, but rather of mixtures of polysaccharides together with varying amounts of non-carbohydrate material. Some properties have also been inferred from examination of microbial polysaccharides, on the assumption that such materials are likely to be produced in considerable amounts in soil by the microbial population. The properties of soil polysaccharides are important both in developing methods for separation from non-carbohydrate material, for resolving polysaccharide mixtures, and also in explaining their behaviour in soil.

Appearance and solubility

Isolated soil polysaccharides have been variously described as white, creamy whity or creamy yellow solids (Bernier, 1958; Mortensen, 1960; Cheshire et al., 1974). They yield colourless to pale yellow opalescent solutions in water or dilute acid, but are insoluble in acetone or ethanol. Freeze-dried preparations are light and fluffy (Mortensen, 1960) and dissolve much more readily than those dried by other means). Insolubility in acetone or ethanol has been used to separate polysaccharides from other acid-soluble components, and also for fractional precipitation to obtain separation of polysaccharide fractions of varying composition (Bernier, 1958).

Polydispersion and molecular weight

Studies with ultracentrifuge, gel filtration and electrophoretic methods have demonstrated that soil polysaccharide isolates are polydisperse (Bernier, 1958; Ogston, 1958; Finch et al., 1968; Swincer et al., 1968a). Various estimates of average molecular weights (MW) of polysaccharide preparations have been made, based on viscosimetric, sedimentation and gel filtration methods. Bernier (1958) reported MW values ranging from 10,000 for a soil under mor humus, up to 124,000 for a mull sample. MW values around

50,000 have been reported for single preparations by Finch et al. (1966) and Mingelgrin and Dawson (1973), based on ultracentrifuge and gel filtration methods, respectively. Mortensen (1960) reported a value of 450,000 for a soil polysaccharide extracted with hot water. However, Saini and Salonius (1969) reported that MW values for polysaccharides from forest humus were much higher than those reported above, and were unable to obtain measurements with the gel filtration technique. Swincer et al. (1968a) using gel filtration on Sephadex, observed a bimodal MW distribution for extracted polysaccharides, with the main components exhibiting MW above 100,000 and below 40,000. In subsequent work, differences in MW distribution were observed (Swincer et al., 1968b) between materials extracted with dilute HCl (51—55% <4,000 MW) and those obtained in a subsequent NaOH extraction (22—23% <4,000 MW), indicating generally larger polysaccharide polymers in the alkaline extract. Based on ultracentrifuge studies, Ogston (1958) suggested that a polysaccharide extracted from a forest soil consisted of elongated, stiff and little solvated rods, with an axial ratio of an average particle of about 80.

Viscosity and optical rotation

Aqueous solutions of soil polysaccharides show considerable viscosity, a property related to molecular weight, shape and chemical constitution. The increase in viscosity with MW has been used, as indicated earlier, to provide estimates of MW. Saini and Salonius (1969) showed that viscosity (in terms of the Staudinger index) was inversely related to the precipitability of polysaccharide solutions with organic solvents. The viscosity of higher MW fractions must also be of some importance in determining their physical behaviour within the soil system.

Another characteristic property of a polysaccharide solution is its optical rotation. Bernier (1958) has used this property as an index of heterogeneity in fractionation studies. Both dextro- and laevo-rotatory fractions have been isolated (Whistler and Kirby, 1956; Bernier, 1958). Specific optical rotation can give information on structural features of polysaccharides. Strong dextro-rotation is associated with the predominance of α-glycosidic linkages, whereas laevo-rotation may indicate predominance of β-glycosidic linkages (Whistler and Smart, 1953; Gorin and Spencer, 1968).

Lucas (1970) observed dextro-rotation for a polysaccharide fraction of a peat, but laevo-rotation for major polysaccharides of higher plants growing in the peat. This suggested not only that the peat polysaccharide had predominantly α-glycosidic linkages, but also that the peat polysaccharide could not be directly derived from the plant polysaccharides.

Functional groups, charge and equivalent weight

As expected for polymers containing neutral sugars, uronic acids and amino sugars, infrared spectra of soil polysaccharides indicate the presence of hydroxyl and carboxyl groups as well as C—C, and C—O bonds typical of sugar rings and glycosidic linkages. The presence of N—H bonds has also been indicated, presumably due to amino sugars or amino acids (Mortensen, 1960). The presence of acidic carboxyl groups is also revealed by titration with bases. For example, Mortensen (1960) and Whistler and Kirby (1956) have reported equivalent weights for isolated soil polysaccharides of 3,050 and 1,945, respectively, based on titrations. Somewhat lower values were found by Parsons and Tinsley (1961) for fractions obtained by formic acid extraction. Charge characteristics arising from the ionization of carboxyl groups, and increased charge with increasing pH, have been utilized in attempts to fractionate soil polysaccharides; either by electrophoresis (Mortensen, 1960) or by ion exchange chromatography (Thomas et al., 1967; Finch et al., 1968). Thomas et al. (1967) reported that electrophoretic mobility could be related to uronic acid content of polysaccharide fractions. Mingelgrin and Dawson (1973) used an anion exchange resin to remove acidic polysaccharides, during isolation of a neutral polysaccharide fraction. The presence of adjacent reactive groups (e.g. COOH, OH, NH_2) will in many cases permit the formation of complexes with metal ions. Insoluble "polysaccharide—copper" complexes are widely used to fractionate polysaccharide mixtures (Jones and Stoodley, 1965). Such reactions also have significance for the behaviour of metals in soil, as do the numerous sites available for H-bonding.

ORIGIN OF SOIL POLYSACCHARIDES

Polysaccharides are clearly added to the soil system from a variety of sources. While the greatest inputs are probably made by plants, in the form of surface additions of leaves, twigs etc. and of roots and root exudates below ground, there is no doubt that micro-organisms and soil animals make a significant contribution on death. Greenland and Oades (1975) have described the diversity of carbohydrate materials that can be expected to be contributed to the soil from plant, microbial or animal sources. Cheshire and Anderson (1975) have recently provided a brief but excellent review of present knowledge of the origins of polysaccharides persisting in the soil.

In essence, no consensus has been reached with regard to the main source of soil polysaccharides. Indeed the different persistent soil fractions are likely to have diverse origins, including (a) residual plant components that have resisted (or escaped) decomposition, (b) direct microbial synthesis from non-carbohydrate substrates, and (c) microbial transformations of monomers released from plant sources during degradation.

Most of the evidence cited in support of a particular origin has been based on monomer composition. Unfortunately such data, indicating similarity or dissimilarity to composition of polysaccharides of known origin, is by no means conclusive, partly because of lack of information on the rates of decomposition of different types of material in the soil. For example, accumulation of a pentose-rich fraction in soil could equally well indicate an origin from a pentose-rich source, or a greater resistance to decomposition of pentose-containing components from a source that was relatively low in pentose.

The relative abundance of methylated sugars in soil polysaccharides has at times been attributed to microbial sources (Duff, 1961). However, similar sugars are known to occur in plant tissue (Bacon and Cheshire, 1971), and accordingly cannot be taken as unambiguous indicators of microbial origin. Similarly, the higher proportions of mannose commonly found in soil polysaccharides, as compared to plant materials, has also been taken to indicate a predominantly microbial origin (Parsons and Tinsley, 1961; Swincer et al., 1969). Again, however, there is evidence that mannose is both a widespread and quantitatively important constituent of some plant cell wall polysaccharides (Albersheim, 1965). Furthermore, mannose is a major polysaccharide constituent of the woody tissue of some conifers (Nikitin, 1966).

The cellulose in soil (Gupta and Sowden, 1964) may well be mainly of direct plant origin, and the predominance of glucose in light fractions (Oades, 1972; Whitehead et al., 1975), in which recognizable plant debris is abundant, is consistent with this view. However, the presence of cellulose in some fungal cell walls is also established (Gorin and Spencer, 1968). In incubation studies with ^{14}C-glucose, Cheshire et al. (1973) demonstrated that a small proportion of added glucose was incorporated in cellulose-like material requiring a 24 N H_2SO_4 hydrolysis to release the glucose. Thus, even in the case of cellulose, the assumption of direct plant origin may not be entirely justified.

In contrast, Cheshire et al. (1973), in incubation studies with ^{14}C-labelled rye straw, showed that plant polysaccharides can persist in soil for considerable periods of time. It was suggested that this might account for much of the xylose in soil organic matter, particularly since earlier studies with ^{14}C-glucose (Cheshire et al., 1969) showed little incorporation of label into pentoses of polysaccharides synthesized during the incubation. This does not of course exclude the possibility of pentose synthesis from other precursors. Amino sugars are known to be important constituents of cell wall material in micro-organisms, and of insect-derived chitin, but are rarely encountered as plant constituents. Bremner (1967) concluded that soil amino sugars are largely, if not entirely, derived from microbial sources. Since most studies of amino sugars in soils have been carried out on hydrolysates of soils rather than those of extracted polysaccharide fractions, little information is available on the association of amino sugars with other polysaccharide constituents.

In conclusion, it is probably safe to assume that the accumulation of polysaccharides in soil arises from a variety of sources, and by a variety of pathways. We are not yet in a position to adequately assess the relative importance of the various alternatives. It is also likely that differences in environmental conditions would lead to the predominance of different processes.

BEHAVIOUR AND SIGNIFICANCE OF SOIL CARBOHYDRATES

Behaviour

Whereas the two previous sections focussed on particular carbohydrate constituents of soils, namely the extractable polysaccharides on which direct chemical studies could be made, this section will attempt to discuss the behaviour and significance of all soil carbohydrates. However, much of this discussion must be to some extent speculative, partly because of the difficulties in studying dynamic features of the behaviour of soil components in an intact soil, and partly because reference must be made to carbohydrate constituents that have been neither successfully isolated nor fully characterized.

Discussion of such a complex topic can be approached in a number of ways. In this section the salient features of the chemical, physical and biological behaviour of carbohydrates will be summarized first. Then, the possible or probable significance of this behaviour will be discussed, under the somewhat arbitrary headings of (a) plant growth, (b) soil genesis and (c) environmental problems. These headings have been selected to emphasize the diversity of implications of soil carbohydrate behaviour.

The chemical behaviour of monosaccharides or polysaccharides is largely a function of their reactive groups, especially hydroxyl and (where uronic acids are present) carboxyl. The amino groups of hexosamines may also be significant. In polysaccharides especially, the abundance of such groups and the linear configuration provide ample opportunity for interactions with metals and with inorganic colloids. In the case of metals, both salt formation and complex formation are possible. The most important feature of such interactions is that they may modify the behaviour of both the polysaccharide and of the inorganic moiety. The magnitude of polysaccharide contributions to cation exchange capacity is unknown, but must be related to the number of uronic acids present, and will be pH-dependent. Interactions with metal ions can result in changes in the solubility and rate of decomposition of polysaccharides added to soil (Martin et al., 1966). Interactions with clays or hydrous oxides also arise from the opportunities for multiple H-bonding, as well as charges arising from carboxyl dissociation. Adsorption of monosaccharides and of soil polysaccharides on clay minerals has been demonstrated (Greenland, 1956; Finch et al., 1966; Parfitt and Greenland, 1970). Polysaccharide adsorption is markedly influenced by type of clay, pH and exchangeable cations (Finch et al., 1966; Parfitt and Greenland, 1970). Adsorption of polysaccharides in acid clays may involve complex formation through exchangeable aluminium. Multiple bonding of a polysaccharide molecule to more than one mineral particle would influence aggregation and hence a range of physical properties of soil. Polysaccharides are generally strongly hydrophilic, and will contribute to water retention in soils. Swelling of soil polysaccharides under wet conditions may possibly contribute to pore-clogging. Free sugars and polysaccharides have long been known to form stable complexes with borate ions, and this property has been widely used in analytical separations. It is possible that such reactions may be of significance in modifying boron availability to plants in soils of high organic matter content. There is as yet, however, no direct evidence for this.

In the context of biological aspects of soil behaviour, the most important feature of soil carbohydrates is their role as substrate for a wide range of soil organisms, although clearly any effects on soil structure and aeration will also have a marked indirect influence on microbial activity. In this regard, the resistance to decomposition will be of particular importance to persistence of polysaccharide materials in soil and to levels of microbial activity.

While simple monosaccharides and most plant and microbially derived polysaccharides, including cellulose, are known to be degraded in soils relatively rapidly under favourable conditions, nevertheless considerable amounts of polysaccharides do accumulate in soils. Martin (1971) has listed the following alternative explanations.

(1) The soil polysaccharide consists of materials of microbial and plant origin which are resistant to decomposition.

(2) Certain plant or microbial products, or their partial degradation products, form salts or complexes with metal ions, clays or humic fractions which are resistant to decomposition.

(3) New complex polysaccharide molecules, peculiar to the soil environment, are produced in soil, and which are resistant to decomposition.

(4) A combination of some or all of the above.

An additional contributing factor suggested by Cheshire et al. (1974) should perhaps be added, namely the possibility that persistance of some materials may be partly a function of inaccessibility without necessarily involving a biologically stable structure. It is also possible that the resistance to decomposition may in some cases be more apparent than real, with degradation to a large extent counterbalanced by synthesis of new microbial polysaccharides.

Martin (1971) has reviewed the factors influencing rates of decomposition of polysaccharides in soil, and stressed the importance of complex formation with metals and with soil colloids in retarding the normally relatively rapid rates of decomposition. For example, complexing with Cu, Fe and Zn has been shown to result in marked reduction in the decomposition of plant and microbial polysaccharides in incubation studies (Martin et al., 1966). Similarly adsorption onto clay, especially montmorillonite, may also reduce decomposition rates (Lynch et al., 1956). There is evidence indicating that some soil polysaccharides are bound through ester linkages to humic fractions (Clark and Tan, 1969; Tan and McCreery, 1970). Such an association may also confer increased resistance to decomposition. Decomposition of cellulose and other polysaccharides can be inhibited in the presence of tannin (Benoit and Starkey, 1968), either as a result of formation of a resistant polysaccharide—tannin complex, or alternatively as a result of inactivation of exocellular enzymes by the tannin. Clearly, a variety of factors, in addition to the normal effects of pH, temperature, moisture and aeration, will influence the decomposition of polysaccharides, making predictions of polysaccharide behaviour, in a particular soil system, very difficult.

Significance in relation to plant growth

Despite the uncertainties of the origin and turnover rates of soil carbohydrates, it is inevitable that such a major, and relatively reactive, component of soil organic matter must have considerable influence on soil as a medium for plant growth. A significant role can be expected in relation to physical properties of soil, nutrient supply to plants, and as a major determinant of both qualitative and quantitative aspects of microbial activity. However, having asserted the overall importance of these materials, it is still not possible to state with any confidence, which effects are peculiar to carbohydrates as compared with other organic constituents, and which particular carbohydrate fractions have the major effect in any given role. Indeed it can be safely assumed that different fractions will vary considerably in their contributions. However, little success has been achieved in defining biologically meaningful fractions of soil organic matter, and soil carbohydrate fractions will vary in their role in relation to chemical composition, environmental factors and location within the soil.

Soil structure, and in particular aggregate stability, is recognized as having considerable influence on plant growth, through effects on such features as aeration, infiltration and drainage characterstics, and seedling emergence. It has long been known that permanent pasture, or the addition of organic residues to soil, results in improved structure involving the formation of aggregates with some degree of stability (Martin et al., 1955). A major stimulus to the study of soil polysaccharides has been the accumulation of evidence suggesting that polysaccharides may play a major role in the development of stable aggregates. The initial evidence took two forms. First, the addition to soil of polysaccharides derived from plants and soil micro-organisms has been shown to improve aggregation (Geoghegan et al., 1948; Clapp et al., 1962; Martin and Richards, 1963, 1968). Concentrations of microbial polysaccharides in the 0.02—0.2% range exerted a marked binding action on soil particles. Secondly, statistical correlations have been reported between soil polysaccharide levels and degree of aggregation (Toogood et al., 1959; Greenland et al., 1962; Acton et al., 1963; Webber, 1965). It has also been demonstrated that the binding action of added polysaccharides can be influenced by such soil properties as pH and the composition of exchangeable cations, as well as by uronic acid content of the polysaccharide (Martin and Aldrich, 1955). However, none of this evidence establishes conclusively that polysacchrides are in fact responsible for aggregation. Other workers have attempted to obtain more direct evidence that polysaccharides are involved in aggregation, by treating stable aggregates with selective chemicals expected to degrade polysaccharides rather than other possible binding agents, and observing any loss in aggregate stability. Periodate treatment has in some cases been shown to produce a marked decrease in aggregation (Greenland et al., 1962; Clapp and Emerson, 1965 a,b; Stefanson, 1971). However, some studies (Mehta et al., 1960) have shown that artificial aggregates, prepared with extracted soil polysaccharides, were more affected by periodate than were natural aggregates, indicating that some binding agents other than polysaccharides played a major role in stabilizing natural soil aggregates. In studies of red-brown earths in Australia, Stefanson (1968, 1971) found that aggregate stability increased with duration of the pasture phase in a rotation, and showed seasonal variation with a maximum in summer. However, these variations could not be explained by variations in "periodate sensitive" materials (presumably polysaccharides). In general, present evidence leads to the conclusion that polysaccharides make a significant contribution to aggregate stability in some soils, particularly cultivated soils of relatively low organic matter Greenland et al., 1962), but that in many soils the role of other binding agents, both organic and inorganic, is at least as important, if not more so. It is possible that polysaccharides are of particular importance in initiating aggregation, but that longer-term stability is more a result of subsequent introduction of other binding agents. This suggestion gains some support from the report of Griffiths and Burns (1972) that the stability of aggregates produced by microbial polysaccharides was markedly enhanced by subsequent treatment with tannic acid or the products of decomposing herbage. Similarly in field studies, tannic acid was found to increase the longevity of stable aggregates in glucose-treated plots.

In addition to effects on physical properties of soil, carbohydrates probably also have a variety of more or less indirect effects on nutrient supply to plants. These effects are complex and have not been thoroughly investigated. Accordingly, the main types of effects that must be considered will be outlined, but without detailed discussion.

Supply of macro-nutrient cations will be influenced by polysaccharide contributions to cation exchange capacity. However, interactions with micro-nutrients and with anions may well be of greater significance. Formation of soluble complexes with micro-nutrient ions may enhance the availability of the latter, in pH ranges in which they would normally precipitate. Conversely, formation of insoluble complexes may reduce availability. It has been suggested that microbial release of simple carbohydrate molecules with chelating properties can accelerate mineral weathering, with consequent release of nutrients (Duff and Webley, 1959). Several effects of carbohydrates on phosphate behaviour have also been indicated. Simple sugars can prevent the precipitation of phosphate by iron and aluminium, depending on pH levels (Bradley and Sieling, 1953), and coating of sesquioxides or clays by polysaccharides can be expected to decrease phosphate fixation. The possibility of an interaction with borate was referred to earlier. The presence of readily metabolizable carbohydrates in soil is likely also to have a profound effect on nitrogen transformations, and under conditions favouring microbial activity, may contribute to increased microbial immobilization of nitrogen, and in some situations, stimulation of denitrification. Alternatively, by stimulating a rapid growth in microbial populations, such materials may lead to accelerated breakdown of more stable humus fractions, leading to eventual increase in nitrogen mineralization. Free monosaccharides, and polysaccharides are presumably at least partly responsible for the rhizosphere effect, i.e., the distinctive population and microbial activity in the region close to plant roots. This region receives inputs of carbohydrates in the form of sloughed off root cell debris, as well as root exudates. The latter are known to contain a variety of monosaccharides and oligosaccharides (Rovira and McDougall, 1967). Webley et al. (1965) have suggested that grasses improve soil structure partly through the action of polysaccharides produced by rhizoplane inhabitants, since bacterial extracellular polysaccharides were more abundant in the rhizoplane than in soil away from the roots.

In conclusion, there is little doubt that carbohydrates in soil have an influence on physical properties, and are also likely to be involved in a wide variety of biochemical processes which influence the soil as a medium for plant growth.

Significance in relation to soil genesis

The role of organic materials in soil formation is not well understood, although a variety of theories have been developed over the years. At present there is very little detailed information on the composition and behaviour of organic fractions on soils that have been thoroughly characterized with respect to inorganic constituents and dominant pedogenetic processes. Thus, few firm conclusions can be drawn with regard to the role of carbohydrates.

Under natural conditions, soils are characterized by continual addition, degradation and synthesis of carbohydrates, mainly as polysaccharides. While carbohydrates are clearly involved in humus formation, in the sense that degradation and synthesis of polysaccharides, through microbial activity, are coincident with humus formation, the nature of this involvement remains uncertain (Kononova, 1961; Martin and Haider, 1971). Several possibilities have been suggested. Polysaccharides are degraded by soil micro-organisms, and may serve only as an energy source; or they may provide simple monomers that are metabolically transformed into complex polymers that contribute to humus. Alternatively, the monomeric degradation products may undergo condensation, oxidation and polymerization, together with amino acid, and phenolic degradation products, to form humic polymers.

Some points seem well established. Some micro-organisms are capable of synthesizing aromatic materials from glucose (Sprinson, 1960), and studies with ^{14}C-labelled glucose, hemicellulose and cellulose have shown incorporation of labelled carbon into humic acid, fulvic acid and humin fractions (Mayaudon and Simonart, 1958—1959). Later studies with labelled glucose and cellulose (Sorensen, 1967, 1972) have also demonstrated the incorporation of carbohydrate-C into amino acids and amino sugars. In one case, 26—30% of the labelled carbon (added as cellulose) remaining in the soil after six years, was present in amino acids. Studies on decomposing plant tissue (Hayes et al., 1968) in the absence of soil indicate that synthesis of new polysaccharides is greatest during early stages of humification, with the development of polyphenolic materials being prominent at a later stage. Other studies on the incorporation of ^{14}C-labelled glucose (Cheshire et al., 1969, 1971) suggest initial transfer of label to other sugars, especially hexoses followed by subsequent appearance of the label in nitrogenous constituents. Quite apart from the information gained in studies of the involvement of carbohydrates in the processes involved in humus formation, the importance of polysaccharides in humus-rich soil horizons is indicated by their presence in all fulvic acid fractions examined for these materials (Forsyth, 1947; Dormaar, 1967; Sequi et al., 1975) as well as their apparent association with polyphenolic materials in the hymatomelanic acid fraction (Clark and Tan, 1969). A relative abundance of polysaccharides in surface horizons (Ah) formed under grassland, may be partly responsible for the characteristically good aggregation observed in such horizons.

Another important feature of the genesis of many soils is the downward translocation of iron and aluminium, or of clays. There are indications that carbohydrates may sometimes have a significant role here also. While simple organic acids and phenolic materials appear to be actively involved in the translocation of iron (Bruckert, 1970; Ellis, 1971), Schnitzer and DeLong (1955) attributed the Fe-complexing and -mobilizing power of poplar leaf extracts to acidic polysaccharides. In contrast Dormaar (1970) found no correlation between iron-dissolving power of extracts from two species of poplar with either polysaccharide or polyphenol content of extracts. He concluded that polysaccharides were not involved in iron mobilization in the soil under investigation. Possibly the significance of polysaccharides in leaf extracts or canopy drip, lies mainly in their production of organic acids on decomposition under wet conditions. A possible role of polysaccharides of leaf extracts in deflocculation and translocation of clays has been suggested by Bloomfield (1956). Polysaccharides appeared to be one of several extract-components capable of causing deflocculation. Such effects might be of importance in the development of luvisols.

A final aspect of carbohydrates in relation to soil genesis and classification that warrants discussion, is the question of how far a knowledge of carbohydrate distribution and composition can be useful in elucidating processes of soil formation, or in defining diagnostic horizons for classification purposes.

While the simple answer is that such an approach currently has little immediate value, because of the lack of adequate detailed information, there are indications that it may justify much more thorough investigation. Recent work by Guidi et al. (1976), Folsom et al. (1974), Oades (1972) and others has indicated considerable quantitative and qualitative differences between carbohydrate fractions in different soils. For example, Dormaar (1967) reported differences in polysaccharide distribution in chernozemic soils that were related to parent material differences, and Folsom et al. (1974) observed variations in relative proportions of hexoses, pentoses and deoxysugars that were related to parent material and to native vegetation. Similarly Guidi et al. (1976) observed considerable variation in hexose : pentose ratios of fulvic fractions for a varied group of soils, with the widest ratio associated with an Ando soil. They also reported variations in individual monosaccharide ratios (e.g., arabinose : xylose) for subfractions of the fulvic acid of the same soil. For a peat soil, Theander (1954) has shown that increased humification is associated with decreasing glucose, xylose and uronic acids, but increasing mannose levels. Bernier (1958) in a comparison of mull and more humus forms, reported marked differences in the mean molecular weights of extractable polysaccharides. It is suggested that relationships between characteristics of soil carbohydrate fractions and both soil morphology and site characteristics may prove to be a fruitful area of study, especially if the more elaborate fractionation methods are employed, along the lines of those developed in recent studies (Oades, 1972; Lowe, 1975; Guidi et al., 1976) and examining polysaccharide distribution in relation to that of nitrogenous and polyphenolic constituents.

Significance in relation to environmental problems.

The soil scientist of today gets increasingly involved (and properly so) in a variety of environmental problems. This seems sufficient justification for drawing attention to a few selected examples, in which knowledge of polysaccharide behaviour in soil may be important.

The retention of metal pollutants entering soil from atmospheric sources or from sewage sludge applications may be influenced by levels and types of polysaccharides present. The degradation of organic pesticides in soil will likely be related to microbial activity, which in turn responds to the level of readily decomposable substrates like carbohydrates. Similarly the rate of degradation of petroleum products spilled on soil, should be accelerated by the presence, or addition, of ample carbohydrate-rich materials, so long as other requirements for sustaining high levels of microbial activity (especially aeration, pH and nutrient supply) are also maintained.

Heavy applications of animal wastes (e.g., cow slurry), which are rich in readily decomposable carbohydrate material, can give rise to a period of highly anaerobic conditions, because free oxygen is consumed by microorganisms faster than it can diffuse into the surface soil layers. Similar effects can arise from temporary flooding of soils rich in carbohydrate materials. Such conditions can give rise to abnormal levels of organic acids, toxic levels of manganese and other problems. A related problem can arise in the use of soil for renovation of sewage effluent wastewater. If sufficient time is not allowed between effluent applications for aerobic conditions to be restored in the surface soil, pore clogging results (De Vries, 1972). This effect has been attributed to the accumulation of bacterial polysaccharides, which leads to a drastic reduction in permeability.

In conclusion carbohydrates are a major component of the organic matter of soils, both in the natural state and as modified by man's activities. They are important in relation to physical, chemical and biological aspects of soil behaviour. However, much still remains to be learned about their origin, composition and behaviour.

REFERENCES

Acton, C.J., Paul, E.A. and Rennie, D.A., 1963. Can. J. Soil Sci., 43: 141—150.
Albersheim, P., 1965. In: J. Bonner and J.E. Varner, (Editors), Plant Biochemistry, Academic Press, New York, N.Y., pp. 298—319.
Alvsaker, E. and Michelsen, K., 1957. Acta Chem. Scand., 11: 1794—1795.
Bacon, J.S.D. and Cheshire, M.V., 1971. Biochem. J., 124: 555—562.
Benoit, R.E. and Starkey, R.L., 1968. Soil Sci., 105: 291—297.
Bernier, B., 1958. Biochem. J., 70: 590—598.
Bloomfield, C., 1956. Trans. Int. Congr. Soil Sci., 6th, Paris, B: 27—32.
Bouhours, J.-F. and Cheshire, M.V., 1969. Soil Biol. Biochem., 1: 185—190.
Bradley, D.B. and Sieling, D.H., 1953. Soil Sci., 76: 175—179.
Bremner, J.M., 1967. Nitrogenous compounds. In: A.D. McLaren and G.H. Peterson (Editors), Soil Biochemistry. Dekker, New York, N.Y. pp. 19—66.
Brink, R.H., Dubach, P. and Lynch, D.L., 1960. Soil Sci., 89: 157—166.
Bruckert, S., 1970. Ann. Agron., 21: 421—452.
Butler, J.H.A. and Ladd, J.N., 1971. Soil Biol. Biochem., 3: 249—257.
Cheshire, M.V. and Anderson, G., 1975. Soil Sci., 119: 356—362.
Cheshire, M.V., Greaves, M.P. and Mundie, C.M., 1974. J. Soil. Sci., 25: 483—498.
Cheshire, M.V., Mundie, C.M. and Shepherd, H., 1969. Soil Biol. Biochem., 1: 117—130.
Cheshire, M.V., Mundie, C.M. and Shepherd, H., 1971. J. Soil Sci., 22: 222—236.
Cheshire, M.V., Mundie, C.M. and Shepherd, H., 1973. J. Soil Sci., 24: 54—68.
Cheshire, M.V., Mundie, C.M. and Shepherd, H., 1974. J. Soil Sci., 25: 90—98.
Clapp, C.E., Davis, R.J. and Waugaman, S.H., 1962. Soil Sci. Soc. Am. Proc., 26: 466—469.
Clapp, C.E. and Emerson, W.W., 1965a. Soil Sci. Soc. Am. Proc., 29: 127—130.
Clapp, C.E. and Emerson, W.W., 1965b. Soil Sci. Soc. Am. Proc., 29: 130—134.
Clark, F.E. and Tan, K.H., 1969. Soil Biol. Biochem., 1: 75—81.
De Vries, J., 1972. J. Water Pollut. Control, 44: 565—573.
Dormaar, J.F., 1967. Soil Sci., 103: 417—423.
Dormaar, J.F., 1970. J. Soil Sci., 21: 105—110.
Duff, R.B., 1961. J. Sci. Food Agric., 12: 826—831.

Duff, R.B. and Webley, D.M., 1959. Chem. Ind. (Lond.), 1376—1377.

Ellis, R.C., 1971. J. Soil Sci., 22: 8—22.

Finch, P., Hayes, M.H. and Stacey, M., 1968. Trans. Int. Congr. Soil Sci., 9th, 3: 193—201.

Finch, P., Hayes, M.H.B. and Stacey, M., 1966. Trans. Comm. II and IV, Int. Soc. Soil Sci., 19—32.

Folsom, B.L., Wagner, G.H. and Scrivner, C.L., 1974. Soil Sci. Soc. Am. Proc., 38: 305—309.

Forsyth, W.G.C., 1947. Biochem. J., 41: 176—181.

Geoghegan, M.J. and Brian, R.C., 1948. Biochem. J. (Lond.), 43: 5—13.

Gorin, P.A.J. and Spencer, J.F.T., 1968. Adv. Carbohyd. Chem., 23: 367—417.

Graveland, D.N. and Lynch, D.L., 1961. Soil Sci., 91: 162—165.

Greenland, D.J., 1956. J. Soil Sci., 7: 329—334.

Greenland, D.J., Lindstrom, G.R. and Quirk, J.P., 1962. Soil Sci. Soc. Am. Proc., 26: 366—371.

Greenland, D.J. and Oades, J.M., 1975. In: J.E. Gieseking (Editor), Soil Components, 1, Organic Components. Springer-Verlag, New York, N.Y., pp. 213—261.

Griffith, S.M. and Schnitzer, M., 1975. Soil Sci. Soc. Am. Proc., 39: 861—867.

Griffiths, E. and Burns, R.G., 1972. Plant Soil, 36: 599—612.

Grov, A., 1963. Acta Chem. Scand., 17: 2301—2306.

Guidi, G., Petruzzelli, G. and Sequi, P., 1976. Can. J. Soil. Sci., 56: 159—166.

Gupta, V.C. and Sowden, F.J., 1963. Soil Sci., 96: 217—218.

Gupta, V.C. and Sowden, F.J., 1964. Soil Sci., 97: 328—333.

Gupta, V.C., Sowden, F.J. and Stobbe, P.C., 1963. Soil Sci. Soc. Am. Proc., 27: 380—382.

Hayes, M.H.B., Stacey, M. and Standley, J., 1968. Trans Int. Congr. Soil Sci., 9th, 2: 247—255.

Ivarson, K.C. and Sowden, F.J., 1962. Soil Sci., 94: 245—250.

Jones, J.K.N. and Stoodley, R.J., 1965. In: R.L. Whistler (Editor), Methods in Carbohydrate Chemistry, 5. Academic Press, New York, N.Y., pp. 36—38.

Klinka, K. and Lowe, L.E., 1975. Brit. Columbia For. Service Res. Note, 74, Vict., B.C., 16 pp.

Kononova, M.M., 1961. Soil Organic Matter. Pergamon Press, Oxford, pp. 118—163.

Lowe, L.E., 1968. Can. J. Soil Sci., 48: 215—217.

Lowe, L.E., 1969. Can. J. Soil Sci., 49: 129—141.

Lowe, L.E., 1974. Can. J. For. Res., 4: 446—454.

Lowe, L.E., 1975. Can. J. Soil Sci., 55: 119—126.

Lucas, A.J., 1970. Geochemistry of Carbohydrates in Some Organic Sediments of the Florida Everglades. Ph.D. Thesis, Penn. State University.

Lynch, D.L. and Cotnoir, L.J., 1956. Soil Sci. Soc. Am. Proc., 20: 367—370.

Martin, J.P., 1945. Soil Sci., 59: 163—174.

Martin, J.P., 1946. Soil Sci., 61: 157—166.

Martin, J.P., 1971. Soil Biol. Biochem., 3: 33—41.

Martin, J.P. and Aldrich, D.G., 1955. Soil Sci. Soc. Am. Proc., 19: 50—54.

Martin, J.P., Ervin, J.O. and Shepherd, R.A., 1966. Soil Sci. Soc. Am. Proc., 30: 196—200.

Martin, J.P. and Haider, K., 1971. Soil Sci., 111: 54—63.

Martin, J.P., Martin, W.P., Page, J.B., Raney, W.A. and De Ment, J.D., 1955. Adv. Agron., 7: 1—37.

Martin, J.P. and Richards, S.J., 1963. J. Bacteriol., 85: 1288—1294.

Martin, J.P. and Richards, S.J., 1969. Soil Sci. Soc. Am. Proc., 33: 421—423.

Mayaudon, J. and Simonart, P., 1958—1959. Plant Soil, 9: 375—380; Plant Soil, 11: 170—175; Plant Soil, 11: 181—192.

Mehta, N.C., Streuli, H., Müller, M. and Deuel, H., 1960. J. Sci. Food Agric., 11: 40—47.

Mingelgrin, V. and Dawson, J.E., 1973. Soil Sci., 116: 36—43.

Mortensen, J.L., 1960. Trans. Int. Congr. Soil Sci., 7th, 2: 98—104.

Nagar, B.R., 1962. Nature. 194: 896—897.

Nikitin, N.I., 1966. The Chemistry of Cellulose and Wood. Israel Program for Scientific Translations, Jerusalem.

Oades, J.M., 1967. Aust. J. Soil Res., 5: 103—115.

Oades, J.M. and Swincer, G.D. 1968. Trans. Int. Congr. Soil Sci., 9th, 3rd, 183—192.

Oades, J.M., 1972. Aust. J. Soil Res., 10: 113—126.

Ogston, A.G., 1958. Biochem. J., 70: 598—599.

Parfitt, R.L. and Greenland, D.J., 1970. Soil Sci. Soc. Am. Proc., 34: 862—865.

Parsons, J.W. and Tinsley, J., 1961. Soil Sci., 92: 46—53.

Parsons, J.W. and Tinsley, J., 1975. In: J.E. Gieseking (Editor), Soil Components, I. Organic Components. Springer-Verlag, New York, N.Y., pp. 263—304.

Pigman, W. and Horton, D., 1972. The Carbohydrates Chemistry and Biochemistry, 2nd ed. Academic Press, New York, N.Y., 2 vols.

Remezov, N.P. and Pogrebnyak, P.S., 1965. Forest Soil Science. Translated from Russian, Israel Program for Scientific Translations, Jerusalem, 1969.

Rovira, A.D. and McDougall, B.M., 1967. In: A.D. McLaren and G.H. Peterson (Editors), Soil Biochemistry, Dekker, New York, N.Y., pp. 417—463.

Saini, G.R. and Salonius, P.O., 1969. Soil Sci. Soc. Am. Proc., 33: 693—695.

Schnitzer, M. and DeLong, W.A., 1955. Soil Sci. Soc. Am. Proc., 19: 363—368.

Sequi, P., Guidi, G. and Petruzzelli, G., 1975. Can. J. Soil Sci., 55: 439—445.

Sorensen, L.H., 1967. Soil Sci., 104: 234—241.

Sorensen, L.H., 1972. Soil Biol. Biochem., 4: 245—255.

Sprinson, D.B., 1960. Adv. Carbohyd. Chem., 15: 235—269.

Sowden, F.J. and Ivarson, K.C., 1962. Plant Soil, 16: 389—400.

Sowden, F.J., 1959. Soil Sci., 88: 138—143.

Stefanson, R.C., 1968. Trans. Congr. Int. Soil Sci. Soc., 9th, Adelaide, 2: 395—402.

Stefanson, R.C., 1971. Aust. J. Soil Res., 9: 33—41.

Stevenson, F.J. and Braids, O.C., 1968. Soil Sci. Soc. Am. Proc., 32: 598—600.

Stevenson, F.J., 1965. In: C.A. Black (Editor), Methods of Soil Analysis, 2. Am. Soc. Agron., Madison, Wisc.

Swincer, G.D., Oades, J.M. and Greenland, D.J., 1969. Adv. Agron., 21: 195—235.

Swincer, G.D., Oades, J.M. and Greenland, D.J., 1968a. Aust. J. Soil. Res., 6: 211—224.

Swincer, G.D., Oades, J.M. and Greenland, D.J., 1968b. Aust. J. Soil Res., 6: 225—235.

Tan, K.H. and McCreery, R.A., 1970. Commun. Soil Sci. Pl. Analysis, 1: 75—84.

Theander, O., 1954. Acta Chem. Scand., 8: 989—1000.

Thomas, R.L., Mortensen, J.L. and Himes, F.L., 1967. Soil Sci. Soc. Am. Proc., 31: 568—570.

Toogood, J.A. and Lynch, D.L., 1959. Can. J. Soil Sci., 39: 151—156.

Waksman, S.A., 1938. Humus: Origin, Chemical Composition and Importance in Nature, 2nd ed. Williams and Wilkins, Baltimore.

Waksman, S.A., Tenney, F.G. and Stevens, K.R., 1928. Ecology, 9: 126—144.

Webber, L.R., 1965. Soil Sci. Soc. Am. Proc., 29: 39—42.

Webley, D.M., Duff, R.B., Bacon, J.S.D. and Farmer, V.C., 1965. J. Soil Sci., 16: 149—157.

Whitehead, D.C., Buchan, H. and Hartley, R.D., 1975. Soil Biol. Biochem., 7: 65—71.

Whistler, R.L. and Kirby, 1956. J. Am. Chem. Soc., 78: 1755—1759.

Whistler, R.L. and Smart, C.L., 1953. Polysaccharide Chemistry. Academic Press, New York, N.Y.

Chapter 3

ORGANIC NITROGEN, PHOSPHORUS AND SULFUR IN SOILS

C.G. KOWALENKO

INTRODUCTION

N, P and S are three important nutrient elements for plant growth. Soil organic matter plays an important role in the soil–plant system, and organic constituents containing these elements plus their chemical, physical and biochemical reactions are critical to plant growth. Reviews are available on each of these elements in soil, but most usually consider each element as a separate entity. It is not uncommon to have two or more of these elements in the same compound (e.g., amino acids, phospholipids, nucleic acids), so that considerable qualitative information can be gained by an integrated orientation. This chapter will attempt, as far as available data permit, to present information using a unified approach. Emphasis will be placed on both quantitative and qualitative information. The importance of a constituent is often more related to its function than to its relative quantity.

SOME CHARACTERISTICS OF SOURCE MATERIALS

During the development and maintenance of soils there is usually addition of organic matter. Sources of the organic matter are plant, animal and microbial materials. Amounts and types of materials that are added vary greatly, depending on the situation. A forest system differs from a grassland system and a grassland system is different from a cultivated system (Bartholomew et al., 1953). Table I illustrates the range of concentrations of N, P and S in the aerial parts of cereals. Concentrations of elements vary with plant part as well as environmental factors during growth. The return of organic matter depends on the situation; for example, in cereal production the straw is often returned to the soil while the grain is removed. Roots of plants also contribute to the organic N, P and S in the soil. Exact amounts vary with a variety of factors. There is a range of concentrations of these elements in plant roots (Table II). The contribution of root materials depends on whether the plant is a perennial or an annual. With perennials there is a turnover of dying plant root material through the years (Wilkinson, 1973), whereas with annuals, the entire root system is available for decomposition after harvest.

The contribution of animal wastes can be significant under certain condi-

TABLE I

N, P and S of selected above ground plant parts
(Adapted from Miller, 1958)

Plant material		% N	% P	% S
Barley:	grain	2.1	0.47	0.19
	straw	0.7	0.09	0.19
Corn:	grain	1.7	0.32	0.12
	straw	1.0	0.09	0.17
Oats:	grain	2.1	0.39	0.23
	straw	0.7	0.10	0.24
Rice:	grain	1.4	0.26	0.05
	straw	0.7	0.09	—
Sorghum:	grain	2.0	0.35	0.18
	straw	0.8	0.11	—
Wheat:	grain	2.3	0.41	0.19
	straw	0.6	0.08	0.19

tions, and types and amounts of additions vary with management practices. In grazing systems, animal wastes are not spread uniformly on soils and types and amounts of N, P and S constituents added vary between the faeces and urine (Barrow and Lambourne, 1962), among other factors (Richards and Wolton, 1976). In other systems, such as feed lots, animal wastes are often spread as a fertilizer and the concentrations of N, P and S added vary (Table III). Human wastes (sewage) are applied to soils as disposal alternatives, and these materials can contribute to the soil organic matter. Sewage sludges are complex materials (Peterson et al., 1971) because industrial sources introduce a great variety of widely differing substances.

TABLE II

N, P and S concentrations in some root materials

Plant	% N	% P	% S	Reference
White clover	3.77	0.22	0.39	Whitehead, 1970
Red clover	2.79	0.26	0.47	Whitehead, 1970
Lucerne	2.47	0.26	0.24	Whitehead, 1970
Perennial grass	1.75	0.13	0.21	Whitehead, 1970
Cocksfoot	1.54	0.14	0.20	Whitehead, 1970
Timothy	1.47	0.14	0.19	Whitehead, 1970
Prairie grassland:				
roots	0.90	0.09	—	Coupland, 1974
rhizomes	—	0.15	—	Coupland, 1974
shoot bases	—	0.12	—	Coupland, 1974
Wheat (maturiry)	1.01	0.097	—	Campbell et al., 1976;
		(0.059 as organic)		Kaila, 1949

TABLE III

N, P and S concentrations in manures

Manure type	% N total [2]	% N organic	% P total	% P organic	% S total	Reference
Cattle:						
general [1]	0.40	0.21	—	—	—	Byrne and Power, 1974
general	—	—	0.455	0.221	—	Kaila, 1949
dairy	0.71	—	0.12	—	0.06	Peterson et al., 1971
dairy (slurry) [1]	0.22	0.11	0.056	—	—	Phillips et al., 1975
fattening	0.88	—	0.25	—	0.10	Peterson et al., 1971
Hogs:						
general	0.67	—	0.19	—	0.18	Peterson et al., 1971
dung	0.44	0.20	—	—	—	Byrne and Power, 1974
Horse:						
general	1.15	—	0.17	—	0.12	Peterson et al., 1971
dung	—	—	0.485	0.175	—	Kaila, 1949
Poultry:						
fresh droppings	6.05	—	—	—	—	Giddens and Rao, 1975
litter	3.37	—	—	—	—	Giddens and Rao, 1975
broiler	6.80	—	3.24	—	—	Peterson et al., 1971
hen	3.51	—	3.24	—	—	Peterson et al., 1971
general	0.51	0.26	—	—	—	Byrne and Power, 1974
Sheep:						
general	3.51	—	3.24	—	—	Peterson et al., 1971

[1] Wet basis. All others on dry basis.
[2] Most of the inorganic N was NH_4-N.

Not all of the materials discussed above are totally organic. In most cases there is some inorganic N, P and S associated in a more or less equilibrium state with the organics. This equilibrium varies with many conditions, and information on this equilibrium is sometimes used for diagnosing the nutrient status of crops (Chapman, 1966; Ensminger and Freney, 1966) or the suitability of plant tissues as feeds (Wright and Davison, 1964) etc. However, once the organic constituents are added to the soil, microorganisms transform these constituents and the result is a new equilibrium between organic and inorganic forms by mineralization and immobilization processes. Microorganisms (and larger fauna) in the soil are also sources of organic N, P and S. Concentrations of the elements in question in microorganisms are illustrated in Table IV.

Although a very great variety of types and quantities of N, P and S containing organic constituents reach the soil, one can simplify the picture by looking at the major compounds containing these elements in plant and

TABLE IV

N, P and S concentrations in microbial material *[1]

Microbial material	% N	% P	% S	Reference
Fungi:				
Aspergillus niger	—	0.24—0.40	—	Kaila, 1949
12 cultures	2.37—4.03 (3.24) *[2]	—	—	Pinck and Allison, 1944
6 cultures	3.53—5.01 (4.50) *[2]	—	—	Jensen, 1932
material	—	0.485 *[3]	—	Kaila, 1949
mycelia	—	0.5—1.0	—	Wilkinson, 1973; Alexander, 1961
Streptomyces	—	0.27—0.63	—	Kaila, 1949
Actinomyces griseus	—	0.24—0.40	—	Kaila, 1949
Bacteria:				
all	—	—	0.04—80.0	Alexander, 1961; Zobell, 1963
excluding S bacteria	—	—	0.04— 8.9	Alexander, 1961; Zobell, 1963
most bacteria	—	—	0.1—1.0	Alexander, 1961; Zobell, 1963
2 soil	5.51—6.69	—	—	Jensen, 1932
soil	—	1.5—2.5	—	Wilkinson, 1973; Alexander, 1961
rumen	—	5.0	—	Wilkinson, 1973
Mixed soil fauna	—	0.16—0.36	—	Kaila, 1949

*[1] See Spector (1956) for some analysis of individual plant fungi and bacteria.
*[2] Average.
*[3] 47% was organic.

TABLE V

Major N, P and S containing compounds in plant and microbial tissue

N	P	S
Amino acids and protein	Phosphoprotein	Amino acids and proteins
Sugar amines	Sugar phosphates	
Nucleic acids	Nucleic acids	
Phospholipids	Phospholipids	Sulfolipids
Vitamins	Vitamins	Vitamins
Teichoic acids	Teichoic acids	

microbial tissues (Table V). Numerous other compounds also occur (Miller, 1957; Freney, 1967), but these are quantitatively of a minor nature. Proportions of these compounds change with the type of plant and microbe and should be considered in the context of their functions within the organism. Functions may include structure where protein, teichoic acid or chitin may play dominant roles or roles in storage, metabolic or transfer functions. Straws of cereals contain significantly lower concentrations of N and P, because stems contain higher proportions of structural material (cellulose), which effectively dilutes out the N and P content (Table I). Seeds, however, have storage functions, so that their amino acid, phosphoprotein and nucleic acid contents are higher. Similarities in S concentrations in grain and straw may be due to lower mobility of this element in the plant as well as to similarities in functions. Microorganisms that do not have elaborate structures tend to contain higher concentrations of N and P, because there is a greater proportion of compounds having metabolic rather than structural functions.

Although there appear to be similarities between the three nutrient elements in living tissue (Table V), there are some distinct differences also. Uptake of N, P and S by plants and microorganisms is most often in the form of the respective anions, NO_3^-, PO_4^{2-} and SO_4^{2-}. Phosphate does not go through an extensive chemical reduction prior to metabolic incorporation but rather energy is conserved and transferred as phosphate bond energy. Both NO_3^- and SO_4^{2-} undergo 8-electron reductions prior to incorporation into organic compounds (Bandurski, 1965). Organic compounds containing nitrate are not normally present in plant or microbial tissues; organic N is present in the —3 redox state. S, on the other hand, can be found in organic combinations in several redox states, but the most dominant form is the —2 state (lowest reduction state) such as in sulfur amino acids. Hence, oxidation-reduction reactions are important in the biochemistry of N and S, whereas the high energy of the phosphate bond is important in phosphorus biochemistry of living organisms.

TOTAL ORGANIC N, P AND S IN SOILS

Quantities of N, P and S in soil organic matter vary with a number of factors, including properties of the soil, vegetation, climate, and management practices. Table VI illustrates ranges of concentrations of these elements in surface samples of mineral soils. These values are averages; individual soils may have quantities above or below those shown. Organic soils contain much higher concentrations of these elements; concentrations in subsurface horizons are, in most cases, considerably lower. The concentrations of these constituents in soils (at least in mineral samples) are smaller than the source materials discussed in the previous section because of the presence of mineral portions of soils. Ratios of these constituents appear to be remarkably similar. They do, however, vary significantly within a group of soil samples. Similarities in ratios of these elements should not be too surprising when one compares these with the sources of the organic constituents. In general, N is the dominant component of the three elements in both soil and source materials, with S and P being lower in quantity. During the decomposition of source materials in an aerobic soil, C is released as CO_2 as a metabolic by-product, however, N, P and S remain in the system as inorganic anions and/or cations. This means that during the decomposition of source materials, there would be a narrowing of C to N, P and S gross ratios. Losses of N, P and S also occur from the system by plant uptake and leaching. Leaching of N is potentially greater than of P or S because the NO_3^- anion is more mobile than PO_4^{2-} and/or SO_4^{2-}. Gaseous losses of N and S are possible in anaerobic situations. Hence, the ratios of C : N : P : S in soil organic matter vary depending on the interaction of sources, losses, and transformations within soil systems. Usually the amounts of these organic components decrease with depth; however, there are exceptions such as in Spodosol humus accumulation horizons or where surface horizons become buried.

As was the case with soil organic matter source materials, N, P and S in the soils are present in both organic and inorganic forms. In most surface soils the majority (95% or more) of the N is present in organic forms. Nitrate, nitrite and ammonium are usually present in ppm quantities in some equilibrium, hence they constitute only small proportions of the total. Ammonium can be present in a nonexchangeable form within certain clay lattices (fixed ammonium). Amounts of fixed ammonium appear to be related to the type of clay (degraded illite and vermiculite) among other factors (Bremner, 1967). Although amounts of fixed ammonium in some soils can be quite significant (>100 ppm), they account in general for only small percentages of the total N of surface samples. Percentages fixed ammonium as % of the total N can be high in subsurface samples (Bremner, 1967).

Organic S is the major form of S in most surface samples. Metal sulfides of plutonic rocks (sulfides of iron, nickel and copper) usually oxidize to sulfates during the formation of soils, so that sulfides usually do not constitute a

TABLE VI

Concentrations and relative proportions of N, P and S in soils

Source	Total N (%)	P (ppm)	S (ppm)	Organic C : N : P : S	Reference
India	0.27	482	101	100 : 12.5 : 1.3 : 1.2	Bhardwaj and Pathak, 1969
Iowa	0.24	300	304	110 : 10 : 1.4 : 1.2	Neptune et al, 1975
Brazil	0.12	145	147	194 : 10 : 1.2 : 1.4	Neptune et al, 1975
New Zealand:					
weakly weathered	0.36	500	400	172 : 10 : 1.4 : 1.1	Walker and Adams, 1959
moderately weathered	0.47	580	570	180 : 10 : 1.2 : 1.2	Walker and Adams, 1959
strongly weathered	0.30	240	380	206 : 10 : 0.8 : 1.3	Walker and Adams, 1959
Scotland:					
granite	0.27	687	400	169 : 10 : 2.4 : 1.4	Williams et al., 1960
basic	0.33	791	470	148 : 10 : 2.4 : 1.4	Williams et al., 1960
old red sandstone	0.29	722	410	130 : 10 : 2.4 : 1.4	Williams et al., 1960
basic	0.34	1056	480	113 : 10 : 1.3 : 1.3	Williams et al., 1960
calcareous	0.24	334	330	140 : 10 : 2.3 : 1.4	Williams et al., 1960

significant portion of the total S (Whitehead, 1964; Blair, 1971; Melville et al., 1971). As with nitrate and extractable ammonium, inorganic sulfate is present in soils in a dynamic equilibrium among the organic and inorganic constituents (Ensminger and Freney, 1966). Amounts of inorganic sulfates in soils are usually in low ppm ranges, making the organic S the major component. Inorganic sulfate is abundant in certain cases such as in arid regions of "white alkali" (Whitehead, 1964).

In contrast to N and S, the inorganic P of soils can be a major component of the total P present. Organic P has been shown to account for from a few percentages to as much as 75% of the total soil P (Black and Goring, 1953). In most mineral surface soils between 1/2 and 2/3 of the total P is organic; P in organic soils is almost entirely organic (Cosgrove, 1967).

Correlations between C, N and S are fairly high for most soils (Barrow, 1961), but correlations between C and organic P are not as high nor as consistent (Black and Goring, 1953, Walker and Adams, 1959; Williams et al., 1960; Barrow, 1961; Neptune et al., 1975). It has been suggested that organic P, or a significant portion of it, derives from an accumulation of resistant P-containing compounds (such as inositol phosphate) that are not intimate components of the remainder of the soil organic matter, which puts organic P "out of phase" with the soil organic C (Walker and Adams, 1959; Williams et al., 1960). A further discussion of organic P in relation to other soil constituents is found elsewhere (Halstead and McKercher, 1975).

ANALYTICAL LIMITATIONS

Organic N is usually determined by measuring total N and subtracting inorganic N (exchangeable and nonexchangeable NH_4^+, NO_2^- and NO_3^-). Two methods that are widely used for total N determinations are wet-oxidation (Kjeldahl) and dry combustion (Dumas). The Kjeldahl method probably has the widest use since it does not require highly specialized equipment and the method also lends itself more readily to routine work. Good reviews of the methods have been published (Bremner, 1965a; Bremner and Tabatabai, 1971). It was mentioned in the previous section that in most surface soils, the majority of the soil N was organic and that at levels of N shown in Table VI (0.12–0.47%) subtraction of ppm levels of NH_4^+- and NO_3^--N would not change the values significantly. But for situations requiring a great degree of accuracy, or where the inorganic N content is high, these corrections should be made. Caution should be exercised in making the correction (subtracting inorganic N from total N), because some methods for measuring total N will not include NO_3-N (Kjeldahl procedures without suitable pre-treatments) and in many cases not all of the fixed NH_4^+-N (Bremner and Tabatabai, 1971; Meints and Peterson, 1972).

Estimations of organic P can be classified into two general groups: those that measure organic P directly and those that do so indirectly by determining total P and subtracting inorganic P. In choosing a method, a fairly simple procedure is desired to facilitate routine analyses aside from giving accurate values. The direct method of Anderson and Black (1965) has not been widely used because the procedure includes use of a column of activated charcoal. Indirect methods for determining total organic P have been much more widely used. Within this general grouping of methods, two subgroups can be separated, namely, extraction and ignition types of methods. In extraction methods total P is extracted from the soil, and inorganic and total P is measured (total P after suitable oxidation of organic to inorganic P); organic P is calculated by difference. A variety of extraction procedures has been proposed to overcome problems associated with the accuracy of the methods. These problems include: (1) incomplete extraction of organic P from the soil; (2) hydrolysis of organic P during extraction (both acid and base labile organic P compounds are known to exist in the soil).

Ignition methods involve extraction and measurements of inorganic P on concurrent ignited and unignited soil samples. A variety of ignition temperatures and extraction procedures have been proposed. However, shortcomings of these methods may include: (1) incomplete conversion of organic to inorganic P; (2) loss of P by volatilization; (3) changes in the solubility of native inorganic P by ignition; (4) incomplete extraction of mineralized organic P; (5) hydrolysis of labile organic P during extraction of inorganic P.

A more complete discussion of these points is available elsewhere (Anderson, 1975).

Organic S, similar to organic N and most methods for organic P, is estimated by subtracting inorganic S from total S. There has not been wide acceptance of any particular method for the analysis of total S in soils, but several methods have been shown to be in close agreement (Tabatabai and Bremner, 1970). Inorganic S in soils may be present in a reduced form (e.g. elemental S or metal sulfide) or in the oxidized SO_4^{2-} form. The reduced form of inorganic S is not expected in soils in abundance and is not normally determined quantitatively. Methods for the estimation of inorganic forms in soils are not entirely satisfactory because they are not specific (Barrow, 1968, 1970; Melville et al., 1971; Fliermans and Brock, 1973). As discussed earlier, a significant amount of inorganic S can be found in soils as inorganic SO_4^{2-}. Numerous methods have been proposed for extracting sulfate-S from soils and these methods extract variable proportions of soluble, adsorbed and organic SO_4. Estimation of sulfate-S is done by one of two groups of methods, namely, methods involving precipitation with barium and methods involving reduction of sulfate with hydriodic acid solution and measurement of the resulting hydrogen sulfide (Beaton et al., 1968). Methods using barium are usually assumed to measure only inorganic sulfate; however, this is not true for certain soil extracts. These methods are subject to interferences from other anions and organic colloids. Hydriodic acid reduction methods do not discriminate between organic and inorganic sulfates (Johnson and Nishita, 1952; Freney, 1961). It has been shown that there are significant amounts of organic sulfates in some soil extracts (Lowe, 1966; Kowalenko and Lowe, 1975a). If a satisfactory method were developed for measuring organic sulfates in soil extracts, quantitative results would still be difficult to interpret because factors such as air drying and others alter extractable sulfate concentrations (Williams, 1967; Peverill et al., 1975; Kowalenko and Lowe, 1975a) and hydrolysis of labile organic sulfuate may also occur.

FRACTIONATIONS

Microbial biomass

It is easy to observe that when plant materials are added to soils, they are relatively quickly converted to unrecognizable forms. In soil systems, then, complex mixtures of organic materials exist. One major portion consists of plant and microbial materials, both decaying and living (biomass) and another one which is dark in color, is usually referred to as soil humus. The proportion of the nutrient elements that is included in the biomass depends on the type of ecological community (Wilkinson, 1973; Porter, 1975). Jansson (1971) reports that 10—15% of the total soil N is found in the biomass. It appears that biomass P accounts for a smaller proportion of the total organic P in soil systems (Holm, 1972). Although the biomass and its decaying products appear to constitute a small proportion of the total N in the system, the relatively fast turnover time of this fraction can result in it being a very important constituent in terms of mineral N released (Clark and Paul, 1970). Much less work of this type been attempted for P and S, but knowing something of biomass weights and concentrations of these elements (e.g., Tables I—IV), approximate calculations can be made for a particular situation. For example, from measurements of the fungal, bacterial plus actinomycete biomass in the grassland system reported by Clark and Paul (1970), and assuming the following: soil = 0.24% N, 0.025% organic P and S [1], density = 1.2 g/cm^3; bacteria and actinomycetes = 7% N, 2% P, 0.5% S; fungi = 4% N, 0.5% P, 0.5% S, then the nutrient element content of the soil biomass can be calculated (Table VII). Calculations of this type are difficult to make accurately because of limited pertinent data such as measures of organism numbers in the soil, their weights and average element concentrations, and similar parameters for roots when plant systems are involved (Russell and Adams, 1954; Babiuk and Paul, 1970; Rennie et al., 1971). Divisions between humus and non-humus material is often operationally defined rather than biologically justified (e.g., Jansson, 1971; Martel and Paul, 1974b; McGill et al., 1975).

---

[1] From Mitchell et al. (1944): a ratio of 10 : 1 : 1 for N : org. P : org. S.

TABLE VII

Nutrient elements in biomass of microorganisms calculated for Matador grassland soil
(From Clark and Paul, 1970)

Organism	g/m^2 to 10 cm:			% of total soil organic:		
	N	P	S	N	P	S
Bacteria	14	0.4	0.1	4.9	1.3	0.3
Fungi	2	0.3	0.3	0.7	1.0	1.0

"Free" constituents

Let us now consider the amounts and types of source materials of soil organic matter. Components that are not a part of the living biomass or the soil humus can be observed as identifiable compounds. Besides the release of such components when the living organism dies, viable organisms release a variety of compounds into the soil. This includes plant root exudates (Rovira, 1965) as well as microbial metabolites and wastes, and animal excretions. Numerous attempts have been made to isolate from soil identifiable "free" N, P and S containing organics with varying degrees of success.

Amino acids

"Free" amino acids have been shown to be present in soils, but their qualitative and quantitative evaluation is difficult. Early work on the extraction of "free" amino acids utilized water or ethanol. It has been shown that "free" amino acids are not completely extracted with water or ethanol and that reagents such as Ba(OH)$_2$, ammonium acetate, ether and water sequentially, or water in presence of carbon tetrachloride (or carbon tetrachloride plus a complexing agent) extract larger quantities and a greater variety of amino acids (Paul and Schmidt, 1960; Grov, 1963; Sowden and Ivarson, 1966). The quantitative levels can also be significantly influenced by the physical handling of the sample. Various methods of drying (oven, air, freeze), steaming and freezing can alter the concentrations that are measured (Paul and Tu, 1965; Ivarson and Sowden, 1966). These factors make it difficult to know what these measurements really mean, that is, whether the amino acids are present in the soil solution as such, whether they are associated with metals, clays or humus in the soil or whether they are living microbial cell constituents. "Free" amino acid levels appear to be fairly dynamic in the soil system, where studies have shown that soil and management factors can influence quantitative and qualitative values. Several factors which have been studied are: (1) cultivation (Mamchenko, 1970); (2) soil horizon (Grov and Alvsaker, 1963; Umarov and Aseyeva, 1971); (3) rhizosphere (Putnam and Schmidt, 1959; Paul and Schmidt, 1961; Ivarson and Sowden,

1969; Ivarson et al., 1970); (4) frost action (Ivarson and Sowden, 1970); (5) application of fungicides (Wainwright and Pugh, 1975); (6) land use (Verstraeten et al., 1972).

Physical and nutritional factors influence exudates of plant roots (e.g., Ivarson et al., 1970) and it is difficult in a soil—plant system to know whether the amino acids originate from plants or rhizosphere microorganisms.

"Free" amino acid N accounts for only a very small fraction of the total N in soils; indeed, it is usually much smaller than the amount of extractable inorganic N. A full range of normal amino acids has been shown to occur as "free" amino acids and some workers have had limited success in showing the presence of more unusual types (e.g., D-amino acids, which are nonprotein in nature and are present in bacteria and soil fauna but not in plants) but these are present in very small quantities (Aldag et al., 1971). S containing "free" amino acids have been shown to be present in some cases (Grov, 1963; Paul and Tu, 1965) but not in others (Sowden and Ivarson, 1966). It is not known whether this is related to extraction or to unusual stability; however, sufficient results are available to show that only a minute proportion of the total S in the soil is accounted for by this fraction.

Besides "free" amino acids, it is probable that "free" peptides or proteins are also present in the soil. Paul and Tu (1965) report the presence of peptides in some of their extracts, but not much more is known. Enzymes, which are protein molecules, are known to be present in the soil in "free" forms outside living organisms (extracellular enzymes) or when released by cell rupture. Most information on enzymes relates to their activities because extraction of intact enzymes is extremely difficult (Skujins, 1967). Adsorption of enzymes on organic and clay colloids probably results in stabilization toward degradation (Ladd and Bulter, 1975), and this association also makes it difficult to extract them from soils. Some success has been achieved in extracting enzymes, but it is difficult to estimate completion of the extraction and what proportion of the soil N they may account for (Lloyd, 1975). S is present in most enzymes. The —S—H groups in amino acids play an important role in enzyme action, but the proportion of total soil S this type of S accounts for is not known. In the same way, free protein has not been isolated from soils and, if present would probably be considered part of the humus fraction. P is also known to be associated with protein in plant and animal tissues and there is some evidence that phosphoprotein may exist in soils (Anderson, 1967), but this has not been confirmed nor quantitatively estimated.

Sugars

Amino sugars have not been detected in the "free" state in soils and it may be that they are readily broken down or are associated chemically and physically with organic or clay colloids. Little is known of the association of S with sugars. Lowe (1968) presents some evidence for the presence of sulfated polysaccharides, but little is known about their actual forms, quantities, distribution or origin. Whether they are "free" sugar sulfates or a part of the humus material is not known. Phosphated sugars have been shown to be present in soils and these forms of phosphate esters appear to account for a significant portion of the total soil organic P. Cheshire and Anderson (1975) present a recent review on carbohydrate phosphates. Robertson (1958) provided evidence for the presence of glucose —1— phosphate, which, however, is not expected to be quantitatively significant (at least in the free state) because it is not very stable. Reactions with other soil components such as clays may increase its stability. Other phosphate esters of sugars include inositol phosphates, nucleic acids and unknown but partially identified soil compounds. The identified compounds (inositol phosphates and nucleic acids) are known to be adsorbed by clays and sesquioxide materials (Goring and Bartholomew, 1952; Anderson and Arlidge, 1962; Anderson, 1963; Greaves and Wilson, 1969) and this complicates distinguishing between "free" and humus forms. These compounds are discussed more fully in following sections.

Nucleic acids

Nucleic acids are composed of pentose sugars (ribose and deoxyribose), purines (adenine and guanine), pyrimidines (cystosine, thymine, 5-methylcytosine, uracil) and phosphates. A nucleoside includes only a purine or pyrimidine plus a sugar molecule, a mononucleotide is a phosphate ester of a nucleoside, and a nucleic acid is a polynucleotide of several mononucleotides joined by a phosphate group. Nucleic acid plus protein forms nucleoprotein. Purines and pyrimidines contain nitrogen in ring structures and a few also contain amino ($-NH_2$) groups. Nucleic acids contain both N and P, but most soil studies have been directed toward the P portion.

Early studies of nucleic acid materials involved purified crystals of derivatives as evidence for their occurrence in soils and it was postulated that a large portion of the soil organic P was in this form (Anderson, 1967). Subsequent studies using chromatographic and spectrophotometric techniques on hydrolyzed soil materials showed that only small percentages of the total soil organic P and very small percentages of total soil N were in the form of ribose nucleic acids (RNA) (Adams et al., 1954) and deoxyribose nucleic acids (DNA) (Anderson, 1961) (Table VIII). These studies did not confirm the actual presence of nucleic acids, but showed all constituents to be present in soil hydrolysates (pentose sugar, purines or pyrimidines and phosphates) and permitted quantitative and qualitative estimates of purine and pyrimidine bases. Other workers have attempted to measure nucleic acids in soils and soil extracts and although there was evidence that these were present, only very small quantities of nucleic acid P were found (Forsyth, 1947; Waldron and Mortensen, 1962; Dormaar, 1963; Martin, 1964; Grindel' and Zyrin, 1965; Wild and Oke, 1966; Kowalenko, 1970). More recently, nucleo-

TABLE VIII

Nucleic acid P in some soil materials

Source	Component	% of org. P	Reference
U.S.A.	RNA	0.17—1.83	Adams et al., 1954
Scotland	DNA	0.60—2.40	Anderson, 1961
Bangladesh	RNA	0.22—1.30	Islam and Ahmed, 1973

side diphosphates were isolated from soils (Anderson, 1970); it is not known if these were artifacts of the extraction (derivatives of polynucleotides) or the actual forms. Nucleic acids are known to be adsorbed on clay minerals (Goring and Bartholomew, 1952) which contribute to difficulties in extractions and measurements as well as to uncertainties in confirming whether these compounds are "free" or in intimate association with the soil humus.

The occurrence of purines and pyrimidines has been used as evidence that a large part of nucleic acid materials in soils is probably of microbial origin (Anderson, 1967). Nucleic acids are a vital part of living systems, hence, it should not be unusual to expect these constituents in soils, at least in the biomass. More recent studies have used this material as an index for microbial biomass in soil and sediment systems (Lee et al., 1971; Sparrow and Doxtader, 1973; Christian et al., 1975). This more recent activity is directed more toward understanding the function of nucleic acids rather than to considering them as a source of the nutrient elements N and P. Although quantities of nucleic acids in soils appear to be fairly small, their association with the soil biomass suggests a relatively fast turnover time, therefore on a nutritional basis they could be very significant.

Lipids

Lipids, which are naturally occurring substances consisting of higher-fatty acids and substances found naturally in chemical combination with them, may contain N, P and S. Phospholipids are lipids which contain ester phosphates. These may also include N when choline, ethanolamine and serine are integral components. Sulfolipids, as the name suggests, are lipids containing S. Most of the work on soil phospholipids has been with reference to organic P. Hance and Anderson (1963a) showed the presence of a glycerophosphate, choline and ethanolamine in soils after hydrolysis of a lipid-solvent extract. The products provided evidence that the P occurred probably as phospholipid, primarily phosphatidyl choline (Hance and Anderson, 1963b). Kowalenko and McKercher (1970) examined extraction methods and as a result isolated and identified phosphatidyl choline and phosphatidyl ethanolamine as the major components (Kowalenko and McKercher, 1971b). Dormaar (1970) and Simoneaux and Caldwell (1965) showed the presence of these materials but did not quantitatively estimate each component. Their extracts showed the presence of a glycerophosphate which was suggested to be a decomposition product of phospholipids. Anderson and Malcom (1974) provided evidence for the possible presence of phosphoinositide lipids in a soil extract, but this was not confirmed conclusively. The origin (microbial and/or plant) of these components cannot be determined from the types of phospholipids extracted, but a clue to the origin may be derived from an examination of the fatty acid constituents of the phospholipids (Kowalenko and McKercher, 1971b).

Quantitatively phospholipids represent small percentages of the total organic P (Table IX). Kowalenko and McKercher (1971a) estimated that 10—20% of the phospholipid P measured in their soils could be accounted for by microbial biomass, so that there is some accumulation of "free" phospholipids. It was also shown that phospholipids are probably stabilized by the clay minerals, because treatment with hydrofluoric acid solution resulted in increased values for phospholipids extracted from air dried soils. N present in the phospholipids accounts for only a very small proportion of the total organic N. Although the quantities of these components appear to be very low, they may be important sources of P and C to plant and microbial life if they turn over rapidly. Phospholipid P levels changed during a laboratory incubation study (Kowalenko and Lowe, 1975b) and during a growing season (Yaskin, 1968) suggesting a dynamic status in the soil. The combined hydrophobic-hydrophylic properties of phospholipids can have significant physical action even with minute quantities.

Not much is known about the sulfolipid content of soils. Unpublished results (Kowalenko, 1972) suggest the possibility that sulfolipids occur in soils, since sulfur in both oxidized and reduced forms was present in chloroform extracts of two organic soils using the Bligh and Dyer method (Kowalenko and McKercher, 1970) (Table X). The extracts were not examined further,

TABLE IX

Phospholipid content of surface soils

Source	Lipid-P ($\mu g/g$)	% of total organic P	Reference
U.S.S.R.	0.6 — 9.0	—	Burangulova, 1959
Germany	5.7 —10.1	—	Wenzel, 1961
Great Britain	3.0 — 7.0	0.6 —0.9	Hance and Anderson, 1963a
U.S.A.	0.24— 1.70	0.06—1.02	Simoneaux and Caldwell, 1965
Canada (S. Alta.)	0.32—13.0	0.1 —7.0	Dormaar, 1970
Canada (Sask.)	0.6 —14.6	0.4 —4.6	Kowalenko and McKercher, 1971a
Bangladesh	0.5 —11.0	0.5 —7.0	Islam and Ahmed, 1973
New Zealand	1.6 — 2.1	—	Baker, 1975

TABLE X

S in a chloroform extract of two organic soils of British Columbia, Canada

Soil	(1) HI-S (μg/g)	(2) Raney Ni-reducible (μg S/g)	Choroform S (1 + 2) as % of total soil S	Lipid-P (μg/g)
Lulu	18	56	0.3	4.7
Metchosin	26	18	0.6	44.5

but chromatography could probably separate and assist in the identification of the components extracted. In these soils at least, S in lipid extracts represents only a small fraction of the total S but a comparison with lipid P shows the S to be equivalent to or greater than P. Plant sulfolipids identified so far in other materials are sulfonic acids (Gordon and Hunter, 1969), which would react as C-bonded forms, but little is known about other sulfolipids especially in microorganisms.

Lipid materials appear to occur in soils in "free" forms or as part of the biomass and not part of the humus. Correlations of lipid P with total organic C of soils is fairly high, whereas they are lower with total and organic P (Kowalenko and McKercher, 1971a), suggesting that it is an organic P component that accumulates and decomposes apart from the remainder of the organic P.

Vitamins

Vitamins are important constituents of living systems that reach the soil in plant materials or are formed by certain microorganisms. Vitamins may contain N and S. Two vitamins which have been detected in soils (biotin and thiamine) contain these two elements. Biotin has been extracted from New Zealand soils (Jones et al., 1962), but the quantities did not represent very large proportions of the total organic N and S of the soils. From 4—15 μg biotin/g of soil was found in three soils and the N and S contribution were fractions of that.

Hagedorn (1969) studied thiamine in Spodosols and showed that levels of this vitamin did not appear to be related to microbial numbers or root exudates, suggesting some accumulation in the "free" state apart from the biomass. However, the proportions of N and S represented by this vitamin as proportions of the total N and S were very small. Probably the physiological effects of vitamins (growth stimulation, etc.) are more important to soil systems than are their N and S contents.

Others

Other N containing compounds such as creatinine and allantoin have been detected in soils but do not appear to be quantitatively significant (Bremner,

1967). A green organic matter fraction has been extracted and examined and if porphyrin derived (Sato and Kumada, 1967; Lowe and Tsang, 1970), may contain N. Goodman and Cheshire (1973) provide evidence that porphyrin derivatives are present in soil HA and estimate that 0.1% of the humic acid N could be in this form. Porphyrins, if present would have chelating ability due to the presence of heterocyclic N. Schnitzer and Skinner (1968) separated a component from a FA extract whose properties resembled those of the "green humic acid", but this material contained very little N. Other transient "free" components are probably present in the soil, such as soil microbial antibiotics (Roy and Chandra, 1976), but these singly account for only a small proportion of the total organic N.

The relative instability of many organic P compounds in acids and bases makes it difficult to isolate specific compounds. Inorganic pyrophosphate has been detected in soils in small quantities (Anderson and Russell, 1969; Anderson and Malcolm, 1974), but it is not known whether it is organically derived or not. If the pyrophosphate is not decomposed during extraction and analysis, it will be included among organic P in many analytical procedures. Organic S compounds, such as trithiobenzaldehyde (Freney, 1967) as well as volatile organic S compounds (Lovelock et al., 1972; Banwart and Bremner, 1975) have been detected but these are considered to occur quantitatively in very small amounts only.

Numerous reports are available which show that organic N, P and S components are present in water or dilute salt solution extracts. In some cases these organics have been identified, such as the "free" amino acids, in others P containing compounds were not identified (Wild and Oke, 1966; Sekhon and Black, 1969; Martin, 1970) while compounds containing S were determined as ester sulfates (Lowe, 1966; Kowalenko and Lowe, 1975a). These materials isolated may include biomass and "free" forms of organic compounds as well as material of humic nature. Other studies indicate that "free" and microbial organic compounds enhance the movement of elements in a soil system, particularly in the case of P, where free inorganic PO_4^{2-} ion is quite reactive (and immobile) in a soil (Hannapel et al., 1964; Bowman et al., 1967; Rolston et al., 1975). Studies which apply chemical analyses of plant or microbial material to soils in an attempt to follow how certain source materials may change during incubation (e.g., Bartholomew and Goring, 1948) are difficult to interpret because of incomplete extraction due to adsorption on the soil and difficulties in differentiating biomass, "free" and humus forms. The use of tracer isotopes may help to elucidate this problem.

Chemical fractionations characteristic for specific elements

The previous discussion has indicated the difficulty of separating biomass, "free" and humus materials in the soil; in cases where this has been attempted it was operationally rather than actually defined. This difficulty has led to other approaches, and some useful methods involve chemical fractionation. Because many studies usually center around one particular element (C, N, P or S), procedures which provide useful information on organic components have evolved for each of the three elements in question. Characteristic chemical fractionations will be discussed for each element to illustrate the type of information derived so far on whole soils and various organic extracts.

N

(a) Whole soil. An early view of the nature of soil organic N was that a major portion of it was protein. This resulted in applying methods for protein analysis to soils. Strong acid hydrolysis techniques have resulted in a greater understanding of the soil N. The actual hydrolysis procedure is not standardized and many variations are used. Because of the presence of a large amount of potentially interfering substances in soils, as compared to plant material, modifications are necessary to increase the completeness of the hydrolysis with minimal destruction of measurable components. In general, the hydrolysis involves refluxing the soil (or an extract of it) with 6N acid for a period between 6 to 24 h. Some of the variables include the following.

(i) Type and concentration of acid. Usually 6N HCl or H_2SO_4 are employed, but higher and lower concentrations have been used for specific studies. Mixtures of acids have been employed for soil extracts to facilitate the analysis of hexosamines (Parsons and Tinsley, 1961). Alkaline hydrolysis has been applied to minimize the destruction of tryptophan (Bremner, 1949). Partial hydrolysis has been used to check for the presence of peptides (Sowden, 1966). Superheated water has been compared to 6N HCl hydrolysis (Cheng and Ponnamperuma, 1974).

(ii) Ratio of acid to soil. Ratios from 3 : 1 to 25 : 1 (acid to soil) have been used; less degradation of amino acids results when the wider ratio is used (Sowden, 1955).

(iii) Time and temperature of hydrolysis. The most frequently used temperatures of hydrolysis are 100°C or reflux temperatures for usually between 12 and 24 h (Bremner, 1965b, 1967; Parsons and Tinsley, 1975). Most amino acids are fairly stable toward acid hydrolysis, and increasing the reaction time beyond 6 h results in little further destruction. Hexosamines are much more senstive to degradation in this medium and if quantitative results are desired, this must be considered. Sowden (1959) suggested a correction factor for the quantitative estimation of hexosamines during acid hydrolysis. Autoclaving soils with 6N HCl at 120°C and 15 lb/inch^2 is another alternative. Autoclaving for 6 h compares favorably with longer reflux times (Lowe, 1973). Besides the advantage of time, it appears that degradation of hexosamine is not as great when autoclaving is used (Lowe, 1973).

(iv) Pretreatments. Mineral material in soils does interfere with the hydrolysis, and HF treatments have been shown to have some effect (Freney, 1958; Cheng, 1975). S-containing amino acids are known to be sensitive to acid

hydrolysis and oxidation with performic acid is recommended before hydrolysis to obtain better values (Freney et al., 1972).

Once the soil is extracted, several types and variations of analyses can be applied and the method varies with the purpose of the study. Methods are available to determine various groups of compounds quantitatively, e.g., Van Slyke decarboxylation or nitrous acid methods for amino acids, periodic acid and phosphate—borate buffer systems for (serine+threonine)-N and hexosamines (Bremner, 1965b). The identification of individual constituents within amino acid and hexosamine groups can be done by paper or thin layer chromatography or electrophoresis, and quantitative analyses can also be made. In some cases, combinations such as ion exchange and paper chromatography have been useful. Ion exchange chromatography coupled with autoanalyser systems are now available (Sowden, 1967; Lowe, 1973). Variations in methods may include desalting before chromatography (Sowden, 1969; Cheng, 1975), different detector systems (Freney et al., 1972; Finlayson and MacKenzie, 1976) or complex systems for special analyses (Aldag et al., 1971). New developments in gas-liquid (detectors specific to an element) and liquid chromatography are other possibilities, which may have advantages for certain applications. Methods for amino acid analysis on gas chromatographs are available and have been compared with other methods (Shearer and Warner, 1971).

Despite variations in methods of analysis and soils, the distribution of amino acids and proportions of the total N in soils is fairly similar. Bremner (1967) reviews information before that date and concludes that between 20 and 50% of the total N in surface soils is amino acid and that its proportion decreases with depth. Cultivation, climate and fertilization affect both the proportions of amino acids in terms of the total N as well as the distribution of individual amino acids. Amino sugars (hexosamines) usually represent 5—15% of the total N in surface soils, with some results indicating levels from 23 to 40% in arid soils (Parsons and Tinsley, 1961; Bremner, 1967). Amino sugars as percentages of total N appear to vary more in subsurface horizons, where in some instances up to 24% of the N of Podzol B horizons is measured in this form.

The distribution of amino acids in soils are what may be expected in protein materials; up to 30 amino acids have been reported (Bremner, 1967). A number of amino acids that are not normally associated with protein (nonprotein amino acids) have been detected in soils. These amino acids include ornithine, β-alanine, α-amino-n-butyric acid, γ-amino-n-butyric acid, 3:4-dihydroxyphenylalanine and α,ϵ-diaminopimelic acid. The latter amino acid is almost exclusively confined to bacteria and also ornithine is found in significant quantities in bacteria. Thus, it appears that microbial amino acids are possibly more prominent than those from plant residues in mineral surface horizons. Lowe (1973) showed small changes in the distribution of amino acids when analysing samples of varying degrees of humification (comparing L, F, H horizons of forest floor materials). He did not observe any consistent changes in proportions of acidic amino acids with increasing humification. Sowden et al. (1976) showed that amorphous allophanic materials in tropical soils accumulated acidic amino acids.

Hydrolysis with 6N acid may result in degradation of some amino acids (Martin and Synge, 1945), and soil materials are assumed to behave similarly. Asparagine, glutamic acid and S containing amino acids (cystine, methionine and related amino acids) are known to be unstable during acid hydrolysis (Sowden, 1958, 1970b; Cheng and Ponnamperuma, 1974) and special methods must be applied if these amino acids are to be determined.

The dominant amino sugar present in soils is glucosamine, and significant amounts of galactosamine have also been shown to be present. N-acetyl-glucosamine and 2-deoxy 2-amino D-talose have also been detected in soil materials (Parsons and Tinsley, 1975) but these have not been shown to account for a very significant proportion of the amino sugars. It should be recalled that hexosamines are not very stable during 6N acid hydrolysis and corrections are often applied for quantitative evaluations. The N-acetyl group of N-acetylglucosamine is removed by 6N acid hydrolysis, so that much of the soil hexosamine may be present originally as N-acetyl amino sugars. Amino sugars are not normally found in higher plant tissues, but are abundant in tissues of insects and fungi, hence their detection in soil suggests microbial origin. This, together with information on the amino acids, indicates that a large portion of soil N is of microbial origin.

Sowden (1959) compared ratios of glucosamine to galactosamine for a variety of soils and showed that they varied from 1.6 to 4.1, with the highest ratio for a humus layer of a Podzol. Two B horizons analysed had ratios of 1.6 to 2.3. Lowe (1973) did not report any consistent trends for glucosamine to galactosamine ratios with degree of decomposition (L → F → H of forest litter) but some very high (e.g., 76) ratios were evident. Analytical problems (degradation of amino sugars) associated with such measurements have been mentioned but, since galactosamine and glucosamine are about equally stable toward hydrolysis, ratios of constituents should reflect the true situation.

Table XI gives an example of what is known about soil organic N. Thirty-two surface mineral soil samples from across Canada analysed for soil sites displayed to the International Soil Science Congress (1978) are shown. Twenty-one amino acids were identified and quantitatively measured in addition to amino sugars glucosamine and galactosamine. The amino acids are grouped into acidic, basic, neutral and sulfur-containing types. The two acidic amino acids (aspartic and glutamic) and glycine and alanine are quantitatively dominant. The amino acids as a group account for 40% of the total N and amino sugars for 7%, which is within ranges discussed earlier. This means that 47% of the total N of these soils has been identified, leaving 53% unidentified. S-containing amino acids did not account for a large portion of

TABLE XI

Quantitative and qualitative N distribution in 32 surface mineral soil samples from across Canada

Fraction		Quantity % of total N
Unhydrolysed N		15.0 ± 6.1
Hydrolysed N		85.0
Unidentified		19.1
NH_3-N		20.7 ± 4.5
Amino acid N		39.6 ± 7.1
Amino sugar N		6.6 ± 1.6
Amino acids		Molar ratio
Acidic:	aspartic	11.9 ± 1.6
	glutamic	8.4 ± 0.6
Basic:	arginine	2.6 ± 0.3
	histidine	1.2 ± 0.4
	lysine	3.6 ± 0.3
	ornithine	0.8 ± 0.3
Neutral:	phenylalanine	2.4 ± 0.2
	tyrosine	1.1 ± 0.3
	glycine	11.8 ± 0.9
	alanine	8.5 ± 0.5
	valine	5.2 ± 0.5
	leucine	5.1 ± 0.6
	isoleucine	3.0 ± 0.3
	serine	5.6 ± 0.6
	threonine	5.6 ± 0.4
	proline	4.1 ± 0.5
	hydroxyproline	0.4 ± 0.2
Sulfur containing:	methionine	0.4 ± 0.2
	cystine	0.4 ± 0.2
	cysteic acid	0.8 ± 0.3
	methionine sulfoxide	0.3 ± 0.1
Miscellaneous		2.1 ± 0.9
Amino sugars:	glucosamine	10.9 ± 2.6
	galactosamine	5.9 ± 1.5
		Ratio:
Glucosamine/galactosamine		1.8

the total N, and for only a small part of the total organic S. A portion of the unidentified N is solubilized by acid hydrolysis and represents about 40% of the total N. Some of this solubilized N appears as NH_4^+-N due to degradation of amino acid amides, amino sugars and some amino acids, e.g., serine (Greenland, 1972) and some of the NH_4^+-N of organic origin would be accounted for in making corrections for sugar amine quantities. Clay fixed NH_4^+, if present, would be included in the NH_4^+-N fraction. Jones and Parsons (1972) suggest that, in certain cases, some of the NH_4-N released into the acid during hydrolysis is organically fixed NH_3. Fifteen percent of the total organic N remains insoluble during acid hydrolysis. Freney and Miller (1970) showed that part of the non-hydrolysable N came from organic N attached to clay minerals which stabilized these compounds.

The nature and origin of unidentified acid soluble N plus insoluble N plus NH_3-N (the source of which is largely unknown), which in Table XI accounts for 53% of the organic N is largely unknown. Some of the unknown N may be in the form of purines, pyrimidines, and vitamins, but as suggested in the previous discussion, these would not account for a very high percentage. Hydroxamic acids have interesting functions in plants and microorganisms, but they have not been examined in soil systems (Waid, 1975). This means that about one-half of the soil organic N is largely unknown. Studies applying alkaline hydrolysis to the acid insoluble materials report differing results. Piper and Posner (1968, 1972) reported significant yields of amino acids by alkaline hydrolysis of the acid insoluble portion and suggested that N-phenoxy amino acid structures account for this. Griffith et al. (1976) found only a small additional release of amino acids (1—3%) by alkaline hydrolysis, and it appeared that the additional NH_3-N due to alkaline hydrolysis was not primarily a result of the destruction of amino acids. Freney (1968) extracted $6N$ HCl insoluble material with alkali and found that this material could be fractionated into soluble and insoluble portions by acidification. He showed that some of the alkali-soluble N was of low molecular weight.

The previous discussion shows that most of the amino acids in soils are bound fairly strongly to the soil matrix. Extractions of "free" amino acids yield very low amounts, but $6N$ acid hydrolysis results in the release of much greater quantities. This is taken as evidence that the amino acids are proteinaceous. Partial hydrolysis has shown that peptide bonds are probably present (Sowden, 1966; Piper and Posner, 1968), and proteolytic enzymes have released amino acids from soil materials (Scharpenseel and Krause, 1962; Ladd and Brisbane, 1967; Sowden, 1970b). Biederbeck and Paul (1973) review work on the proteinaceous nature of some soil extracts and fractions designated "lignoproteins" and "humoproteins" that have been isolated in soils. Their experimental study shows that humate fractions contain proteinaceous material.

(b) Extracted-fractionated soil. Numerous extraction and fractionation procedures have been done on soils where $6N$ acid hydrolysis has been applied on fractions to determine amino acid and amino sugar N qualitatively and quantitatively. The classical fractionation of soil organic matter is extraction with a base, then fractionation into acid soluble (FA), and insoluble (HA) portions and material not extracted by NaOH (humin). A large number of variations of this extraction and fractionation scheme are reported in the literature, including variations in concentration, type and pH of the extractant,

ratio of soil to solution, pretreatments and shaking times, all of which make comparisons among various reports exceedingly difficult if not impossible. In some cases, the distribution of components such as amino acids and amino sugars appears similar for various extracts, but a wide range of extraction efficiencies is observable (Sowden, 1970a). A recent report (Sowden et al., 1976) provides an example of the type of N distribution using the "classical" extraction and fractionation procedures. Averages are presented only for soils 2 and 5 of that study to simplify the information and to make the values comparable (Table XII). The results illustrate differences in nitrogen components among the fractionated humus materials. Individual amino acids vary as much as two-fold from one fraction to the other e.g., aspartic acid in HA and FA. Molar ratios are presented to facilitate comparisons of the various amino acids among the fractions. The authors note that the acidic amino acid composition in these soils was unusually high. Most of the N in soils is in the HA and humin fractions and only a small amount (6.5% of the total N) is in the FA fraction. The ratio of the amino sugars also varies widely from one fraction to the other.

More detailed studies of HA's and FA's are available; in each case the fractions were further subdivided. Biederbeck and Paul (1973) used phenol, ultrafiltration and polyvinylpyrrolidone columns to separate "humoproteins" from soil HA. Recently Sequi et al. (1975) and Guidi et al. (1976) fractionated FA by polyamide and Sephadex chromatography into several fractions. One of these was rich in N (proteinaceous compound), containing a high proportion of the total N as amino acids.

There is controversy on whether or not polypeptides are truly part of the humic components (Felbeck, 1971; Schnitzer, et al., 1974), but most workers seem to agree that N is an important component associated in some way with humic materials. The differences in opinion arise from differences in definition. A "core" of humus, which is N-, P- and S-free, has been suggested to which peptides, etc. are relatively loosely attached (Haworth, 1971). Material resembling HA's has been shown to be formed by fungi, and fungal humic materials have been compared with soil humic substances (De Serra et al., 1973). Analysis of the amino acids after acid hydrolysis shows that fungal and soil humic acid amino acids differ, suggesting that these materials must undergo considerable modification to be similar to amino acids in HA's. However, despite controversies, acid hydrolysis of soils and soil materials has yielded valuable information on soil organic N. Molecular weight fractionations of FA's show several distinct fractions and in general the N content decreases as the molecular weight decreases (Piper and Posner, 1968; Schnitzer and Skinner, 1968). The amino acids appear to be tightly bonded to the fulvic material, show variations in individual amino acid distribution in the fractions and contain peptide bonds (Piper and Posner, 1968).

P

(a) *Whole soil.* Inositol phosphates, nucleic acids and phospholipids are

TABLE XII

Quantitative and qualitative N distribution in a classical extraction-fractionation of two tropical soils (soils 2 and 5 from Sowden et al., 1976)

Fraction		Humic acid	Fulvic acid	NaOH insoluble
% total N		4.68	2.28	0.72
% N unhydrolysed		15.6	4.4	16.3
% N hydrolysed		84.4	95.6	83.7
Unidentified (%)		15.4	30.5	10.4
NH_3-N (%)		14.6	20.3	16.4
Amino acid N (%)		50.5	42.7	48.4
Amino sugar N (%)		3.7	2.2	8.6
Amino acids (molar ratio)				
Acidic:	aspartic	12.4	24.6	17.5
	glutamic	8.6	18.0	9.2
Basic:	arginine	2.2	0.4	1.9
	histidine	1.9	0.6	1.8
	lysine	3.4	1.6	2.6
	ornithine	0.7	0.8	0.8
Neutral:	phenylalanine	3.0	1.1	2.3
	tyrosine	1.4	0.7	1.4
	glycine	11.2	13.0	12.2
	alanine	8.0	8.2	8.4
	valine	5.6	3.8	4.9
	leucine	5.1	2.4	4.8
	isoleucine	3.4	1.8	2.6
	serine	5.0	4.8	5.0
	threonine	5.0	4.2	4.6
	proline	4.6	3.9	3.4
	hydroxyproline	0.7	trace	0.6
Sulfur containing:	methionine	0.6	0.4	0.5
	cystine	0.2	0.2	0.1
	cysteic acid	0.5	2.6	0.6
	methionine sulfoxide	0.4	0.6	0.2
Miscellaneous [1]		1.6	1.9	1.8
Glucosamine/galactosamine ratio		1.3	1.3	3.0
% N of total soil N		41.6	6.5	(51.9) [2]

[1] Includes allo-isoleucine, γ-NH_2-butyric acid, α,4-disaminobutyric acid, diaminopimelic acid, β-alanine, ethanolamine, and unidentified compounds.
[2] Calculated by difference, therefore may include material in humic and fulvic acids lost during their purification.

the three organic P compounds that have been identified and shown to occur in measurable quantities in soils. Previous discussion in this chapter showed that nucleic acids and phospholipids accounted for only small percentages of the total organic P. Inositol phosphate is the largest group of organic P compounds that has been identified, and studies on the nature of soil organic P should include an analysis of inositol phosphates. Halstead and McKercher (1975) determined quantities of inositol phosphates in soils and found amounts ranging from 1 to 460 μgP/g, constituting between 3 and 77% of the organic P. Islam and Ahmed (1973) reported on one soil containing 83% of the total organic P as inositol phosphates. This concentration falls within the range given by Halstead and McKercher (1975). A few reports include measurements of all three major organic components (inositol, nucleic acid and phospholipid) on the same soils (Table XIII). Inositol phosphates are quantitatively the most significant organic compounds. Numerous quantitative reports are now available on analyses of these organic P compounds. This phosphate is quite interesting chemically in that a range of inositol phosphates from mono- to hexa-phosphates are possible, depending on the number of ester phosphates. They can also occur as sterioisomers, with 9 possible polyphosphate forms (Halstead and McKercher, 1975). Penta- and hexa-phosphates of inositol dominate the forms identified in soils although lower forms have also been shown to be present but in very small quantities. Identification of five (of a possible nine) isomers in soils has been used as evidence of the probable origin of these components, but with limited success. The myo-, scyllo-, neo- and chiro-isomers have been detected in soils and to date only the myo- and chiro-inositol phosphates are reported to be present in plant materials (Halstead and McKercher, 1975), suggesting that microbial and chemical action in the soil results in the formation of the other isomers. The chiro- form is an optically active form with D- and L-rotations. Microbial degradation pathways of myo-, scyllo- and DL-chiro-inositol hexaphosphates have been studied, showing the various inositol phosphates produced (Halstead and McKercher, 1975). McKercher (1968) reviews the information on inositol phosphates and shows that myo + DL:

TABLE XIII

Identified organic P constituents of soils

Source	% of total organic P				Reference
	Phospholipid	Nucleic acid	Inositol phosphates	Total identified	
Canada	1	$\ll 1$ [*1]	16	17	Kowalenko and McKercher, 1971a
Bangladesh	2	1	45	48	Islam and Ahmed, 1973
Scotland	1	3	50	54	Anderson and Malcolm, 1974

[*1] Estimated from Kowalenko, 1970.

scyllo isomer ratios and hexa : penta ester ratios range from 0.8 to 8 : 1 and 0.9 to 4.3 : 1, respectively. The myo isomer and hexaphosphate ester is usually predominant. Realization of the quantitative significance of inositol phosphates in soils has generated a fair amount of activity in the development of methods. Anderson (1967) reviews methods developed up to 1967. Two new approaches, using either alkaline extraction and gel permeation chromatography or acid-complexing extraction and ion-exchange chromatography have subsequently been described (Cheshire and Anderson, 1975).

(b) Extracted-fractionated soil. Most studies have been directed toward the quantitative and qualitative evaluation of compounds discussed above but much less is known about how these compounds are related to the remainder of the soil organic matter. Most of the methods for determining inositol phosphates require removal of "extraneous" organic matter. This may be done by precipitating inositol with iron or aluminum, oxidation of "extraneous" organic matter with hypobromite (Anderson, 1967), treatment with hot 6N HCl (Tinsley and Özsavasci, 1974) or using Sephadex column separation (Steward and Tate, 1971). Some of these treatments may be strong enough to rupture phosphate ester or other bonds, resulting in the appearance of these compounds as unbound organic substances. Application of the method of Steward and Tate (1971), however, suggests that a significant portion of inositol polyphosphates are moieties separate from other soil organic constituents. Moyer and Thomas (1970) showed that organic matter removed by mild extraction (chelating resin), could be fractionated into three molecular weight ranges over Sephadex G-75. The two lower molecular weight fractions were found to contain inositol polyphosphates, whereas no evidence for inositol polyphosphates in the higher molecular weight fraction was found. Veinot and Thomas (1972) subsequently found inositol in a similar high-molecular weight fraction and postulated the presence of inositol polyphosphates. These studies suggest that inositol polyphosphates occur in both "free" and "bound" forms in relation to the remainder of the organic matter. Omotoso and Wild (1970) and Halstead and Anderson (1970) provided further evidence to confirm the bound status of inositol P in soils and the latter workers also showed that the nature of the extractant was important in this type of evaluation. Inositol phosphates are known to react with sesquioxides and calcium, which may increase the stability of these organic phosphorus compounds in soils.

Present information shows that one-half or more of the organic P in most soils has not yet been identified. Several workers (McKercher, 1968; Halstead and Anderson, 1970; Omotoso and Wild, 1970; Steward and Tate, 1971; Anderson and Malcolm, 1974; Tinsley and Özsavasci, 1974) have reported studies on a distinct fraction (or fractions) that account for significant portions of the organic P and which could be separated on exchange resin or by chromatography. In each case it is difficult to know for certain whether or not the material is a single component or a group of chromatographically similar components. These components appear to have relatively high molecular weights and are not usually adsorbed by ion exchange resins (possibly due to their large molecular weight). Although their general chromatographic reactions suggest similarities, comparisons are difficult because different soils, extractions, fractionations and analyses were used in each case. Halstead and Anderson (1970) compared several extractants and showed that only mild extraction (0.2 M acetylacetone at pH 8.0) produced an organic P fraction that was not adsorbed by an ion exchange resin. But, after NaOH hydrolysis, adsorption of organic P containing fractions occurred. With this in mind, it is difficult to evaluate the work of Anderson and Malcolm (1974) who fractionated a hot 3 M NaOH extract. Three reports where attempts were made to identify unknown P-rich fractions of soil extracts are compared in Table XIV. The results show that some of the soil organic P may consist of phosphorylated polysaccharides (McKercher, 1968). Bacterial lipopolysaccharides or bacterial enzyme-induced reaction products have also been suggested as possible sources (Anderson and Malcolm, 1974). Chemical (non-enzymatic) reactions in soils will contribute to the formation of complex products.

It has been suggested that not all of the organic P is intimately connected with the remainder of the soil organic matter. The relative ease of extracting soil organic P in comparison to C and N has been used to support this idea (Barrow, 1961). Theoretical structures of HA molecules do not include P (Felbeck, 1965). Even recent studies assume that HA's and FA's contain negligible amounts of P since the O content of these materials is calculated by subtracting %C + %H + %N + %S from 100 (Schnitzer and Vendette, 1975). This may depend on how one defines the purity of the substances in question and the nature of bonds between substances such as fatty acids with "building blocks" (Neyroud and Schnitzer, 1974). Inorganic P has been shown to interact with HA's and FA's through an iron or aluminum linkage and properties of such complexes have been studied (Weir and Soper, 1963; Lévesque and Schnitzer, 1967; Lévesque, 1969; Sinha, 1971). The complexes have relatively strong bonds between phosphates and humic components. Alkaline and acid hydrolysis are required to break these bonds, showing that the complexes behave as a unit during soil extraction and fractionation procedures. It is desirable then, that complexes of this nature should be included in estimations of organic P although according to strict definitions of organic vs. inorganic P, this type of P would be inorganic rather than organic P. Veinot and Thomas (1972) showed that a high molecular weight, P-containing soil organic extract could be partially degraded by mild treatments and also showed the presence of measurable quantities of iron and aluminum. They concluded that iron and aluminum were at least partially responsible for the formation of high molecular weight complexes in these soils. Dormaar (1972) showed that bonding of phosphate to humic substances by calcium was important in the system he studied, where seasonal fluctuations of organic P were noted. Part of the problem in studying soil organic P is de-

TABLE XIV

Information of unknown, P-rich fractions reported by several workers

	Unidentified PO$_4$ ester (McKercher, 1968)	SF$_1$ (Omotoso and Wild, 1970)	"Fulvic" fraction (Steward and Tate, 1971)
Basis of separation	Unadsorbed by ion exchange resin	Sephadex G-25	Sephadex G-50, separation into "fulvic" acid, desalted with Sephadex G-10, adsorbed on ion exchange resin
Chemical data	25% C, 0.75% N, 1.4% P, 25% anthrone sugars, 4% uronic acids (carbazole) Structural hydroxyl and carboxyl groups present	—	—
Hydrolysis: (a) method (b) identified products	acid galactose, glucose, mannose, arabinose, xylose, fucose, rhamnose and 3 methyl sugars	hypobromite tetra- and hexa-inositol PO$_4$; myo- and dl-inositols, product Pa contained: 807 ppm total organic P, 433 ppm inositol P, 1,370 ppm, reducing sugar, 660 ppm total N, 370 ppm amino-N, 4,170 ppm org. C, no 260 mμ absorption max. (no nucleic acid)	acid reducing sugars (uronic acids), series of non-reducing PO$_4$ (resembling α and β glycerophosphates), inorganic PO$_4$
Other	Not dephosphorylated with alkaline phosphatase	—	—

termining precisely what is actually measured as organic P and further elucidation of this situation would be helpful.

S

(a) Whole soil. A characteristic analytical approach to studying soil organic S is to distinguish between two fractions: HI-reducible (SO$_4$-S, reducible S or O-bonded S) and C-bonded S. Determination of the HI-reducible S involves

reduction of S with a reducing agent containing hydriodic, formic and hypophosphorous acids. This reagent reacts with inorganic sulfate, organic ester sulfate (O-bonded S) as well as with N-sulfate (as in heparin) and produces hydrogen sulfide (Johnson and Nishita, 1952; Freney, 1958; Spencer and Freney, 1960). To determine total soil organic HI-reducible S, inorganic sulfate must be subtracted from this value. Various reagents to extract soluble and adsorbed inorganic sulfate are available (Beaton et al., 1968). It should be recalled from a previous discussion that these reagents may also measure easily soluble organic sulfate, which may influence the results. However, amounts of inorganic sulfate and easily soluble organic sulfate in most surface soils are relatively small (compared to organic HI-reducible S), so that a reasonably good estimate of total organic sulfate S can be made. Lowe and DeLong (1963) proposed that reaction of soil organic S with Raney nickel was specific for most forms of S directly bonded to C. Aliphatic sulfones and sulfonic acid (both C-bonded forms) are not converted to hydrogen sulfide by this reaction, and are not included in the analysis. Freney et al. (1970) showed that HI-reducible S plus C-bonded S (by Raney nickel) did not account for all of the organic S in some soils and went on to demonstrate that iron and manganese interfered with the Raney nickel reaction for C-bonded S. They suggested that a better estimate of C-bonded S resulted from the difference between total and HI-reducible S.

Little is known about the nature of soil organic S components that make up the HI-reducible S and C-bonded S groups. Sulfated polysaccharides, sulfated esters of phenols, choline sulfate and sulfated lipids have been suggested as components of HI-reducible S (Freney, 1967; Beaton et al., 1968), but these may not account for all of this fraction. Evidence for sulfated polysaccharides and some other fractions, e.g., sulfated lipids, have been discussed in an earlier section, but qualitative and quantitative information is lacking. S-containing amino acids are present in soils, and this was discussed in connection with amino acids. The S of amino acids does not appear to account for all of the C-bonded S, and little is known of the nature of the unknown S-component. A wide variety of components may make up these fractions (HI-reducible and C-bonded S), coming from a variety of sources as well as from a variety of reaction products (Gilbert, 1965; Freney, 1967; Behrman, 1967; Suter, 1970).

About one-half of the total S is in the C-bonded form, and slightly less than one-half is in the HI-reducible form in most surface soils (Table XV). A small fraction of the HI-reducible form is inorganic sulfate. The table also shows differences in values for the C-bonded form using the direct method (Raney nickel) and the difference method (total minus HI-reducible S). In some cases (Iowa, Brazil and Scottish soils) recovery of C-bonded S with Raney nickel was very low, which may be due to iron and manganese interferences mentioned earlier. Soils in these studies showed a relatively narrow range of HI-reducible S. HI-reducible S also includes inorganic sulfate. Few

TABLE XV

S fractions in surface mineral soils

Source	Total S (ppm)	HI-reducible S [1]		C-bonded S (difference)		C-bonded S (Raney Ni)	
		(ppm)	(% of total)	(ppm)	(% of total)	(ppm)	(% of total)
Saskatchewan (Bettany et al., 1973)	284	127	45	157	55	—	—
Alberta gleysols (Lowe, 1969b)	2422	962	40	1460	60	1018	42
Quebec (Lowe, 1964)	2653	1362	51	1291	49	1258	47
Iowa (Neptune et al., 1975)	319	172	54	147	46	35	11
Brazil (Neptune et al., 1975)	166	85	51	81	49	12	7
Australia (Freney et al., 1970)	426	200	47	226	53	49	12
Scotland [2] (Scott and Anderson, 1976)	480	276	58	204	42	95	20

[1] Some of the HI-reducible S includes soluble and adsorbed inorganic SO_4.
[2] Averages calculated from Table V of reference. HI-reducible S corrected for inorganic SO_4.

analyses are available that distinguish S variability as a function of soil characteristics. Bettany et al. (1973) showed a trend with respect to soils of chernozemic, forested and transitional zones and Scott and Anderson (1976) attempted to show the influence of drainage and acidity. Freney et al. (1971) showed small changes in these fractions during an incubation experiment.

(b) Extracted-fractionated soil. Very little work has been reported in which HI-reducible S as well as total S have been determined on fractionated extracts of soil organic matter. More values are available for total S, and these measurements are often done to enable calculations of O in an organic fraction by difference. Freney et al. (1969) compared the proportion of the total S that is extracted by a variety of extractants, using different times of extraction. They did analyse HI-reducible S in the extracts but did not determine HI-reducible S in HA and FA after fractionation. In a subsequent study (Freney et al., 1971) some information is presented on total, HI-reducible and Raney nickel C-bonded S for one soil and this is summarized in Table XVI. The proportion of HI-reducible S (48% of total S) in the whole soil is

TABLE XVI

Fractionation of S in an Australian soil
(From Freney et al., 1971).

Fraction	Total S (ppm)	HI-reducible S		C-bonded S (difference)	
		(ppm)	(% of total)	(ppm)	(% of total)
Whole soil	102	49	48	53	52
NaOH extract:					
humic acid	33	30	91	3	9
fulvic acid	17	5	29	12	71
NaHCO$_3$ extract:					
humic acid	14	9	64	5	36
fulvic acid	13	6	46	7	54

close to the average shown for a variety of soils (Table XV) and most of the
HA-S is ester sulfate. The two extractants, however, showed different percentages of HI-reducible S in HA's and FA's. Sodium hydroxide extracted HA's
more completely (shown by S values) than did sodium bicarbonate. No
further information was presented on these fractions. Freney (1961) also
applied 6N NCL hydrolysis to a NaOH extract and showed the release of
inorganic sulfate, but not much more is known about the nature of S components that are involved. Lowe (1966) presents a few Raney nickel and
HI-reducible S values for a HA preparation, with an average of 61% being C-
bonded and the remainder HI-reducible-S. Recently, Houghton and Rose
(1976) showed that organic sulfate esters in HA fractions were quite resistant to sulfohydralase enzyme action and suggested on the basis of this information that sulfated macromolecules were present in these materials.

Other fractions

Multiple element analysis of fractions

Several extraction-fractionation approaches provide detailed data on more
than one of the three elements (N, P and S) in the same fraction. Only a partial survey will be presented here to illustrate some of the information that is
available. A relatively mild extraction system (acetylacetone with ultrasonic
dispersion) was shown to be relatively efficient for extracting organic P and
S (extracting almost all present in eight soils tested), but was less efficient
for N (48—68%) (Halstead et al., 1966). Some use this as evidence that organic P as well as organic S are not intimate components of humic materials.
Milder reagents are unable to extract highly humified, stable organic N components. There is a lack of comparisons of various extractants with regard to
their relative abilities to extract organic N, P and S. Also few characteristic
analyses for each of these elements have been done on the same soil mate-

rials, analyses such as 6N acid hydrolysis for N, inositol phosphate tests for P and HI-reducible and C-bonded S for S components. Lowe (1975) fractionated a FA extract by polyvinyl pyrrolidone (PVP) adsorption and membrane diafiltration into seven subfractions. Material adsorbed on PVP accounted for 55% of the organic matter recovered, and for 28%, 72% and 4% of the N, S and P, respectively. Molecular size fractionations varied slightly in N, S and P, but no trends were apparent. The seven subfractions ranged from 0.67—3.31% in N, 0.61—2.46% in S and 0.02—1.49% in P. One subfraction, designated B, was of interest in that it contained the highest percentages of N and P; 75.5% of the original P was in this fraction and inositol may have been an important constituent. The percentage sulfate S in the fractions was always less than one-half of the total S (17.7—36.4%), showing that C-bonded S was uniformly distributed through all fractions. The B fraction accounted for 40.8% of the original FA-S, with most (82.3%) in the C-bonded form, making this a very quantitatively significant S fraction. Polysaccharides were also measured and the largest percentages were found in PVP-unadsorbed fractions. Polysaccharides are of interest in that linkages to N, P and S are all possible.

Swift and Posner (1972a) fractionated HA from two soils with agar gel to prepare subfractions of different molecular weights. The higher molecular weight fractions had the highest N and P contents and their concentrations decreased with decreasing molecular weight. The S content of the subfractions remained essentially constant throughout the molecular weight range. Only the N was examined further and the amino acid content was shown to decrease with decreasing molecular weight of the HA fractions.

More studies are available in which only N and S analyses were done rather than N, S and P. This probably reflects the opinion that organic P is not an intimate part of humic components and if present, is a "contaminant". As mentioned earlier, some studies assume soil organic fractions to be P-free. Griffith and Schnitzer (1975) extracted HA's and FA's from six tropical soils and found that N ranged from 3.47 to 5.51% and S from 0.60% to 1.06% in the HA's and N from 2.02 to 3.29% and S from 1.25 to 3.56% in the FA's. A comparison of the effect of 6N HCl hydrolysis on N and S content of soil HA's showed a decrease of both these elements in the same samples, showing that N and S containing organics were unstable toward this treatment (Riffaldi and Schnitzer, 1973). It was suggested that 6N HCl hydrolysis "purified" HA, so that N and S components were considered "contaminants". Pyrolysis of soil organic matter extracts showed that some of the N and S was recoverable at temperatures up to 540°C, that is, some N and S constituents were very stable (Schnitzer and Hoffman, 1964). Lowe (1969a) gave some detail on both N and S in a HA fraction. An average of 61% of the total S in this fraction was C-bonded, and 39% of the C-bonded S could be accounted as amino acid S. This meant that slightly less than one-half of the S was in the HI-reducible form. About 39% of the total N in this fraction was accounted for by amino acids. In a comprehensive study of HA fractions of forest humus layers, Lowe and Godkin (1975) showed correlations among N, HI-reducible S, total S and ash content.

Clay-organic and particle size association

It has long been known that there is an intimate association between clay particles and organic matter in soils. Mortland (1970) summarizes theoretical bonding mechanisms; amino and amide-N groups appear to be important in these interactions. It has been suggested frequently that clays enhance the stability of proteinaceous organic constituents. Clay—organic complexes have been extracted from a variety of soils. The N content of these complexes increases and the C : N ratio decreases as the particle size decreases (Chichester, 1969; McKeague, 1971; Swift and Posner, 1972b; Anderson et al., 1974; Watson and Parsons, 1974). This increased concentration of N may be a result of an accumulation of amino acid and amino sugars (Anderson et al., 1974; Watson and Parsons, 1974; Whitehead et al., 1975). Clay-fixed inorganic NH_4^+ may also be a significant component of these clay complexes, but was not considered in any of these studies.

Reports on organic P measurements in particle size fractions are few, but the limited evidence available shows that it is concentrated in the silt and clay fractions, especially in the clays (Williams, 1959). Probable reasons for this lack of information are: (a) that most researchers have been primarily interested in inorganic P and have assumed that organic P is but a minor contributor to plant nutrition; and (b) analytical problems. As discussed previously in relation with organic-matter—metal—phosphate complexes and total organic P measurements, it is difficult to do accurate measurements of soil organic P and to properly interpret these measurements. Although some work has been done on molecular weight fractions of organic P with concurrent measurements of sesquioxides and inositol (Veinot and Thomas, 1972), one suspects that measurements of organic P also include inorganic phosphate—sesquioxide/Ca—humus complexes. Inositol phosphates can behave similar to inorganic phosphate ions toward iron, aluminum, and calcium so that interactions between organic P and these constituents are possible.

Little, if any, information is available on organic S concentrations of various particle size fractions or organo—clay complexes. This is probably so because of the rather limited interest in soil organic S as compared to C and N. It is not known whether organic S or fractions of it (e.g., HI-reducible or C-bonded S) are concentrated in finer particles as are N and P. This type of information would be interesting and useful.

BIOLOGICALLY "MEANINGFUL" FRACTIONS

Stability

Several theories have been proposed for the stability of organic N compounds in the soil, and although each has some supporting evidence, the situation has not been completely resolved. The theories are summarized below; some have been discussed in detail in the past (Bremner, 1965c; Bartholomew, 1965).

(1) Reaction of N-containing constituents with other soil organic components such as polyphenolics, humic substances, etc.

(2) Adsorption by clays (some metals may be involved) and adsorption of enzymes capable of degrading N-compounds.

(3) Limitation of energy or proper balance of nutrient elements for microbial degradation.

Similar causes may also be responsible for the stability of organic S and P compounds. The apparent stability coupled with a desire to be able to predict the N supplying power of soils has led to a number of studies involving the dynamics of soil organic N components. It has been postulated that various organic N constituents turn over at varying rates (active and passive phases), resulting in the release of inorganic N from various fractions for plant growth (Jansson, 1958). This theory of cycling N has also been extended to S (Van Praag, 1973) and could also apply to P. The controversy as to whether organic P is an intimate constituent of the rest of the organic matter makes it a different case.

Correlation approaches

Numerous approaches have been used to determine the relative availability of N, S and P in chemically prepared organic matter fractions. These studies involve correlations of chemical fractionations with mineralization during incubations and/or plant growth. Such studies have been done with N (Keeney and Bremner, 1966; Cornforth, 1968; Stanford, 1968; Verstraeten et al., 1970), P (Bromfield and Jones, 1970; Dormaar, 1972; Stewart et al., 1973; Adhikari and Ganguly, 1973) and S (Spencer and Freney, 1960; Bardsley and Lancaster, 1960; Ensminger and Freney, 1966; McLachlan and De Marco, 1975), but no single extraction technique has been universally accepted for any of the nutrients.

Tracer approaches

"Meaningful" fractions can be examined more precisely with isotopes. Numerous isotope studies have been conducted with the stable isotope ^{15}N and a variety of approaches have been used to examine this problem. Materials labelled with ^{15}N have been added and the distribution of the isotope in various fractions has been determined after a period of time (e.g., Hauck and Bystrom, 1970; Moore and Russell, 1970; Legg et al., 1971; Van Praag and Brigode, 1973; Kai et al., 1973; Allen et al., 1973; Chicester et al., 1975; Knowles, 1975). Variations in natural ^{15}N abundance have been demonstrated, and these are a result of biological discrimination of the N isotope in microbial metabolic pathways. This is being examined for possible use in evaluating qualitative and quantitative aspects of N cycle processes (Shearer et al., 1974; Meints et al., 1975; Rennie et al., 1976). This technique has been oriented more toward environmental evaluations than organic N fractionation. The method requires high precision ^{15}N analyses and information on isotope discrimination of the biological pathways involved. The latter has yet to be fully evaluated, making this approach difficult to apply and interpret. An approach utilizing two labelled isotopes shows a great deal of promise. This approach has included C and N labelled materials (McGill et al., 1975). Radiocarbon dating has been used to evaluate the relative stability of various soil organic matter fractions, and these measurements together with N measurements can provide information on the biodynamics of various fractions (Martel and Paul, 1974a,b; Rennie et al., 1974). One can probably summarize that, although considerable progress has been made, these approaches (chemical correlations and isotope methods which involve examinations of the biological significance of various fractionations of soil organic N) show that no single extraction-fractionation system is entirely satisfactory. It appears that refinement of the fractionation procedures must go hand in hand with biological evaluations for "meaningful" fractions. Biological evaluations with one or more isotopes appear to be the most fruitful.

Several S isotope studies have been done which evaluate the usefulness of S fractionations for biological systems (Scharpenseel, 1962; Till and May, 1971; Freney et al., 1971, 1975; Bettany et al., 1974; Van Praag, 1973). Again, the fractionations that were done were not entirely satisfactory in that all fractions contributed toward the available or mineralizable S. Probably chemical fractionation procedures must be further refined to establish "meaningful" fractions. The limited information on variations of natural S isotopes in soil fractions appears promising for this type of evaluation, but much more work would be needed for more complete interpretations (Lowe et al., 1971).

Few, if any, reports of a comparable nature on soil organic P-containing fractions using isotope tracers are available. This is probably due to the assumption that organic P plays a small role in plant nutrition, to the availability of limited information on chemical extraction-fractionation procedures and to the nature of P isotopes that are available. It is probably unfortunate that N, P and S measurements are not available concurrently for all the extraction-fractionation studies, particularly for radiocarbon dating studies, so that comparative information would be available on all three of these important nutrient elements. However, biological evaluations of various fractions do show that large quantities of an element in a particular fraction are perhaps not the best criterion for evaluating its importance. A fraction which is quantitatively small, but turns over rapidly may be equally or more important to plant nutrition than a quantitatively large fraction that turns over slowly. This is analogous to constituents within plant systems such as vitamins, etc. that occur in small amounts but which are very important metabolically.

CONCLUDING REMARKS

Progress has been made in determining the nature of N-, P- and S-containing organics in soils, yet significant portions of these components have yet to be identified. Useful work has been done on various fractionation schemes to determine whether the fractions so produced have biological significance. A double thrusted effort of this type is needed to understand the complex and dynamic soil system. New analytical techniques are being developed which may provide us with better tools. Examples of these include detectors on gas chromatographs specific for N, P and S, gas chromatographic-mass spectrometric systems, liquid chromatography, improved automated systems and new analytical methods such as molecular emission cavity analysis for specific S and organophosphorus species. Older analytical techniques should be critically evaluated to discover precisely what the measurements mean (for example, are methods determining total N, P and S accurate, etc.); otherwise many assumptions will be misleading.

REFERENCES

Adams, A.P., Bartholomew, W.V. and Clark, F.E., 1954. Soil Sci. Soc. Am. Proc., 18: 40—46.
Adhikari, M. and Ganguly, T.K., 1973. Indian J. Agric. Res., 7: 1—10.
Aldag, R.W., Young, J.L. and Yamamoto, M., 1971. Phytochemistry, 10: 267—274.
Alexander, M., 1961. An Introduction into Soil Microbiology. Wiley, New York, N.Y., 472 pp.
Allen, A.L., Stevenson, F.J. and Kurtz, L.T., 1973. J. Environ. Quality, 2: 120—124.
Anderson, C.A. and Black, C.A., 1965. Soil Sci. Soc. Am. Proc., 29: 255—259.
Anderson, D.W., Paul, E.A. and St. Arnaud, R.J., 1974. Can. J. Soil Sci., 54: 317—323.
Anderson, G., 1961. Soil Sci., 91: 156—161.
Anderson, G., 1963. J. Sci. Food Agric., 14: 352—359.
Anderson, G., 1967. In: A.D. McLaren and G.H. Peterson (Editors), Soil Biochemistry. Dekker, New York, N.Y., pp. 67—90.
Anderson, G., 1970. J. Soil Sci., 21: 96—104.
Anderson, G., 1975. In: J.E. Giesking (Editor), Soil Components, 1. Organic Components. Springer-Verlag, New York, N.Y., pp. 305—331.
Anderson, G. and Arlidge, E.Z., 1962. J. Soil Sci., 13: 216—224.

Anderson, G. and Malcolm, R.E., 1974. J. Soil Sci., 25: 282—297.

Anderson, G. and Russell, J.D., 1969. J. Sci. Food Agric., 20: 78—81.

Babiuk, L.A. and Paul, E.A., 1970. Can. J. Microbiol., 16: 57—62.

Baker, R.T., 1975. J. Soil Sci., 26: 432—436.

Bandurski, R.S., 1965. In: J. Bonner and J.E. Varner (Editors), Plant Biochemistry. Academic Press, New York, N.Y. pp. 467—490.

Banwart, W.L. and Bremner, J.M., 1975. Soil Biol. Biochem., 7: 359—364.

Bardsley, C.E. and Lancaster, J.D., 1960. Soil Sci. Soc. Am. Proc., 24: 265—268.

Barrow, N.J., 1961. Soils Fertilizers, 24: 169—173.

Barrow, N.J., 1968. J. Sci. Food Agric. 19: 454—456.

Barrow, N.J., 1970. J. Sci. Food Agric., 21: 439—440.

Barrow, N.J. and Lambourne, L.J., 1962. Aust. J. Agric. Res., 13: 459—471.

Bartholomew, W.V., 1965. In: W.V. Bartolomew and F.E. Clark (Editors), Soil Nitrogen. Am. Soc. Agron., Madison, Wisc., pp. 285—308.

Bartholomew, W.V. and Goring, C.A.I., 1948. Soil Sci. Soc. Am. Proc., 13: 238—241.

Bartholomew, W.V., Meyer, J. and Laubelout, H., 1953. Institut National Pour L'Etude Agronomique du Congo Belge (I.N.E.A.C.) Ser. Sci., 57.

Beaton, J.D., Burns, G.R. and Platou, J., 1968. Tech. Bull., 14. The Sulphur Institute, Washington, 56 pp.

Behrman, E.J., 1967. J. Am. Chem. Soc., 89: 2424.

Bettany, J.R., Stewart, J.W.B. and Halstead, E.H., 1973. Soil Sci. Soc. Am. Proc., 37: 915—918.

Bettany, J.R., Stewart, J.W.B. and Halstead, E.H., 1974. Can. J. Soil Sci., 54: 309—315.

Bhardwaj, S.P. and Pathak, A.N., 1969. J. Soil Water Conserv. India, 17: 28—29. In: Soils Fertilizers, 34: 359 [2686] (1971).

Biederbeck, V.O. and Paul, E.A., 1973. Soil Sci., 115: 357—366.

Black, C.A. and Goring, C.A.I., 1953. Agronomy, 4: 123—152.

Blair, G.J., 1971. J. Aust. Inst. Agric. Sci., 37: 113—121.

Bloomfield, C., Brown, G. and Catt, J.A., 1970. Plant Soil, 33: 479—481.

Bowman, B.T., Thomas, R.L. and Elrick, D.E., 1967. Soil Sci. Soc. Am. Proc., 31: 477—481.

Bremner, J.M., 1949. J. Agric. Sci., 39: 183—193.

Bremner, J.M., 1965a. In: C.A. Black, D.D. Evans, J.L. White, L.E. Ensminger, F.E. Clark and R.C. Dinauer (Editors), Methods of Soil Analysis, Part 2. Chemical and Microbiological Properties. Am. Soc. Agron., Madison, Wisc. Agron Ser., 9: 1149—1237.

Bremner, J.M., 1965b. In: C.A. Black, D.D. Evans, J.L. White, L.E. Ensminger, F.E. Clark and R.C. Dinauer (Editors), Methods of Soil Analysis, Part 2. Chemical and Microbiological Properties. Am. Soc. Agron., Madison, Wisc., Agron. Ser., 9: 1238—1255.

Bremner, J.M., 1965c. In: W.V. Bartholomew and F.E. Clark (Editors). Soil Nitrogen. Am. Soc. Agron., Madison, Wisc., pp. 285—308.

Bremner, J.M., 1967. In: A.D. McLaren and G.H. Peterson (Editors), Soil Biochemistry. Dekker, New York, N.Y., pp. 19—66.

Bremner, J.M. and Tabatabai, M.A., 1971. In: Instrumental Methods for Analysis of Soils and Plant Tissue. Soil Sci. Soc. Am., Madison, Wisc., pp. 1—15.

Bromfield, S.M. and Jones, O.L., 1970. Aust. J. Agric. Res., 21: 699—711.

Burangulova, M.N., 1959. Biol. nucein Obmen. Rast., 177—182 R.Zh. (Biol.) 1961. Soil Fertilizers, 24: 261 [1783] (1961).

Byrne, E. and Power, T., 1974. Comm. Soil Sci. Plant Anal., 5: 51—65.

Campbell, C.A., Cameron, D. and Davidson, H.R., 1976. Plant Soil. In press.

Chapman, H.D., 1966. Diagnostic Criteria for Plants and Soils. Univ. of Calif., Berkeley, Calif., 793 pp.

Cheng, C.N., 1975. Soil Biol. Biochem., 7: 319—322.
Cheng, C.N. and Ponnamperuma, C., 1974. Geochim. Cosmochim. Acta, 38: 1843—1848.
Cheshire, M.V. and Anderson, G., 1975. Soil Sci., 119: 356—362.
Chichester, F.W., 1969. Soil Sci., 107: 356—363.
Chichester, F.W., Legg, J.O. and Stanford, G., 1975. Soil Sci., 120: 455—460.
Christian, R.R., Bancroft, K. and Wiebe, W.J., 1975. Soil Sci., 119: 89—97.
Clark, F.E. and Paul, E.A., 1970. Adv. Agron., 22: 375—435.
Cornforth, I.S., 1968. Expl. Agric., 4: 193—201.
Cosgrove, D.J., 1967. In: A.D. McLaren and G.H. Peterson (Editors), Soil Biochemistry. Dekker, New York, N.Y., pp. 216—228.
Coupland, R.T., 1974. Canadian Committee for the International Biological Program (Matador Project), Tech. Rep., 41. University of Sask., Saskatoon.
Dormaar, J.F., 1963. Can. J. Soil Sci., 43: 235—241.
Dormaar, J.F., 1970. Soil Sci., 110: 136—139.
Dormaar, J.F., 1972. Can. J. Soil Sci., 52: 107—112.
Ensminger, L.E. and Freney, J.R., 1966. Soil Sci., 101: 283—290.
Felbeck Jr., G.T., 1965. Adv. Agron., 17: 327—368.
Felbeck Jr., G.T., 1971. Soil Sci., 111: 42—48.
Finlayson, A.J. and MacKenzie, S.L., 1976. Anal. Biochem., 70: 397—402.
Fliermans, C.B. and Brock, T.D., 1973. Soil Sci., 115: 120—122.
Forsyth, W.C.G., 1947. Biochem. J., 41: 176—181.
Freney, J.R., 1958. Soil Sci., 86: 241—244.
Freney, J.R., 1961. Aust. J. Agric. Res., 12: 424—432.
Freney, J.R., 1967. In: A.D. McLaren and G.H. Peterson (Editors), Soil Biochemistry. Dekker, New York, N.Y., pp. 229—259.
Freney, J.R., 1968. Trans. 9th Int. Congr. Soil Sci., 3: 531—539.
Freney, J.R., Melville, G.E. and Williams, C.H., 1969. J. Sci. Food Agric., 20: 440—445.
Freney, J.R., Melville, G.E. and Williams, C.H., 1970. Soil Sci., 109: 310—318.
Freney, J.R., Melville, G.E. and Williams, C.H., 1971. Soil Biol. Biochem., 3: 133—141.
Freney, J.R., Melville, G.E. and Williams, C.H., 1975. Soil Biol. Biochem., 7: 217—221.
Freney, J.R. and Miller, R.J., 1970. J. Sci. Food Agric., 21: 57—61.
Freney, J.R., Stevenson, F.J. and Beavers, A.H., 1972. Soil Sci., 114: 468—476.
Giddens, J. and Rao, A.M., 1975. J. Environ. Quality, 4: 275—278.
Gilbert, E.E., 1965. Sulfonation and Related Reactions. Interscience Publishers, New York, N.Y., pp. 529.
Goodman, B.A. and Cheshire, M.V., 1973. Nature New Biol., 244: 158—159.
Gordon, M. and Hunter, A.S., 1969. J. Heterocycl. Chem., 6: 739—744.
Goring, C.A.I. and Bartholomew, W.V., 1952. Soil Sci., 74: 149—164.
Greaves, M.P. and Wilson, M.J., 1969. Soil Biol. Biochem., 1: 317—323.
Greenland, L.G., 1972. Plant Soil, 36: 191—198.
Griffith, S.M. and Schnitzer, M., 1975. Soil Sci. Soc. Am. Proc., 39: 861—867.
Griffith, S.M., Sowden, F.J. and Schnitzer, M., 1976. Soil Biol. Biochem., 8: 529—531.
Grindel', N.M. and Zyrin, N.G., 1965. Sov. Soil Sci., 12: 1393—1401.
Grov, A., 1963. Acta Chem. Scand., 17: 2316—2318.
Grov, A. and Alvsaker, E., 1963. Acta Chem. Scand., 17: 2307—2315.
Guidi, G., Petruzzelli, G. and Sequi, P., 1976. Can. J. Soil Sci., 56: 159—166.
Hagedorn, H., 1969. Plant Soil, 31: 161—178.
Halm, B.J., 1972. Ph.D. Thesis, Univ. of Sask., Saskatoon.
Halstead, R.L. and Anderson, G., 1970. Can. J. Soil Sci., 50: 111—120.
Halstead, R.L., Anderson, G. and Scott, N.M., 1966. Nature, 211: 1430—1431.

Halstead, R.L. and McKercher, R.B., 1975. In: E.A. Paul and A.D. McLaren (Editors), Soil Biochemistry, 4. Dekker, New York, N.Y., pp. 31—63.

Hance, R.J. and Anderson, G., 1963a. Soil Sci., 96: 94—98.

Hance, R.J. and Anderson, G., 1963b. Soil Sci., 96: 157—161.

Hannapel, R.J., Fuller, W.H., Bosma, S. and Bullock, J.S., 1964. Soil Sci., 97: 350—357.

Hauck, R.L. and Bystrom, M., 1970. ^{15}N. A Selected Bibliography for Agricultural Scientists. The Iowa State Univ. Press, Ames, Iowa.

Haworth, R.D., 1971. Soil Sci., 111: 71—79.

Houghton, C. and Rose, F.A., 1976. Appl. Environ. Microbiol., 31: 969—976.

Islam, A. and Ahmed, B., 1973. J. Soil Sci., 24: 193—198.

Ivarson, K.C. and Sowden, F.J., 1966. Can. J. Soil Sci., 46: 115—120.

Ivarson, K.C. and Sowden, F.J., 1969. Can. J. Soil Sci., 49: 121—127.

Ivarson, K.C. and Sowden, F.J., 1970. Can. J. Soil Sci., 50: 191—198.

Ivarson, K.C., Sowden, F.J. and Mack, A.R., 1970. Can. J. Soil Sci., 50: 183—189.

Jansson, S.L., 1958. Kungl. Lantbrukshögsk. Ann., 24: 101—361.

Jansson, S.L., 1971. In: A.D. McLaren and J. Skujiņs (Editors), Soil Biochemistry, 2. Dekker, New York, N.Y., pp. 129—166.

Jensen, H.L., 1932. J. Agric. Sci., 22: 1—25.

Johnson, C.M. and Nishita, H., 1952. Anal. Chem., 24: 736—742.

Jones, M.J. and Parsons, J.W., 1972. J. Soil Sci., 23: 128—134.

Jones, P.D., Graham, V., Segal, L., Baillie, W.J. and Briggs, M.H., 1962. Life Sci., 1: 645—648.

Kai, H., Ahmad, Z. and Harada, T., 1973. Soil Sci. Plant Nutr., 19: 275—286.

Kaila, A., 1949. Soil Sci., 68: 279—289.

Keeney, D.R. and Bremner, J.M., 1966. Soil Sci. Soc. Am. Proc., 30: 714—719.

Knowles, R., 1975. In: B. Bernier and C.H. Winget (Editors), Forest Soils and Forest Land Management. Proc. N. Am. Forest Soils Conf., 4th. Les Presses de l'Université, Laval, Que., pp. 53—65.

Kowalenko, C.G., 1970. M. Sc. Thesis, Univ. of Sask., Saskatoon, 106 pp.

Kowalenko, C.G., 1972. Unpublished data.

Kowalenko, C.G. and McKercher, R.B., 1970. Soil Biol. Biochem., 2: 269—273.

Kowalenko, C.G. and McKercher, R.B., 1971a. Soil Biol. Biochem., 3: 243—247.

Kowalenko, C.G. and McKercher, R.B., 1971b. Can. J. Soil Sci., 51: 19—22.

Kowalenko, C.G. and Lowe, L.E., 1975a. Can. J. Soil Sci., 55: 1—8.

Kowalenko, C.G. and Lowe, L.E., 1975b. Can. J. Soil Sci., 55: 9—14.

Ladd, J.N. and Brisbane, P.G., 1967. Aust. J. Soil Res., 5: 161—171.

Ladd, J.N. and Butler, J.H.A., 1975. In: E.A. Paul and A.D. McLaren (Editors), Soil Biochemistry, 4. Dekker, New York, N.Y., pp. 143—194.

Lee, C.C., Harris, R.F., Williams, J.D.H., Syers, J.K. and Armstrong, D.E., 1971. Soil Sci. Soc. Am. Proc., 35: 86—91.

Legg, J.O., Chicester, F.W., Stanford, G. and DeMar, W.H., 1971. Soil Sci. Soc. Am. Proc., 35: 273—276.

Levesque, M., 1969. Can. J. Soil Sci., 49: 365—373.

Levesque, M. and Schnitzer, M., 1967. Soil Sci., 103: 183—190.

Lloyd, A.B., 1975. Soil Biol. Biochem., 7: 357—358.

Lovelock, J.E., Maggs, R.J. and Rasmussen, R.A., 1972. Nature, Lond., 237: 452—543.

Lowe, L.E., 1964. Can. J. Soil Sci., 44: 176—179.

Lowe, L.E., 1966. Can. J. Soil Sci., 46: 92—93.

Lowe, L.E. 1968. Can. J. Soil Sci., 48: 215—217.

Lowe, L.E., 1969a. Can. J. Soil Sci., 49: 129—141.

Lowe, L.E., 1969b. Can. J. Soil Sci., 49: 375—381.

Lowe, L.E., 1973. Soil Sci. Soc. Am. Proc., 37: 569—572.

Lowe, L.E., 1975. Can. J. Soil Sci., 55: 119—126.
Lowe, L.E. and DeLong, W.A., 1963. Can. J. Soil Sci., 43: 151—155.
Lowe, L.E. and Godkin, C.H., 1975. Can. J. Soil Sci., 55: 381—393.
Lowe, L.E. and Tsang, W.-C., 1970. Can. J. Soil Sci., 50: 456—457.
Lowe, L.E., Sasaki, A. and Krouse, H.R., 1971. Can. J. Soil Sci., 51: 129—131.
Mamchenko, O.A., 1970. Sov. Soil Sci. 2: 125.
Martel, Y.A. and Paul, E.A., 1974a. Can. J. Soil Sci., 54: 419—426.
Martel, Y.A. and Paul, E.A., 1974b. Soil Sci. Soc. Am. Proc., 38: 501—505.
Martin, A.J.P. and Synge, R.L.M., 1945. Adv. Protein Chem., 2: 1—84.
Martin, J.K., 1964. N.Z.J. Agric. Res., 7: 736—749.
Martin, J.K., 1970. Soil Sci., 109: 362—375.
McGill, W.B., 1971. Ph.D. Thesis, Univ. of Sask., Saskatoon, 269 pp.
McGill, W.B., Shields, J.A. and Paul, E.A., 1975. Soil Biol. Biochem., 7: 57—63.
McKeague, J.A., 1971. Can. J. Soil Sci., 51: 499—505.
McKercher, R.B., 1968. Trans. 9th Int. Congr. Soil Sci., 3: 547—553.
McLachlan, K.D. and DeMarco, D.C., 1975. Aust. J. Soil Res., 13: 169—176.
Meints, V.W., Boone, L.V. and Kurtz, L.T., 1975. J. Environ. Quality 4: 486—490.
Meints, V.W. and Peterson, G.A., 1972. Soil Sci. Soc. Am. Proc., 36: 434—436.
Melville, G.E., Freney, J.R. and Williams, C.H., 1971. Soil Sci., 112: 245—248.
Miller, D.F., 1958. Composition of Cereal Grains and Forages. National Academy of
    Sciences, National Research Council, Washington, 661 pp.
Miller, E.V., 1957. The Chemistry of Plants. Reinhold, New York, N.Y., pp. 229—259.
Mitchell, J., Moss, H.C. and Clayton, J.S., 1944. Soil Surv. Rep., 12. Univ. of Sask.,
    Saskatoon, 186 pp.
Moore, A.W. and Russell, J.S., 1970. Aust. J. Soil Res., 8: 21—30.
Mortland, M.M., 1970. Adv. Agron., 22: 75—117.
Moyer, J.R. and Thomas, R.L., 1970. Soil Sci. Soc. Am. Proc., 34: 80—83.
Neptune, A.M.L., Tabatabai, M.A. and Hanway, J.J., 1975. Soil Sci. Soc. Am. Proc., 39:
    51—55.
Neyroud, J.A. and Schnitzer, M., 1974. Can. J. Chem., 52: 4123—4132.
Omotoso, T.I. and Wild, A., 1970. J. Soil Sci., 21: 224—232.
De Serra, M.I., Sowden, F.J. and Schnitzer, M., 1973. Can. J. Soil Sci., 53: 125—127.
Parsons, J.W. and Tinsley, J., 1961. Soil Sci., 92: 46—53.
Parsons, J.W. and Tinsley, J., 1975. In: J.E. Giesking (Editor), Soil Components, 1.
    Organic Components. Springer-Verlag, New York, N.Y., pp. 263—304.
Paul, E.A. and Schmidt, E.L., 1960. Soil Sci. Soc. Am. Proc., 24: 195—198.
Paul, E.A. and Schmidt, E.L., 1961. Soil Sci. Soc. Am. Proc., 25: 359—362.
Paul, E.A. and Tu, C.M., 1965. Plant Soil, 22: 207—219.
Peterson, J.R., McCalla, T.M. and Smith, G.E., 1971. In: R.A. Olsen, T.J. Army, J.J.
    Hamway and V.J. Kilmer (Editors), Fertilizer Technology and Use. 2nd ed. Soil Sci.
    Soc. Am. Proc., Madison, Wisc., pp. 557—596.
Peverill, K.I., Briner, G.P. and Douglas, L.A., 1975. Aust. J. Soil Res., 13: 69—75.
Philips, P.A., McLean, A.J., Hore, F.R., Sowden, F.J., Tennant, A.D. and Patni, N.K.,
    1975. Eng. Res. Serv., Agric. Canada, Ottawa, Rep., 7043—540.
Pinck, L.A. and Allison, F.E., 1944. Soil Sci., 57: 155—161.
Piper, T.J. and Posner, A.M., 1968. Soil Sci., 106: 188—192.
Piper, T.J. and Posner, A.M., 1972. Plant Soil, 36: 595—598.
Porter, L.K., 1975. In. E.A. Paul and A.D. McLaren (Editors), Soil Biochemistry, 4.
    Dekker, New York, N.Y., pp. 1—30.
Putnam, H.D. and Schmidt, E.L., 1959. Soil Sci., 87: 22—27.
Rennie, D.A., Paul, E.A. and Johns, L.E., 1974. In: Effects of Agricultural Production on
    Nitrates in Food and Water with Particular Reference to Isotope Studies. IAEA,
    Vienna, pp. 77—90.

Rennie, D.A., Paul, E.A. and Johns, L.E., 1976. Can. J. Soil Sci., 56: 43—50.
Rennie, D.A., Warembourg, F. and Paul, E.A., 1971. In: International Symposium of Use of Isotopes and Radiation in Agriculture and Animal Husbandry Use, New Delphi, pp. 597—615.
Richards, I.R. and Wolton, K.M., 1976. J. Sci. Food Agric., 27: 426—428.
Riffaldi, R. and Schnitzer, M., 1973. Soil Sci., 115: 349—356.
Robertson, G., 1958. J. Sci. Food Agric., 9: 288—294.
Rolston, D.E., Rauschkolb, R.S. and Hoffman, D.L., 1975. Soil Sci. Soc. Am. Proc., 39: 1089—1094.
Rovira, A.D., 1965. In: K.F. Baker and W.C. Snyder (Editors), Ecology of Soil-borne Plant Pathogens. Univ. of Calif. Press, Berkley, Calif., pp. 170—186.
Roy, M.K. and Chandra, A.L., 1976. Folia Microbiol., 21: 50—53.
Russell, R.S. and Adams, S.N., 1954. Plant Soil, 5: 223—225.
Sato, O. and Kumada, K., 1967. Soil Sci. Plant Nutr., 13: 121—122.
Scharpenseel, H.W., 1962. Radioisotopes in Soil—Plant Nutritional Studies. IAEA, Vienna.
Scharpenseel, H.W. and Krausse, R., 1962. Z. Pflanzenernähr. Düng. Bodenk., 96: 11—34.
Schnitzer, M. and Hoffman, I., 1964. Soil Sci. Soc. Am. Proc., 28: 520—525.
Schnitzer, M. and Skinner, S.I.M., 1968. In: Isotopes and Radiation in Soil Organic-matter Studies. IAEA, Vienna, pp. 41—54.
Schnitzer, M. and Vendette, E., 1975. Can. J. Soil Sci., 50: 93—103.
Schnitzer, M., Sowden, F.J. and Ivarson, K.C., 1974. Soil Biol. Biochem., 6: 401—407.
Scott, N.M. and Anderson, G., 1976. J. Sci. Food Agric., 27: 358—366.
Sekhon, G.S. and Black, C.A., 1969. Plant Soil., 31: 321—327.
Sequi, P., Guidi, G. and Petruzzelli, G., 1975. Can. J. Soil Sci., 55: 439—445.
Shearer, D.A. and Warner, R.M., 1971. Int. J. Environ. Anal. Chem., 1: 11—21.
Shearer, G., Duffy, J., Kohl, D.H. and Commoner, B., 1974. Soil Sci. Soc. Am. Proc., 38: 315—322.
Simoneaux, B.J. and Caldwell, A.G., 1965. Agron. Abst., Annual Mtg. Am. Soc. Agron., La. Agric. Ex. Sta., pp. 77.
Sinha, M.K., 1971. Plant Soil, 35: 485—493.
Skujiņś, J.J., 1967. In: A.D. McLaren and G.H. Peterson (Editors), Soil Biochemistry. Dekker, New York, N.Y., pp. 371—414.
Sowden, F.J., 1955. Soil Sci., 80: 181—188.
Sowden, F.J., 1958. Can. J. Soil Sci., 38: 147—154.
Sowden, F.J., 1959. Soil Sci., 88: 138—143.
Sowden, F.J., 1966. Soil Sci., 102: 264—271.
Sowden, F.J., 1967. In: Automation in Analytical Chemistry. Technicon Symposium, New York, N.Y., pp. 129—132.
Sowden, F.J., 1969. Soil Sci., 107: 364—371.
Sowden, F.J., 1970a. Can. J. Soil Sci., 50: 227—232.
Sowden, F.J., 1970b. Can. J. Soil Sci., 50: 233—241.
Sowden, F.J. and Ivarson, K.C., 1966. Can. J. Soil Sci., 46: 109—114.
Sowden, F.J., Griffith, S.M. and Schnitzer, M., 1976. Soil Biol. Biochem., 8: 55—60.
Sparrow, E.B. and Doxtader, K.G., 1973. U.S. I.B.P. Tech. Rep. 224.
Spector, W.S., 1956. Handbook of Biological Data. Wright Air Development Center Tech. Rep. 56-273. Wright-Patterson Air Force Base, Ohio, p. 89.
Spencer, K. and Freney, J.R., 1960. Aust. J. Agric. Res., 11: 948—959.
Stanford, G., 1968. Soil Sci., 106: 345—351.
Steward, J.H. and Tate, M.E., 1971. J. Chromatogr., 60: 75—82.
Stewart, J.W.B., Halm, B.J. and Cole, C.V., 1973. Can. Comm. Int. Biol. Proj. (Matador Project), Tech. Rep. 40.

Suter, C.M., 1970. Organic Chemistry of Sulfur. (Reprint edition, 1970). Intra-Science Research Foundation, Santa Monica, Calif.

Swift, R.S. and Posner, A.M., 1972a. J. Soil Sci., 23: 50—57.

Swift, R.S. and Posner, A.M., 1972b. Soil Biol. Biochem., 4: 181—186.

Tabatabai, M.A. and Bremner, J.M., 1970. Soil Sci. Soc. Am. Proc., 34: 417—420.

Till, A.R. and May, P.E., 1971. Aust. J. Agric. Res., 22: 391—400.

Tinsley, J. and Özsavasci, C., 1974. Trans. Int. Congr. Soil Sci., 10th, Moscow, 2: 332—340.

Umarov, M.M. and Aseyeva, I.V., 1971. Sov. Soil Sci., 3: 639.

Van Praag, H.J., 1973. Plant Soil, 39: 61—69.

Van Praag, H.J. et Brigode, N., 1973. Plant Soil, 39: 49—59.

Veinot, R.L. and Thomas, R.L., 1972. Soil Sci. Soc. Am. Proc., 36: 71—73.

Verstraeten, L.M.J., Vlassak, K. and Livens, J., 1970. Soil Sci., 110: 299—305.

Verstraeten, L.M.J., Vlassak, K. and Livens, J., 1972. Soil Sci., 113: 200—203.

Waid, J.S., 1975. In: E.A. Paul and A.D. McLaren (Editors), Soil Biochemistry, 4. Dekker, New York, N.Y., pp. 65—101.

Wainwright, M. and Pugh, G.J., 1975. Soil Biol. Biochem., 7: 1—4.

Waldron, A.C. and Mortensen, J.L., 1962. Soil Sci., 93: 286—293.

Walker, T.W. and Adams, A.F.R., 1959. Soil Sci., 87: 1—10.

Watson, J.R. and Parsons, J.W., 1974. J. Soil Sci., 25: 1—8.

Weir, C.C. and Soper, R.J., 1963. Can. J. Soil Sci., 43: 393—399.

Wenzel, W., 1961. Z. Pflanzenernähr. Düng. Bodenk., 93: 148—164.

Whitehead, D.C., 1964. Soils Fertilizers, 27: 1—8.

Whitehead, D.C., 1970. J. Br. Grasslands Soc., 25: 236—241.

Whitehead, D.C., Buchan, H. and Hartley, R.D., 1975. Soil Biol. Biochem., 7: 65—71.

Wild, A. and Oke, O.L., 1966. J. Soil Sci., 17: 356—371.

Wilkinson, S.R., 1973. In: G.W. Butler and R.W. Bailey (Editors), Chemistry and Biochemistry of Herbage, 2. Academic Press, New York, N.Y., pp. 247—315.

Williams, C.H., 1967. Plant Soil, 26: 205—223.

Williams, C.H., Williams, E.G. and Scott, N.M., 1960. J. Soil Sci., 11: 334—346.

Williams, E.G., 1959. Agrochimica, 3: 279—309.

Wright, M.J. and Davidson, K.L., 1964. Adv. Agron., 16: 197—247.

Yaskin, A.A., 1968. Dokl. Soil Sci., A Supplement to Sov. Soil Sci., 13: 1823—1831.

Zobell, C.E., 1963. In: E. Ingerson (Editor), Organic Geochemistry. Earth Sci. Ser. Monograph, 16, Pergamon Press, New York, N.Y., pp. 543—578.

Chapter 4

THE INTERACTION OF ORGANIC MATTER WITH PESTICIDES

S.U. KHAN

INTRODUCTION

The chemicals used as pesticides represent many different classes of compounds. They are grouped according to the purpose for which they are used. In agriculture, herbicides, insecticides and fungicides are used for controlling weeds, insects, and plant pathogens, respectively. The interaction of these chemicals with organic matter is an important factor affecting the fate of pesticides in the soil environment. Several comprehensive reviews have been made recently indicating that persistence, degradation, bioavailability, leachability, and volatility of pesticides bear a direct relationship to the nature and content of organic matter in soil (Hayes, 1970; Wolcott, 1970; Adams, 1972; Khan, 1972; Stevenson, 1972; Weed and Weber, 1974). The role of organic matter in studying the organic matter—pesticide interactions in soil can be considered from two principal aspects.

(1) Adsorption of pesticides by organic matter. This may exert the most profound influence of the several processes operating to determine the fate of a pesticide in soil. Adsorption will control the quantity of a pesticide in soil solution, and thus determine its persistence, leaching, mobility and bioavailability. The extent of adsorption of a pesticide depends upon the nature and properties of the chemical itself, the kind and amount of organic matter present, and the environment provided in the soil. Once adsorbed by the organic matter surface, a pesticide may be easily desorbed, or it may be desorbed with difficulty, or not at all. There are conflicting reports in the literature regarding the reversible nature of adsorption. Complete desorption of the pesticide from organic surfaces (Nearpass, 1969, 1971; Dunigan and McIntosh, 1971), partial desorption (Talbert and Fletchall, 1965; Hance, 1967; Tompkins et al., 1968; Coffey and Warren, 1969), as well as nearly complete irreversibility (Coffey and Warren, 1969; Deli and Warren, 1971) have been observed.

(2) Nonbiological degradation of pesticides by organic matter. The organic fraction of soil can be exceptionally important in nonbiological degradation of pesticides (Crosby, 1970; Armstrong and Konrad, 1974; Stevenson, 1976). The behavior and fate of such compounds in soil would be greatly influenced by this process.

Only recently have attempts been made to study the interactions between

TABLE I

Common and chemical names of pesticides

Common name	Chemical name
Aldrin	1,2,3,4,10,10-hexachloro-1,4,4a,5,8,8a-hexahydro-exo-1,4-endo-5,8-dimethanonaphthalene
Amitrole	3-amino-s-triazole
Atrazine	2-chloro-4-ethylamino-6-isopropylamino-1,3,5-triazine
Benefin	N-butyl-N-ethyl-α,α,α-trifluoro-2,6-dinitro-p-toluidine
Bromacil	5-bromo-3-sec-butyl-6-methyluracil
Butralin	4-(1,1-dimethylethyl)-N-(1-methylpropyl)-2,6-dinitrobenzenamine
Cacodylic acid	hydroxydimethylarsine oxide
Carbaryl	1-naphthalenyl methylcarbamate
Chlordane	1,2,4,5,6,7,8,8-octachloro-2,3,3a,4,7,7a-hexahydro-4,7-methano-1H-indene
Chlorpropham	isopropyl 3-chlorophenylcarbamate
2,4-D	2,4-dichlorophenoxyacetic acid
DDT	1,1,1-trichloro-2,2-di(chlorophenyl)ethane
Diazinon	OO-diethyl O-2-isopropyl-6-methylpyrimidin-4-yl phosphorothioate
Dicamba	3,6-dichloro-2-methoxybenzoic acid
Dichlobenil	2,6-dichlorobenzonitrile
Dieldrin	1,2,3,4,10,10-hexachloro-6,7-epoxy-1,4,4a,5,6,7,8,8a-octahydro-exo-1,4-endo-5,8-dimethanonaphthalene
Dimefox	tetramethylphosphorodiamidic fluoride
Dinoseb	2-sec-butyl-4,6-dinitrophenol
Diphenamide	NN-dimethyldiphenylacetamide
Diquat	1,1'-ethylene-2,2'-bipyridylium ion
Diuron	3-(3,4-dichlorophenyl)-1,1-dimethylurea
DSMA	disodium methanearsonate
Endrin	1,2,3,4,10,10-hexachloro-6,7-epoxy-1,4,4a,5,6,7,8,8a-octahydro-exo-1,4-exo-5,8-dimethanonaphthalene
Fonofos	O-ethyl S-phenyl ethylphosphonodithioate
Heptachlor	1,4,5,6,7,8,8-heptachloro-3a,4,7,7a-tetrahydro-4,7-methanoindene
Heptachlor epoxide	1,4,5,6,7,8,8-heptachloro-2,3-epoxy-3a,4,7,7a-tetrahydro-4,7-methanoindene
Isocil	5-bromo-3-isopropyl-6-methyluracil
Lindane	r-1,2,3,4,5,6-hexachlorocyclohexane

TABLE I (continued)

Common name	Chemical name
Linuron	3-(3,4-dichlorophenyl)-1-methoxy-1-methylurea
Monuron	3-(4-chlorophenyl)-1,1-dimethylurea
Oryzalin	3,5-dinitro-N^4N^4-dipropylsulphanilamide
Paraquat	1,1'-dimethyl-4,4'-bipyridylium ion
Parathion	OO-diethyl O-4-nitrophenyl phosphorothioate
PCP	Pentachlorophenol
Phanacridane chloride	9-(p-n-hexyloxyphenyl)-10-methylacridinium chloride
Phosphon	tributyl-2,4-dichlorobenzyl phosphonium chloride
Phorate	OO-diethyl S-ethylthiomethyl phosphorodithioate
Picloram	4-amino-3,5,6-trichloropicolinic acid
Profluralin	N-cyclopropylmethyl-2,6-dinitro-N-propyl-4-trifluoromethyl-aniline
Prometone	2,4-di(isopropylamino)-6-methoxy-1,3,5-triazine
Prometryne	2,4-di(isopropylamino)-6-methylthio-1,3,5-triazine
Propazine	2-chloro-4,6-di(isopropylamino)-1,3,5-triazine
Propham	isopropyl phenylcarbamate
Pyridyl pyridinium chloride	N-(4-pyridyl)-pyridinium chloride
Simazine	2-chloro-4,6-di(ethylamino)-1,3,5-triazine
Terbacil	3-tert-butyl-5-chloro-6-methyluracil
Terbutryne	2-tert-butylamino-4-ethylamino-6-methylthio-1,3,5-triazine

soil organic matter and pesticides. Stevenson (1976) pointed out that information on the nature of organic matter—pesticide interactions may provide a more rational basis for their effective use, thereby reducing undesirable side effects due to carry-over and contamination of the environment. However, a proper understanding of the precise nature of these interactions is hindered due to the complexity of organic matter and the numerous other interactions in the soil environment all operating simultaneously. In recent years careful studies with simplified systems involving well-defined organic matter components have resulted in the elucidation of some of the mechanisms of interactions. The merits of using organic matter components, such as humic acids (HA) and fulvic acid (FA), are: (1) they can be readily extracted from soil organic matter in relatively pure form; (2) they have been thoroughly characterized by various techniques; and (3) they are the major and common constitutents of soil organic matter.

A limitation that needs to be considered in organic matter—pesticide interactions is that in most mineral soils, organic matter and clay minerals are intimately associated in the form of clay—organic matter complexes. Thus, organic matter may not function as a separate entity and its relative contribution in pesticide adsorption will depend upon the extent to which the clay is coated with organic matter (Stevenson, 1976). However, it should be realized that the association of organic matter with clay still provides an organic surface for adsorption (Stevenson, 1976).

Properties of soil systems, such as clay-mineral composition, pH, exchangeable cations, moisture, and temperature are also influential to some degree depending upon the pesticide involved. However, under any given set of soil conditions specified above, the organic matter—pesticide interactions could better be evaluated if adequate information was available about the adsorbent (organic matter) and the nature of the pesticide. The physical and chemical properties of the soil organic matter have been discussed in the previous chapters of this book. For additional information regarding humic substances, the reader is referred to the recent book of Schnitzer and Khan (1972). The nature of pesticides relevant to this subject matter will be briefly discussed first, followed by a discussion on various aspects of our current knowledge concerning the interactions of organic matter with pesticides.

Common and chemical names of pesticides cited in this chapter are listed in Table I.

NATURE OF PESTICIDES

One of the main characteristics of organic pesticides is that most of them are generally low molecular weight compounds with low water solubility. According to Bailey and White (1970) the chemical character, shape and configuration of the pesticide, its acidity (pKa) or basicity (pKb), its water solubility, the charge distribution on the cation, the polarity of the molecule, its molecular size and polarizability all affect the adsorption-desorption by soil colloids. In the following paragraphs only those factors which are particularly relevant to pesticide adsorption by soil organic matter are discussed briefly.

Four structural factors determine the chemical character of a pesticide molecule and thus, influence its adsorption on soil colloids (Bailey and White, 1970).

(1) Nature of functional groups, such as carboxyl ($-\overset{\overset{\text{O}}{\|}}{\text{C}}-\text{OH}$), carbonyl ($\text{C}=\text{O}$), alcoholic hydroxyl ($-\text{OH}$), and amino ($-\text{NH}_2$).

(2) Nature of substituting groups that may alter the behavior of functional groups.

(3) Position of substituting groups with respect to the functional groups

which may enhance or hinder intermolecular bonding.

(4) Presence and magnitude of unsaturation in the molecule, which affects the lyophylic—lyophobic balance.

The charge characteristics of a pesticide are probably the most important property governing its adsorption. The charge may be weak, arising from an unequal distribution of electrons producing polarity in the molecule, or it may be relatively strong, resulting from dissociation.

The pH of a system is also an important factor as it governs the ionization of most of the organic molecules. Acidic pesticides are proton donors, which at high pH (one or more pH unit above the pKa of the acid) become anions due to dissociation. On the other hand, basic compounds, when protonated may behave like organic cations. The adsorption behavior of pesticides which ionize in aqueous solutions to yield cations is different from those that yield anions. Furthermore, nonionic or neutral pesticides behave differently from cationic, basic or anionic pesticides. Neutral pesticides may be subjected to "temporary polarization" in the presence of an electrical field which contributes to adsorption on a charged surface. The availability of mobile electrons, such as π electrons in the benzene ring, influence the polarization of a neutral molecule. Thus, adsorption of neutral pesticides on charged surfaces may increase with molecular size when the increase involves addition of aromatic groups.

There is some controversy in the literature concerning the influence of water solubility on pesticide adsorption. Bailey et al. (1968) suggested that within a chemical family the magnitude of pesticide adsorption is directly related to and governed by the degree of water solubility. The hydrophobic character of a pesticide will increase by a decrease in its water solubility thereby resulting in stronger adsorption on soil organic matter (Hance, 1965b; Leenheer and Ahlrichs, 1971). Several workers have observed an inverse relationship between solubility and adsorption (Leopold et al., 1960; Hilton and Yuen, 1963; Ward and Upchurch, 1965). Weber (1972) noted that the amount of some acidic herbicides adsorbed on a muck soil was inversely related to the water solubilities of these compounds. Carringer et al. (1975) observed that adsorption of certain nonionic pesticides on organic matter was inversely related to the water solubilities of the chemicals. On the other hand, Hance (1965a, 1967) and Weber (1966, 1970) observed no relationship between solubility and adsorption of certain pesticides. It appears that for a particular family of pesticides, several factors may be interacting in determining direct, inverse or no relationship between water solubility and adsorbability. These may include such factors as surface acidity and the relative polarity of the adsorbent (Bailey and White, 1970). In general, pesticides of relatively high water solubility would be expected to be more effective in higher organic matter soils than chemicals of low water solubility.

For detailed information on the chemical properties of pesticides the reader is referred to the work of Metcalf (1971) and Melnikov (1971).

MECHANISMS OF ADSORPTION

Several mechanisms have been proposed for adsorption of pesticides by soil organic matter. Two or more mechanisms may occur simultaneously depending upon the nature of the pesticide and organic matter surface. The mechansims most likely involved in the adsorption of pesticides on organic matter surfaces are outlined below.

(1) Van der Waals attractions. Van der Waals forces are involved in the adsorption of nonionic, nonpolar molecules or portions of molecules. Van der Waals forces result from short range dipole—dipole interactions of several kinds. The additive nature of Van der Waals forces between the atoms of adsorbate and adsorbent may result in considerable attraction for large molecules. The adsorption of carbaryl and parathion on soil organic matter in aqueous systems is considered to be physical involving Van der Waals bonds between the hydrophobic portions of the adsorbate molecules and the adsorbent surface (Leenheer and Ahlrichs, 1971). Recently Nearpass (1976) suggested that the principal adsorption mechanism for picloram by humic materials was molecular adsorption due to Van der Waals forces.

(2) Hydrophobic bonding. Nonpolar pesticides or compounds whose molecules often have nonpolar regions of significant size in proportion to polar regions are likely to adsorb onto the hydrophobic regions of soil organic matter. Water molecules present in the system will not compete with nonpolar molecules for adsorption on hydrophobic surfaces. The potential importance of the hydrophobic fractions of organic matter for the retention of pesticides was cited by Hance (1969b). This type of bonding also may be largely responsible for the strong adsorption by soil organic matter of many pesticides such as DDT and organochlorine insecticides. The primary sites for chlorinated hydrocarbon pesticides in organic matter is lipid. As much as 20% lipid content is not uncommon for some peat and muck soils (Stevenson, 1966). Lipids are also associated with soil humus (Schnitzer and Khan, 1972). Thus, association of nonpolar (chlorinated hydrocarbons) pesticides with the lipid fraction of soil organic matter and humus might be described by hydrophobic bonding (Pierce et al., 1971). This also explains the relative independence of pesticide adsorption on moisture in soils with high organic content (Pierce et al., 1971). Khan and Schnitzer (1971, 1972) suggested that hydrogen bonding between humic polymers could form internal voids which trap hydrophobic molecules. Nonpolar portions of the humic polymer and hydrophobic molecules trapped within the polymer could provide hydrophobic binding sites for DDT (Pierce et al., 1974). The hydrophobic portion of peats such as fats, waxes and resins can be a significant adsorbent of phenylureas (Hance, 1969b; Morita, 1976). The adsorption of pesticides involving this mechanism would be independent of pH (Hance, 1965b; Walker and Crawford, 1968). Methylation of organic matter or humic substances to block hydrophilic-OH groups would increase

the adsorption by this mechanism. In view of this concept, adsorption of pesticides by a soil can be considered to be primarily a matter of partitioning between organic matter and water (Lambert et al., 1965; Lambert, 1968).

(3) Hydrogen bonding. This is a special kind of dipole-dipole interaction in which the hydrogen atom serves as a bridge between two electronegative atoms, one being held by covalent bond and the other by electrostatic forces. There is a parallel between hydrogen bonding and protonation (Hadzi et al., 1968). Protonation may be considered as a full charge transfer from the base (electron donor) to the acid (electron acceptor). The hydrogen bonding interaction is a partial charge transfer.

The presence of oxygen containing functional groups, as well as amino groups, on organic matter indicates that adsorption could occur by formation of a hydrogen bond with organic pesticides containing similar groups. For example, carbonyl oxygens on pesticide molecules may bound to amino hydrogens or hydroxyl groups on the organic matter. Additional sites for hydrogen bonding by soil organic matter includes —SH and —O— linkages (Stevenson, 1972). Hayes (1970) stressed the participation of a hydrogen bonding mechanism in s-triazines and organic matter interactions. Evidence for this type of bonding was obtained from infrared studies by Sullivan and Felbeck (1968). They observed that hydrogen bonding may take place between C=O groups of the humic compounds and the secondary amino groups of s-triazines. The heat of HA-atrazine complex formation was estimated as 8—13 Kcal./mole, which most likely is the heat of formation of one or more hydrogen bonds (Li and Felbeck, 1972). Maslennikova and Kruglow (1975) suggested that binding of simazine by hydrogen bonds with weakly acidic groups of HA may result in the formation of a stable complex.

Anionic pesticide adsorption at pH values below their pKa values can be attributed to adsorption of the unionized form of the molecule on organic surfaces. Thus, hydrogen bonding may take place between the COOH group and C=O or NH group of organic matter (Kemp et al., 1969). Hydrogen bonding would be limited to acid conditions where COOH groups are unionized (Stevenson, 1972).

(4) Charge transfer. In the formation of charge transfer complexes, electrostatic attraction takes place when electrons are transferred from an electron-rich donor to an electron-deficient acceptor. Charge transfer interaction will take place only within short distances of separation between the interacting species. The formation of charge transfer complexes has been postulated as the possible mechanism involved in the adsorption of s-triazines onto soil organic matter (Hayes, 1970). The charge transfer reactions are particularly important when trying to explain the high adsorption of methylthiotriazines onto organic matter (Hayes, 1970). Burns et al. (1973b) postulated the involvement of charge transfer mechanisms in paraquat adsorption by HA. However, ultraviolet spectroscopic studies failed to provide evidence for such a mechanism in paraquat—HA complexes in an aqueous system. Presumably,

the ultraviolet methods are not sufficiently sensitive to detect any charge transfer interactions. Recently, Khan (1973b, 1974a) provided evidence for such interactions using infrared spectrometry. The interaction of bipyridylium herbicides with humic materials resulted in a shift of C—H out-of-plane bending vibrations from 815 cm^{-1} to 825 cm^{-1} for paraquat, and from 729 cm^{-1} to 765 cm^{-1} for diquat (Fig. 1). The observed shifts in the out-of-plane C—H vibration frequencies provide evidence for the charge transfer complex formation between the humic materials and the bipyridylium herbicides.

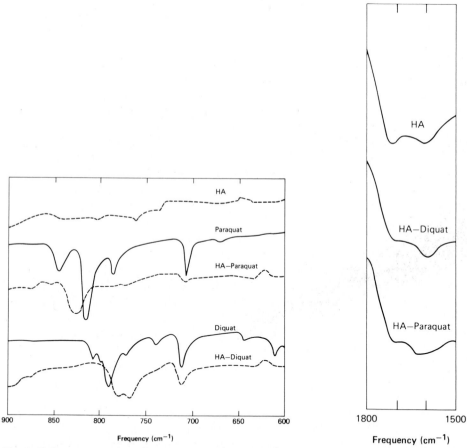

Fig. 1. Infrared spectra in the region 600—900 cm^{-1} on expanded scale (Khan, 1974a). Published by permission of Springer-Verlag, New York Inc.

Fig. 2. Infrared spectra of HA and HA-herbicide complex in the region 1,500—1,800 cm^{-1} (Khan, 1974a). Published by permission of Springer-Verlag New York Inc.

(5) Ion exchange. Ion exchange adsorption takes place for those pesticides which either exist as cations or which become positively charged through protonation. Adsorption of cationic pesticides, such as paraquat and diquat, via cation exchange functions through COOH and phenolic-OH groups associated with the organic matter (Broadbent and Bradford, 1952; Schnitzer and Khan, 1972). The adsorption is always accompanied by the release of a significant concentration of hydrogen ions (Best et al., 1972; Khan, 1974a). According to Stevenson (1976), diquat and paraquat can react with more that one negatively charged site on soil humic colloids, such as through two COO^- ions, a COO^- ion plus a phenolate ion combination, or a COO^- ion (or phenolate ion) plus a free radical site.

Burns et al. (1973c) and Khan (1974a) utilized infrared spectroscopy to demonstrate that ion exchange is the predominant mechanism for adsorption of bipyridylium herbicides by humic substances. Spectra for the HA-herbicide complexes are presented in Fig. 2. It can be seen that upon addition of herbicides the intensity of the $1,720$ cm^{-1} band (carbonyl of carboxylic acid) diminished while that at $1,610$ cm^{-1} (carboxylate) increased. This indicated conversion of COOH to COO^- groups which react with bipyridylium cations to form carboxylate bonds. Notice that the $1,720$ cm^{-1} band did not disappear completely, indicating that a considerable proportion of H^+ in COOH remained inaccessible to the large herbicide cations. It has been observed that HA and FA retained paraquat and diquat in amounts that were considerably less than the exchange capacity of humic materials (Khan, 1973b). The large size of the organic cation seems to have resulted in steric hindrance so that they may not have been exchanged with ionizable H^+ as effectively as the smaller inorganic cations.

Further evidence for the ion exchange mechanism was procured by the potentiometric titrations of HA and herbicides—HA complexes (Fig. 3). The decreases in consumption of alkali for herbicide—HA complexes titration (curves b, c vs curve a) suggest that ionization of acid functional groups are involved in the bipyridylium cations interactions with humic materials (Khan, 1973b).

It was suggested earlier that charge transfer mechanisms are also involved in the adsorption of bipyridylium cations by humic acid. An estimate of the relative importance of charge transfer and ion exchange mechanisms in the adsorption of bipyridylium cations by HA will remain a matter of conjecture until more information is available. However, judging from the available data in the literature it appears that an ion exchange mechanism plays a dominant role in the adsorption processes.

The cationic adsorption mechanism is also responsible for the adsorption on organic matter of less basic pesticides, such as s-triazines (Weber et al., 1969; Gaillardon, 1975). The pesticide may become cationic through protonation, either in the soil solution or during adsorption. Thus, a weakly basic pesticide may be protonated and adsorbed on organic matter according

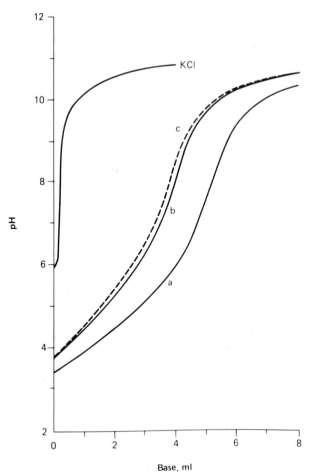

Fig. 3. Potentiometric titration curves of: *a* HA; *b* HA-paraquat; and *c* HA-diquat (Khan, 1974a). Published by permission of Springer-Verlag New York Inc.

to the following series of equation:

$$P + H^+ \rightleftharpoons PH^+ \tag{1}$$

where P = weakly basic organic pesticide. When the solution pH is equal to the pKa of the compound, 50% of the basic pesticide molecules are protonated. In this case, the pKa is derived from the expression:

$$Ka = \frac{[H^+][P]}{[PH^+]} \tag{2}$$

Maximum adsorption of s-triazines by organic soil colloids occurs at pH levels near the pKa of the respective compound (Weber et al., 1969). Thus, the adsorption capacities of organic matter and humic substances for s-triazines were found to follow the order expected on the basis of pKa values for the compounds (Weber et al., 1969; Gilmour and Coleman, 1971). The pH of soil solution will govern the ionization of the acidic functional groups on organic matter which may be available for cation exchange. This also would affect adsorption of weakly basic pesticides (Nearpass, 1965, 1969, 1971). Reduction in solution pH results in an increase in the protonated species. For the subsequent adsorption of PH^+ it should compete with initially adsorbed cation (M^+).

$$PH^+ + MR \rightleftharpoons M^+ + PHR \tag{3}$$

where R is the organic matter cation exchanger. Sullivan and Felbeck (1968) showed that ion exchange could take place between a protonated secondary amine group on s-triazine and a carboxylate anion on the HA. Gilmour and Coleman (1971) also suggested an ion exchange process between protonated s-triazine and Ca-humate. Larger Ca-saturation of HA resulted in less s-triazine adsorption. Adsorption was greater for more strongly basic s-triazines as compared to weakly basic s-triazines under the same conditions because, at a given pH, the proportion of protonated s-triazine was greater.

Protonation may also occur by H^+ already countering the charge on R^- and the protonated pesticide remains on the surface as counter ion:

$$P + HR \rightleftharpoons PHR \tag{4}$$

Thus, the acidity of the organic matter surface will influence the protonation of the adsorbed basic pesticide molecule. According to Hayes (1970), the pH at the surface of soil organic colloids may be as much as two pH units lower than that of the liquid environment. Thus, the protonation of a basic pesticide may occur even though the measured pH of the water-adsorbent system is greater than the pKa of the compound.

Diprotonation of picloram at pH values below 1 was reported by Nearpass (1976). The cation thus formed cannot compete with H^+ for adsorption sites, thereby resulting in a slight decrease in picloram adsorption in this pH region.

Ion exchange adsorption of pesticides by soil organic colloidal constituents will also depend on the Donnan properties of the adsorbent. According to Burns and Hayes (1974), an imaginary boundary can be drawn around spherical or coiled HA macromolecules encompassing a certain volume of solvent. This boundary can behave as a semipermeable membrane. Burns and Hayes (1974) suggested that in order to evaluate completely Donnan effects in ion exchange systems involving HA it would be necessary to know the volume of solution enclosed by the hypothetical membranes which surround the polymer molecules. The approach outlined by Burns and Hayes (1974) warrants further study in its application in the organo-cation-HA adsorption studies. The Donnan effects will be insignificant in the presence of an excess of diffusible electrolytes in the water—polyelectrolyte system (Burns and Hayes, 1974).

(6) Ligand exchange. Adsorption by this mechanism involves replacement of one or more ligands by the adsorbent molecule. The necessary condition is that the adsorbent molecule be a stronger chelating agent than that of the replaced ligands. This type of mechanism may be involved for the binding of s-triazines on the residual transition metals of HA (Hamaker and Thompson, 1972). In ligand exchange partially chelated transition metals may serve as possible sites for adsorption (Hayes, 1970). The pesticide molecule may displace water of hydration acting as a ligand.

Coordination through an attached metal ion (ligand exchanged) was considered to be the main process in the adsorption of linuron by peat samples saturated with different cations (Hance, 1971). However, on the basis of infrared study, Khan (1974c) found it very unlikely that complex formation with cation is a possible mechanism of linuron adsorption on organic matter. It may be pointed out, however, that bridging of pesticide, present in the anionic form, to polyvalent cations associated with organic matter may be possible in the adsorption (Nearpass, 1976).

ADSORPTION ISOTHERMS

Adsorption of pesticides is generally evaluated by the use of adsorption isotherms. An isotherm represents a relation between the amount of pesticide adsorbed per unit weight of adsorbent and the pesticide concentration in the solution at equilibrium. Giles et al. (1960) investigated the relation between solute adsorption mechanisms on solid surfaces and the types of adsorption isotherms obtained. They developed an empirical classification of adsorption isotherms into four main classes according to the initial slope (Fig. 4). The S-type isotherms are common when the solid has a high affinity for the solvent. The initial direction of curvature shows that adsorption becomes easier as concentration increases. In practice, the S-type isotherm usually appears when the solute molecule is monofunctional, has moderate intermolecular attraction, and meets strong competition, for substrate sites, from molecules of the solvent or of another adsorbed species. The L-type curves, the normal or "Langmuir" isotherms, are the best known and represent a relatively high affinity between the solid and solute in the initial stages of the isotherm. As more sites in the substrate are filled, it becomes increasingly difficult for solute molecules to find a vacant site available. The C-type curves are given by solutes which penetrate into the solid more readily than does the solvent. These curves are characterized by the constant

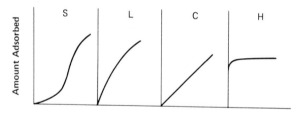

Fig. 4. Classification of adsorption isotherms according to Giles et al. (1960). Reproduced from "Pesticides in Soil and Water", 1974, p. 45, by permission of Soil Science Society of America.

partition of solute between solution and substrate, right up to the maximum possible adsorption, where an abrupt change to horizontal plateau occurs. The H-type curves are quite uncommon and occur only when there is very high affinity between solute and solid. This is a special case of the L-type curves, in which the solute has such high affinity that in dilute solutions it is completely adsorbed, or at least there is no measurable amount remaining in solution. The initial part of the isotherm is therefore vertical. The foregoing four classes of isotherms have been referred to in the literature on many instances concerning pesticide adsorption on organic matter.

In general, the following two mathematical equations have been used for quantitative description of pesticide adsorption on organic matter.

(1) Freundlich adsorption equation. The empirically derived Freundlich eq. 5 has been used to describe the adsorption of pesticides by organic matter in the majority of published reports. The Freundlich equation can be expressed as:

$$\frac{x}{m} = KC^{1/n} \tag{5}$$

where x/m is the ratio of pesticide to organic matter mass, C is the pesticide concentration in solution upon achieving equilibrium, and K and n are constants. The form $1/n$ emphasizes that C is raised to a power less than unity. When eq. 5 is expressed in the logarithmic form, a linear relationship is obtained:

$$\log \frac{x}{m} = \log K + \frac{1}{n} \log C \tag{6}$$

Normally, within a reasonable range of pesticide concentration, the relationship between $\log x/m$ and $\log C$ is linear, with $1/n$ being constant. In comparing adsorptivity of various pesticides by different organic surfaces, the K value may be considered to be a useful index for classifying the degree of adsorption. The necessary conditions are that $1/n$ values be approximately equal and determination be made at the same C value (Hance, 1967). In general, K and $1/n$ values for the adsorption of pesticides on organic matter or humic materials decreases and increases, respectively, with increase in temperature (Haque and Sexton, 1968; Khan, 1973c, 1974c, 1977). Typical Freundlich plots for adsorption of fonofos on H^+ —HA at $0°$ and $23°C$ are shown in Fig. 5. An increase in the value of K with humification of organic matter has been implied in pesticide adsorption (Hamaker and Thompson, 1972; Morita, 1976).

(2) Langmuir adsorption equation. The Langmuir adsorption equation was initially derived from the adsorption of gases by solids using the following assumptions: (1) the energy of adsorption is constant and independent of surface charge; (2) adsorption is on localized sites and there is no interaction between adsorbate molecules; and (3) the maximum adsorption possible is that of a complete monolayer. The Langmuir adsorption equation may be expressed in terms of concentration in the form:

$$\frac{x}{m} = \frac{K_1 K_2 C}{1 + K_1 C} \tag{7}$$

the terms x/m and C have been defined earlier. K_1 is a constant for the system dependent on temperature and K_2 is the monolayer capacity. The

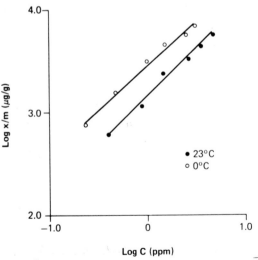

Fig. 5. Freundlich isotherms of fonofos on H^+ — HA (Khan, 1977). Published by permission of the Agricultural Institute of Canada.

reciprocal of eq. 7 gives:

$$\frac{1}{x/m} = \frac{1}{K_2} + \frac{1}{K_1 K_2 C} \tag{8}$$

A plot of $1/(x/m)$ against $1/C$, should give a straight line with an intercept $1/K_2$ and a slope of $1/(K_1 K_2)$ when the Langmuir relation holds. The adsorption of a number of pesticides on organic surfaces was found to conform to an isotherm type which was similar to Langmuir model for adsorption (Weber and Gould, 1966; Li and Felbeck, 1972). Fig. 6 shows Langmuir plots for adsorption of atrazine on HA at two different temperatures.

Under certain conditions both the Freundlich and Langmuir equations may reduce to linear relationship. In the case of the Freundlich eq. 5, if the exponent $1/n$ is 1, the adsorption will be linearly proportional to the solution concentration. In practice it has been found that adsorption of pesticides on soil organic matter do fit the Freundlich equation with an exponent close to unity. In the case of the Langmuir eq. 7 the denominator, $1 + K_1 C$, becomes indistinguishable from 1 at low concentration. In this situation, the

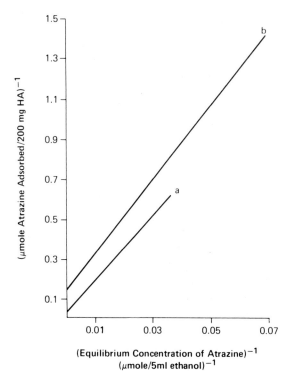

Fig. 6. Langmuir plot for adsorption of atrazine on HA at: a 25°C; and b 3°C (Li and Felbeck, 1972b). Published by permission of the Williams & Wilkins Co.

amount adsorbed becomes directly proportional to the concentration in solution.

Eqs. 5 and 7 will not be obeyed if the adsorption of pesticides on organic matter is predominantly due to an ion-exchange mechanism. Burns et al. (1973b) examined the validities of two ion-exchange isotherm equations for the adsorption of paraquat cation (P^{2+}) on H^+—HA. The Rothmund-Kornfeld equation is given by Burns et al., 1973b:

$$\frac{[\bar{P}^{2+}]}{[\bar{H}]^2} = K\left(\frac{[P^{2+}]}{[H]^2}\right)^{1/n} \qquad (9)$$

where the superimposed bars refer to the ions in the adsorbent. Eq. 9 is reduced to an expression of the law of mass action when $n = 1$. The logarithmic form of eq. 9 can be expressed as:

$$A = \log K + \left(\frac{1}{n}\right)S \qquad (10)$$

where $A = \log [\bar{P}^{2+}] - 2 \log [\bar{H}^+]$ and $S = \log [P^{2+}] - 2 \log [H^+]$. This can be used to test the data in both Rothmund-Kornfeld and mass action equations. Burns et al. (1973b) found that only the Rothmund-Kornfeld eq. 9 satisfactorily fitted the results. However, at low concentrations small deviations were observed, which were attributed to non-exchange adsorption because of deviations from Donnan behavior at low concentrations. Neither Freundlich nor Langmuir plots fitted the data, although some of the data at lower concentration level were in reasonable accord with the Freundlich model for adsorption.

According to Burns and Hayes (1974) it is possible to distinguish between ionic and other mechanisms of adsorption by using the isotherm equation. Thus, carefully controlled adsorption studies at different temperatures can give some idea of the mechanism involved.

ADSORPTION OF SPECIFIC TYPES OF PESTICIDES BY ORGANIC MATTER

Weber (1972) suggested that organic pesticides may be classified as ionic and nonionic. The ionic pesticides include cationic, basic and acidic compounds. The broad group of pesticides classified as nonionic vary widely in their properties and include chlorinated hydrocarbons, organophosphates, substituted anilines and anilides, phenyl carbamates, phenylureas, phenylamides, thiocarbamates, acetamides, benzonitriles and esters.

Ionic pesticides

(1) Cationic. The water solubility of this group of pesticides is generally high and they ionize in aqueous solution to form cations. The herbicides,

diquat and paraquat, are the only compounds of this group that have been studied in any detail concerning the reaction with various organic surfaces. In solution they exist as divalent cations and postive charges are distributed around the molecules (Hayes et al., 1975). Diquat and paraquat are known to become inactivated in highly organic soils (Harris and Warren, 1964; O'Toole, 1966; Calderbank, 1968; Calderbank and Tomlinson, 1969; Damanakis et al., 1970; Khan et al., 1976b). However, due to a slow approach to the adsorption equilibria the inactivation process in the field has been occasionally either very slow or incomplete (Calderbank and Tomlinson, 1969). The adsorption from the solution phase by the organic matter was demonstrated by the reduction in paraquat phytotoxicity to plants grown in media containing organic soils (Scott and Weber, 1967; Coffey and Warren, 1969; Damanakis et al., 1970).

The amount of diquat or paraquat adsorbed by soil organic matter is related to the amount of the herbicide in solution. The plot of the herbicide concentration in solution against the amount adsorbed generally has an L-shaped isotherm which levels off at a certain adsorption maximum (Calderbank, 1968; Calderbank and Tomlinson, 1969; Weber, 1972). A typical adsorption curve for paraquat on fen peat is shown in Fig. 7. The herbicide is completely adsorbed at low levels of application. This region often has been referred to as the strong adsorption capacity of the organic soils (Knight and Tomlinson, 1967). However, the definition of this region depends on the analytical method available (Calderbank, 1968). Tucker et al. (1967, 1969) arbitrarily defined two types of bonding in paraquat and diquat adsorption processes by a muck soil. The "loosely bound" paraquat was classified as adsorbed paraquat that can be desorbed with saturated ammonium chloride. The "tightly bound" paraquat was classified as

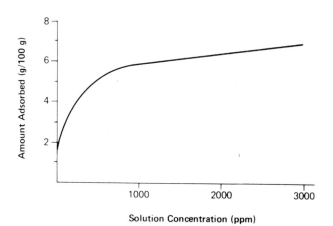

Fig. 7. Adsorption isotherm of paraquat on fen peat (Calderbank and Tomlinson, 1969). Published by permission of Springer-Verlag, New York Inc.

adsorbed paraquat that cannot be desorbed with saturated ammonium chloride, but can only be released from soil by refluxing with 18 N sulphuric acid (Tucker et al., 1967). The "tightly bound" capacity of muck soil for bipyridylium cations was noted to be considerably less than the "loosely bound" capacity. Since high cation exchange capacities are characteristics of organic soil, they would have a high "loosely bound" bipyridylium cation capacity (Tucker et al., 1967). The "tightly bound" paraquat is not available to plants whereas the "loosely bound" paraquat can potentially become available (Riley et al., 1976).

Adsorption of diquat and paraquat on fractionated and well characterized humic substances has been studied in greater detail (Damanakis et al., 1970; Best et al., 1972; Khan, 1973b, 1974a, b; Burns et al., 1973a, b, c, d). Khan (1973b) investigated the binding of diquat and paraquat by HA and FA by using gel filtration technique. Paraquat was complexed by humic materials in greater amounts than was diquat, but the amounts of the two herbicides complexed by HA were higher than those complexed by FA. The adsorption is influenced by the nature of the cation present on HA (Best et al., 1972; Burns et al., 1973a, Khan, 1974a). Thus, the cation order for increasing adsorption for the two herbicides was found to be nearly the same and followed the sequence: $Al^{3+} < Fe^{3+} < Cu^{2+} < Ni^{2+} < Zn^{2+} < Co^{2+} < Mn^{2+} < H^+ < Ca^{2+} < Mg^{2+}$ (Khan, 1974a). The competitive ion effect between diquat and paraquat for sites on HA has been investigated by equilibrating the material with an equal molar mixture of the two herbicides (Best et al., 1972; Khan, 1974b). Table II shows the adsorption of paraquat and diquat in competition on HA. The ratio of paraquat adsorbed to the total paraquat + diquat was also calculated. A value of 0.50 denotes no preference, while larger or smaller values indicate the preference in favor of paraquat or diquat, respectively. It was found that the preference was always slightly in favor of paraquat (Table II). This was attributed to the relationship between surface charge density of the adsorbent and cation charge spacings, as well as steric hindrance due to cation size (Best et al., 1972).

Charcoal and activated carbon adsorb cationic pesticides such as diquat and paraquat, similar to the adsorption which occurs on soil organic matter and humic substances (Weber et al., 1965, 1968; Faust and Zarins, 1969; Best et al., 1972; Parkash, 1974). Bipyridylium cations may also adhere to the surface of grass, lignin and cellulose when present in soil (Damanakis et al., 1970). Adsorption of other cationic pesticides, such as phosphon, phanacridane chloride, pyridyl pyridinium chloride and ethyl pyridinium bromide on organic matter also has been reported (Weber, 1972).

(2) Basic. Basic pesticides, such as s-triazine herbicides, readily associate with hydrogen to form a protonated species and may behave as positive counter ions just as organic cations. The latter may be adsorbed via a negative site on the organic matter (Weber et al., 1969, 1974). Evidence which demonstrates the importance of soil organic matter in adsorbing s-

TABLE II

The adsorption of paraquat and diquat in competition on humic acid (Khan, 1974b). Published by permission of Springer-Verlag, New York Inc.

Ad-sorbent	Herbicide added		Herbicide adsorbed			Ratio [1] of $\frac{P}{P+D}$
	paraquat (meq./ 100 g)	diquat (meq./ 100 g)	paraquat (meq./ 100 g)	diquat (meq./ 100 g)	total (meq./ 100 g)	
HA [2],[3]	80	80	40.8	35.8	76.6	0.53
HA [2]	80	80	44.1	43.1	87.2	0.51
Humin [2]	80	80	42.1	36.1	78.2	0.54
HA [4]	50	50	39.1	39.5	78.6	0.50

[1] Ratio pf paraquat (P) and diquat (D) adsorbed.
[2] Best et al. (1972)
[3] Aldrich commercial.
[4] Khan (1973c).

triazines, in reducing their phytotoxicity, and in affecting their movement in soil, has been reviewed and discussed by Hayes (1970). The adsorption of basic pesticides by soil organic matter is pH dependent (McGlamery and Slife, 1966; Doherty and Warren, 1969; Weber et al., 1969). Maximum adsorption of basic pesticides, such as s-triazines, occurred near the pKa of the compound. The number of protonated molecules decreased at higher pH thereby reducing the adsorption. McGlamery and Slife (1966) observed much greater adsorption of atrazine on HA under acid than neutral conditions. In similar studies by Hayes et al. (1968), the adsorption of atrazine by hydrogen saturated muck was found to be considerably greater than that by calcium saturated muck. Gaillardon (1975) observed that terbutryn is very readily adsorbed by HA in an acid medium.

Other studies have shown that charcoal has a high adsorption capacity for some basic pesticides, such as s-triazines (Robinson, 1965; Weber et al., 1965, 1968). Walker and Crawford (1968) observed that straw and lucerne adsorbed atrazine, propazine, prometone, and prometryne. Synthetic and natural polymers, such as cellulose acetate and nylon, have been shown to adsorb s-triazines (Ward and Holly, 1966). The ether and alcohol extractable components of soil organic matter, i.e., fats, oils, waxes, and resins have negligible capacity to adsorb atrazine (Dunigan and McIntosh, 1971). It has been observed that polysaccharide types of compounds have low affinities, and HA have high affinities for atrazine.

(3) Acidic. The acidity of this class of pesticides is mainly due to carboxylic or phenolic groups which may ionize to produce organic anions. The activity of acidic pesticides has been shown to be related to organic matter content of soil (Upchurch and Mason, 1962; Schliebe et al., 1965;

Hamaker et al., 1966; Herr et al., 1966; Scott and Weber, 1967; Grover, 1968; Keys and Friesen, 1968; O'Connor and Anderson, 1974). The magnitude of adsorption of acidic pesticides by soil organic matter is much lower than that of cationic or basic pesticides (Weber, 1972). The adsorbed pesticides can readily be released to water (Harris and Warren, 1964; Weber et al., 1968). Adsorption of acidic pesticides depends on the pH of the system. At low pH levels most of the weakly acidic herbicides are present in the molecular rather than the anionic form. Thus, they would be adsorbed to a greater extent than stronger acid herbicides. The relative adsorption of different herbicides by organic matter is shown in Fig. 8. The adsorption of acidic herbicides, dinoseb, picloram, 2,4-D, and dicamba on a muck soil was relatively low compared with the basic and cationic herbicides. The stronger acidic herbicides, picloram (pKa = 1.9) were adsorbed in lower amounts than the weakly acidic phenol, dinoseb (pKa = 4.4). The latter is a weaker acid than the other herbicides, suggesting that most of the molecules present at the soil pH were in the nonionic form. Picloram has been shown to be adsorbed on HA and humin largely in the form of uncharged molecules (Nearpass, 1976). Some phenolic pesticides exist as the free acid in acidic soils and may be adsorbed on organic matter. Su and Lin (1971) observed that the efficacy of PCP was strongly influenced by organic matter. PCP efficacy decreased with increase in organic matter. Positive correlations have been observed between PCP and organic matter content of soil (Tsunoda, 1965; Choi and Aomine, 1972). Choi and Aomine (1974) suggested that organic matter plays an important role in adsorption of PCP in soil. They

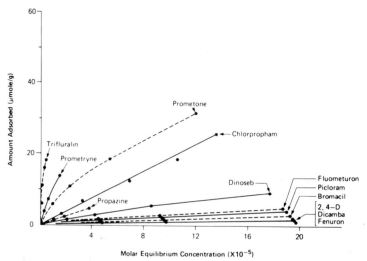

Fig. 8. Adsorption isotherms of some herbicides by soil organic matter according to Weber (1972).

observed that a decrease in organic matter content resulted in a decrease in adsorption of PCP.

Adsorption of acidic pesticides on charcoal also has been reported (Ogle and Warren, 1954; Donaldson and Foy, 1965; Weber et al., 1965; Hamaker et al., 1966; Haque and Sexton, 1968).

(4) Miscellaneous ionic pesticides. Some of the ionic pesticides do not fall into the above described categories of compounds. Included in this group are bromacil, terbacil, isocil, oryzalin, DSMA, and cacodylic acid. They exhibit weak acidic or basic properties and may also possess certain functional groups in the molecule. The latter cause them to behave differently from cationic, basic or acidic pesticides.

The uracil herbicides are partially adsorbed by soil organic matter (Burnside et al., 1969; Rhodes et al., 1970). Fig. 8 shows the adsorption of bromacil by soil organic matter relative to several other herbicides.

Nonionic pesticides

Pesticides included in this category vary widely in their properties and they do not ionize significantly in aqueous or soil system. Adsorption of nonionic pesticides on soil organic matter depends mainly upon the chemical properties of the compounds and the types of organic surface involved. In the following paragraphs the adsorption of the broad group of pesticides classified as nonionic on soil organic matter is discussed.

(1) Chlorinated hydrocarbons. The effect of soil organic matter on the insecticidal activity of several chlorinated hydrocarbons was first observed by Fleming (1950), Fleming and Maines (1953, 1954), and Edwards et al. (1957). Later investigations confirmed the influence of soil organic matter on the bioactivity of both volatile and non-volatile chlorinated hydrocarbons (Barlow and Hadaway, 1958; Bowman et al., 1965; Harris, 1966; Hermanson and Forbes, 1966; Burkhardt and Fairchild, 1967; Harris and Sans, 1967; Beall and Nash, 1969; Adams and Li, 1971; Weil et al., 1973).

Many investigators found that the retention and inactivation of DDT in soil was related to the organic matter content of the soil (Lichtenstein et al., 1960; Porter and Beard, 1968; Beall and Nash, 1969; Shin et al., 1970; Peterson et al., 1971). Shin et al. (1970) observed that DDT adsorption in soil was greater in more humified soil organic matter. Volatilization of degradation products of DDT in soil has been shown to decrease by organic matter application (Farmer et al., 1974; Spencer et al., 1974). The persistence of DDT, lindane, and aldrin was found to be greater in a muck soil than in a mineral soil (Lichtenstein, 1959), and the persistence of aldrin and its epoxidation product, dieldrin, was directly related to the amount of organic matter in soil (Lichtenstein and Schulz, 1960). Pierce et al. (1974) investigated DDT adsorption to a marine sediment, sediment fractions, clay and HA suspended in sea water. The humic fraction was found to have a greater adsorbing capacity than the clay or sediment. Removal of humic fractions from sediment reduced the adsorption capacity to less than 50% of the original sediment sample. Pierce et al. (1974) concluded that suspended humic particulates may be important agents for transporting chlorinated hydrocarbons through the water column and for concentrating them in sediments. Movement of DDT in forest soils has been attributed to its association with HA and FA fractions of soil organic matter (Warshaw et al., 1969; Ballard, 1971). Warshaw et al. (1969) observed that DDT was more soluble in sodium humate than it was in distilled water. The increased solubility of the insecticide was related to the effect of humate on lowering the surface tension of water.

The lipid fraction of soil organic matter has also been implicated in the adsorption of DDT (Pierce et al., 1971). It has been suggested that the adsorption of nonpolar pesticides on soil organic matter is mainly due to pesticide-lipid interaction.

Soil organic matter affects the adsorption of endrin, dieldrin and aldrin by soils (Edwards et al., 1957; Harris and Sans, 1967; Beall and Nash, 1969). Aldrin was more persistent in soil containing higher levels of organic matter (Lichtenstein, 1959; Lichtenstein and Schulz, 1959a). Adsorption and inactivation of other chlorinated hydrocarbons, such as lindane (Edwards et al., 1957; Lichtenstein and Schulz, 1959b; Kay and Elrick, 1967; Adams and Li, 1971), chlordane (Harris and Sans, 1967) and heptachlor (Beall and Nash, 1969) has also been related to the organic matter content of the soil.

Activated carbon added to soils strongly adsorbed aldrin, dieldrin, heptachlor, and heptachlorepoxide and adsorption increased with time (Lichtenstein et al., 1969, 1971).

(2) Organophosphates. Adsorption of organophosphates has been related to the organic matter content of soils (Kirk and Wilson, 1960; Harris, 1966; Swoboda and Thomas, 1968; Khan et al., 1976a). The bioactivity of phorate was found to decrease with an increase in organic matter content of soils (Kirk and Wilson, 1960). Khan et al. (1976a) found that fonofos was persistent for more than two years in an organic soil. Harris (1966) observed that the bioactivity of diazinon and parathion decreased with increased organic matter content of moist soil but this relationship was not observed for dry soil. Soil moisture affects the adsorption of organophosphates similar to that of chlorinated hydrocarbons. Saltzman et al. (1972) observed that in aqueous solution parathion had a greater affinity for organic than for mineral adsorptive surfaces in soils.

Various organic matter fractions were found to adsorb parathion (Leenheer and Ahlrichs, 1971). Furthermore, it was observed that organic matter with H^+ on exchange sites adsorbed significantly larger amounts of the insecticide than with Ca on the exchange sites. In a recent study Khan (1977) investigated adsorption of fonofos on HA saturated with different cations. The amount of the insecticide adsorbed was affected by the cation with which the HA was saturated. It was suggested that the mobility and persistence of fonofos in soils will be partly a function of adsorption on humic materials. Grice et al. (1973a) showed that HA has a high affinity for organophosphorus compounds. Their experiments gave an adsorption capacity of about 30 g for dimefox for monolayer coverage per 100 g HA.

Organophosphates also have been found to be readily adsorbed by charcoal (Sigworth, 1965).

(3) Substituted anilines. The substituted anilines are readily adsorbed by soil organic matter (Lambert, 1967; Hollist and Foy, 1971; Weber et al., 1974). The phytotoxicity of benefin was significantly correlated with the organic matter content of soil (Weber et al., 1974). According to Lambert (1967), the adsorption of some substituted anilines by organic matter is related to the parachor of the compounds; larger molecules are adsorbed more than smaller molecules.

The adsorption of eighteen substituted anilines by nylon and cellulose triacetate was reported to be inversely related to their water solubility (Ward and Upchurch, 1965).

(4) Phenylureas. The herbicidal activity of phenylureas was shown to be related to the organic matter content of the soils (Upchurch and Mason, 1962; Harris and Sheets, 1965; Upchurch et al., 1966; Darding and Freeman, 1968; Doherty and Warren, 1969; Hsu and Bartha, 1974a, b; Savage and Wauchope, 1974; Weber et al., 1974; Carringer et al., 1975; Rahman et al., 1976). Khan et al. (1976b) observed that linuron was persistent in an organic soil and about 15—20% of the herbicide remained in the soil fifteen months after application. The movement of phenylureas in soils has been shown to decrease with an increase in organic matter (Upchurch and Pierce, 1958; Ashton, 1961; Ivey and Andrews, 1965) and the adsorption to soil material was related to the organic matter content (Sherbourne and Freed, 1954; Sheets, 1958; Hance, 1965a; Harris and Sheets, 1965; Doherty and Warren, 1969). The adsorption of linuron by organic soils was found to increase with decomposition (Morita, 1976). The pH of the system did not affect adsorption of phenylureas significantly (Yuen and Hilton, 1962; Hance, 1969a). Hance (1965a) observed competition between water and diuron for adsorption sites, and that diuron was a more effective competitor at soil organic matter surfaces than at soil mineral matter surfaces.

The adsorption of linuron by organic matter or humic substances is affected by the cation with which the adsorbent is saturated. Thus, the adsorption of linuron by peat and HA samples saturated with various cations decreased in the following order:

$$Ce^{4+} > Fe^{3+} > Cu^{2+} > Ni^{2+} > Ca^{2+} \qquad \text{(Hance, 1971)}$$

and:

$$H^+ > Fe^{3+} > Al^{3+} > Cu^{2+} > Ca^{2+} > Zn^{2+} > Ni^{2+} \qquad \text{(Khan, 1974c).}$$

Phenylurea-derived chloroaniline residues in soil were found to be immobilized by adsorption on humic materials (Hsu and Bartha, 1974a, b; Bartha and Hsu, 1976). It was suggested that chemical attachment of chloranilines to humic substances occurs both in a hydrolyzable and in a non-hydrolyzable manner (Hsu and Bartha, 1974a).

Large amounts of phenylureas are adsorbed by charcoal (Sherbourne and Freed, 1954; Leopold et al., 1960; Yuen and Hilton, 1962; Hance, 1965b). Adsorbed diuron and monuron were readily desorbed by water (Yuen and Hilton, 1962). Cellulose and chitin did not adsorb diuron strongly, whereas lignin adsorbed diuron appreciably (Hance, 1965b).

(5) Phenylcarbamates and carbanilates. Chlorpropham and propham inactivation has been related to the organic matter content of the soil (Upchurch and Mason, 1962; Harris and Sheets, 1965; Roberts and Wilson, 1965; Robocker and Canode, 1965). Chlorpropham was adsorbed reversibly by muck (Harris and Warren, 1964; Hance, 1967) and its phytotoxicity was reduced by the organic matter added to the soil (Scott and Weber, 1967). The relative adsorption of chlorpropham by soil organic matter is shown in Fig. 8. Carbaryl, an insecticide, was shown to be adsorbed by various organic matter fractions (Leenheer and Ahlrichs, 1971).

Charcoal has been shown to adsorb chlorpropham and propham in large amounts (Schwartz, 1967; Coffey and Warren, 1969). Adsorption of chlorpropham on nylon and cellulose triacetate in small amounts also has been observed (Ward and Upchurch, 1965).

(6) Substituted anilides. The adsorption of substituted anilides on soil organic matter has not been studied in detail. Recently, butralin and profluralin were shown to be strongly adsorbed by soil organic matter (Carringer et al., 1975). Ward and Upchurch (1965) observed that several acetanilide derivatives were adsorbed from aqueous solutions by powdered nylon and cellulose triacetate. According to these workers adsorption was somewhat related to their water solubilities.

(7) Phenylamides. In leaching experiments it was observed that diphenamide moved less as the organic matter content of the soil was increased (Dubey and Freeman, 1965; Deli and Warren, 1971). It was reported that up to 90% of the 3,4-dichloroaniline released during the biodegradation of several phenylamide herbicides becomes unextractable by solvents due to binding to the soil organic matter (Hsu and Bartha, 1976). Diphenamide was found to be adsorbed in moderate amounts by muck and charcoal (Coffey and Warren, 1969).

(8) Thiocarbamates, carbothioates, and acetamides. Movement of certain thiocarbamates was found to be considerably less in soil as the organic matter content increased (Gantz and Slife, 1960; Fang et al., 1961; Gray and Weierich, 1968; Koren et al., 1968, 1969). Increase in organic matter content resulted in increased adsorption of thiocarbamates and acetamides (Ashton and Sheets, 1959; Deming, 1963; Koren et al., 1968, 1969; Carringer

et al., 1975). Organic matter content of soil was related to the herbicidal activities of thiocarbamate and acetamide (Ashton and Sheets, 1959; Damielson et al., 1961; Jordan and Day, 1962). According to Koren et al. (1969), thiocarbamates and acetamides are readily adsorbed by charcoal.

(9) Benzonitriles. The benzonitrile herbicide, dichlobenil was adsorbed on soil organic matter (Massini, 1961). Lignin also was reported to adsorb dichlobenil from aqueous solution (Briggs and Dawson, 1970).

ADSORPTION OF PESTICIDES BY ORGANIC MATTER—CLAY COMPLEXES

The presence of organic matter—clay complexes in most of the mineral soils needs to be considered in evaluating the importance of organic matter in pesticide adsorption. Recently, Stevenson (1976) quoted Walker and Crawford (1968) indicating that up to an organic matter content of about 6%, both mineral and organic surfaces are involved in adsorption. However, at higher organic matter contents, adsorption will occur mostly on organic surfaces. Stevenson (1976) pointed out that the amount of organic matter required to coat the clay will depend on the soil type and the kind and amount of clay that is present.

The intimate association of organic matter and clay may cause some modification of their adsorptive properties, or they may complement one another in the role of pesticide adsorption (Pierce et al., 1971; Niemann and Mass, 1972). Only recently have attempts been made to study the adsorption of pesticides by organic matter—clay complexes. Burns (1972) pointed out that a humus—clay microenvironment is a site of high biological and non-biological activity and it is here that we need to look for the basic information concerning soil—pesticide interactions.

The adsorptive capacity of sedimentary organomineral complexes for lindane and parathion were found much greater than the corresponding mineral fraction (Graetz et al., 1970). Furthermore, the extent of adsorption was related to the organic carbon content of the complex. Wang (1968) obtained similar results for the adsorption of parathion and DDT on organo—clay fractions. Miller and Faust (1972) investigated the adsorption of 2,4-D by several organo—clay complexes. The latter were prepared by treating dimethylbenzyl octadecylammonium chloride and various benzyl and aliphatic amines with Wyoming bentonite. It should be noted, however, that the nature of the organic matter in soil differs profoundly from the organic compounds used by Miller and Faust (1972). Thus, the adsorption behavior of their organo—clay complexes may differ significantly from those found in soil. Khan (1974d) investigated the adsorption of 2,4-D by a FA—clay complex prepared by treating FA with Na-montmorillonite. This FA—clay complex was similar to the naturally occurring organo—clay complexes found in soil (Kodama and Schnitzer, 1971). Khan (1974d) observed that the FA—clay complex adsorbed about 6.5 and 5.2 μmole of 2,4-D per g of complex at 5° and 25°C, respectively.

Hance (1969a) suggested that in soil, clay and organic matter associate in such a manner that little of the clay mineral surface will be accessible to herbicide molecules. Thus, the contribution to adsorption of the clay fraction in soils would be much less than studies with the isolated mineral would indicate. On the other hand, Mortland (1968) is of the opinion that organic compounds in soil organic matter, upon interaction with clay, may facilitate and stabilize adsorption of pesticides beyond that observed in purely inorganic clay systems. Khan (1973d) also concluded that FA, which is the most prominent humic compound in "soil solution" on interacting with clay minerals may facilitate the adsorption of pesticides on clays in soils.

TECHNIQUES USED IN PESTICIDE—ORGANIC MATTER INTERACTION STUDIES

The slurry technique has been used widely by a number of workers in studying the adsorption of pesticides by soil organic matter and humic substances. This involves shaking several samples of adsorbent in known volumes of different concentrations of aqueous pesticide solution. When equilibrium is established, the suspension is centrifuged and the amount of the pesticide adsorbed is calculated from the decrease in its initial concentration in solution. The results are usually expressed in terms of the amounts of pesticide adsorbed per unit weight of adsorbent. Hayes (1970) pointed out that the slurry technique cannot be satisfactorily applied in adsorption studies with materials which are not sedimented by centrifugation.

Grice et al. (1973b) used the continuous flow technique to study the adsorption of prometone on a hydrogen ion saturated HA. An Amicon Model 12 ultrafiltration cell, with minor alteration, was adopted for studying the adsorption. Fig. 9 shows a comparison of isotherms obtained for adsorption of prometone on H^+—HA by continuous flow technique and by slurry method. These workers concluded that the continuous flow method has advantages over the conventional slurry technique in that it can be utilized to obtain adsorption isotherm data for very low concentrations of adsorbate, such as those which are likely to be present in soil solution. In subsequent studies use of the technique was extended to studies concerning adsorption of paraquat, prometone and dimefox (Grice and Hayes, 1972; Grice et al., 1973a; Burchill et al., 1973). It was observed that the continuous flow technique gives adsorption isotherms comparable to those from the slurry procedure (Grice and Hayes, 1972; Grice et al., 1973b).

Adsorption studies also have been conducted by transferring the humic material to dialysis tubing which was then immersed in pesticide solution of known concentration (Hayes et al., 1968).

Microcalorimetry has been applied in studying the interaction between pesticides and soil constituents (Grice and Hayes, 1970; Lundie, 1971; Hayes et al., 1972). It was observed that the heat exchange which accompanies the adsorption of small organic molecules on HA could be determined readily by means of a microcalorimeter.

Hayes (1970) outlined the potential use of the gel filtration technique for the study of binding of s-triazines by soluble humic materials. The gel filtration technique was used in studying the adsorption of s-triazines and paraquat on sodium humate in aqueous solution (Grice and Hayes, 1970; Burns et al., 1973d). Hummel and Dreyer (1962) developed a gel filtration technique suitable for studying the interactions between macromolecules and substances of low molecular weight. The appearance of a peak followed by a trough in the elution profile is used as the criterion of binding. This procedure could have given more quantitative evaluation of the interaction between sodium humate and herbicides in Grice and Hayes (1970) and Burns et al. (1973d) studies. The method described by Hummel and Dreyer (1962) was successfully employed by Khan (1973b) in studying the binding of bipyridylium cations by HA and FA at pH 6.9 using Sephadex G-10 gel. The results of a typical experiment demonstrating the binding of diquat with HA are shown in Fig. 10. The appearance of a trough at the elution volume of diquat was taken as evidence of binding. The attainment of equilibrium during gel filtration was indicated by the return of the base line concentration of diquat to its initial value after the emergence of the leading HA-diquat peak and after the appearance of a trough in the elution profile

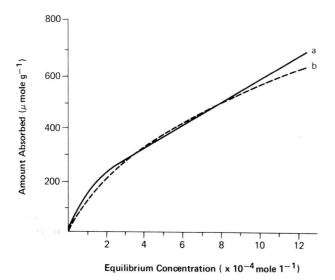

Fig. 9. Adsorption isotherm of prometone by H^+ — HA: *a* flow technique, and *b* slurry technique (Grice et al., 1973b).

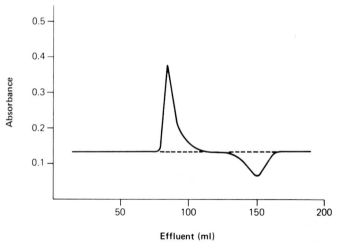

Fig. 10. Elution profile for measurement of binding of diquat by HA (Khan, 1973b). Published by the permission of the Agricultural Institute of Canada.

(Fig. 10). The area of the trough provided a direct measure of the amount of diquat bound or complexed with HA put on the column when it was in equilibrium with free diquat of the concentration in the equilibrium solution.

CHEMICAL ALTERATION AND BINDING OF PESTICIDES

The organic fraction of the soil can be important in producing chemical changes in a wide variety of pesticides. A close correlation between the rate of lindane degradation and the native organic matter content of flooded soil was reported (Yoshida and Castro, 1970). The hydrolysis of organophosphorus esters (Gatterdam et al., 1959) and dehydrochlorination of DDT and lindane (Lord, 1968) may be catalyzed by basic amino acids and similar organic compounds. Chlorinated hydrocarbon insecticides were shown to be reductively dechlorinated by other nitrogenous soil constituents, the reduced porphyrins (Crosby, 1970). Organic matter and the nature of its constituents have an important role in the biodegradation of DDT, heptachlor, endrin, and the four isomers (α, β, γ, and δ) of BHC (Castro and Yoshida, 1974). Other nonbiological transformations brought about by the organic fraction include the decomposition of 3-amino-triazole, the s-oxidation of phorate and the slow conversion of aldrin to dieldrin (Crosby, 1970; Kaufman, 1970).

It has been suggested that degradation of atrazine proceeds more rapidly in soil with higher organic matter content (Harris, 1967). Nearpass (1972) investigated the effect of the ionized surface hydrogen of organic matter on the hydrolysis of propazine in a slurry system. It was observed that hydrolysis of propazine was enhanced by the presence of organic matter. The increase in hydrolysis rate directly reflected the increase proton supply associated with the acidic surface of the organic matter.

Humic materials, by virtue of their acidity, may chemically alter or degrade certain pesticides (Armstrong et al., 1967; Crosby, 1970; Hayes, 1970). They may catalyze the hydroxylation of the chloro-s-triazines (Armstrong et al., 1967; Hayes, 1970; Maslennikova and Kruglow, 1975). Li and Felbeck (1972b) examined the catalytic effects of HA on the chemical hydrolysis of atrazine in aqueous solution. The half-life of atrazine was least at low pH values and increased as pH values increased. The rate of atrazine hydrolysis in the aqueous suspension of HA at pH 4 was found to be first-order in relation to atrazine concentration. The half-life of atrazine, resulting from a first-order plot, varies nonlinearly with the concentration of HA.

The occurrence of stable free radicals in HA and FA may bring about a variety of reduction and other reactions (Steelink and Tollin, 1967). The heterocyclic ring of amitrole is highly susceptible to attack by free radicals (Kaufman, 1970).

Humic substances strongly absorb ultraviolet and visible light and may act as photosensitisers for other non-absorbing compounds. This may affect the rate and the route of breakdown of various pesticides (Burkhard and Guth, 1976).

Goring et al. (1975) suggested that carbon of many pesticides may be converted to natural soil constituents including humic substance. Furthermore, some aromatic pesticides are metabolized to structures having ring systems heavily substituted with such functional groups as NH_2, OH and COOH. These structures may then be incorporated by polymerization, oxidation, and reduction reactions into humic substances (Goring et al., 1975). The amounts of "bound" pesticide metabolites thus incorporated in humic substance will be very small. These metabolites may not be released in their original forms from humic substances in which they have been incorporated (Schnitzer and Khan, 1972). According to Stevenson (1976) pesticides or their decomposition products can form stable chemical linkages with organic matter and such binding greatly increases the persistence of the pesticide residue in the soil. He envisaged two main mechanisms for such processes: (1) the pesticide residues could be directly attached by chemical linkages to the reactive sites on colloidal organic surfaces; and (2) during the humification processes the pesticide residues can be incorporated into the structures of newly formed HA and FA. It has been suggested that the bulk of the chloroanilines produced by partial degradation of the phenylamide herbicides become immobilized in soil by chemical bonding to organic matter (Chiska and Kearney, 1970; Bartha, 1971; Hsu and Bartha, 1974a, b). The chemically bound residues could not be recovered by extraction with organic solvents and inorganic salts (Hsu and Bartha, 1974a). It is rather evident that more research is needed to determine whether or not the bound residue consists of intact pesticides or their degradation products which are immobilized by soil organic matter.

SUMMARY

During the past decade much effort has been directed towards understanding the organic matter—pesticide interactions. The main limitations in our understanding of these interactions have been the complex nature of soil organic matter and the numerous processes in soil environment all operating simultaneously. Recent work with simplified systems involving well defined organic matter components is encouraging in this respect. Pesticides are adsorbed on organic matter surfaces through mechanisms which may include Van der Waals forces, hydrophobic bonding, hydrogen bonding, ion exchange, protonation, charge transfer, coordination complexes, and ligand exchange. The extent and nature of adsorption of a pesticide will depend upon the properties of the compound itself, the kind of organic matter, and the environment provided. The adsorption of pesticides on organic matter is usually expressed by establishing the adsorption isotherms. In most mineral soils the organic matter is intimately bound to the clay, and the relative contribution of organic matter to adsorption will depend upon the extent to which the clay is coated with organic substances. Soil organic matter also has the potential for promoting the nonbiological degradation of many pesticides.

Recent work appearing on soil bound residues is encouraging with respect to our understanding of the fate of pesticides and their decomposition products in soil. Our knowledge of bound residues has only progressed to the point that pesticides become immobilized in soil by chemical bonding to organic matter. The bound residues are not extractable by extensive sequential extraction. However, for a better understanding of the bound residues we need to know whether or not the bound residue consists of intact pesticide or degradation products which are adsorbed, incorporated or entrapped in organic matter. We should be concerned about the release of bound pesticides or bound metabolites and contamination of subsequent crops. On the other hand, it is possible that binding of pesticide residues by organic matter represents the most effective and safe method of decontamination. It is obvious that much work is needed in this area.

REFERENCES

Adams Jr., R.S., 1972. Proc. Trace Subst. Environ. Health, V: 81—93.
Adams Jr., R.S. and Li, P., 1971. Soil Sci. Am. Proc., 35: 78—81.

Armstrong, D.E. and Chesters, G., 1968. Environ. Sci. Technol., 2: 683—689.

Armstrong, D.E. and Konrad, J.G., 1974. In: W.D. Guenzi (Editor), Pesticide in Soil and Water. Am. Soc. Agron., Madison, Wisc., pp. 123—131.

Armstrong, D.E., Chesters, G. and Harris, R.F., 1967. Soil Sci. Soc. Am. Proc., 31: 61—66.

Ashton, F.M., 1961. Weeds, 9: 612—619.

Ashton, F.M. and Sheets, T.J., 1959. Weeds, 7: 88—90.

Bailey, G.W. and White, J.L., 1970. Residue Rev., 32: 29—92.

Bailey, G.W., White, J.L. and Rothberg, T., 1968. Soil Sci. Soc. Am. Proc., 32: 222—234.

Ballard, T.M., 1971. Soil Sci. Soc. Am. Proc., 35: 145—147.

Barlow, F. and Hadaway, A.B., 1958. Bull. Entomol. Res., 49: 315—338.

Bartha, R. and Hsu, T.S., 1976. In: D.D. Kaufman, G.G. Still, G.D. Paulson and S.K. Bandal (Editors), Bound and Conjugated Pesticides Residues. ACS Symp. Ser., 29, pp. 258—271.

Bartha, R., 1971. J. Agric. Food Chem., 19: 385—575.

Beall Jr., M.L. and Nash, R.G., 1969. Agron. J., 61: 571—575.

Best, J.A., Weber, J.B. and Weed, S.B., 1972. Soil Sci., 114: 444—450.

Bowman, M.C., Schechter, M.S. and Carter, R.L., 1965. J. Agric. Food. Chem., 13: 360—365.

Briggs, G.G. and Dawson, J.E., 1970. J. Agric. Food Chem., 18: 97—99.

Broadbent, F.E. and Bradford, G.R., 1952. Soil Sci., 74: 447—457.

Burchill, S., Cardew, M.H., Hayes, M.H.B. and Smedley, R.J., 1973. Proc. Eur. Weed Res. Council Symp. Herbicides—Soil, pp. 70—79.

Burkhard, N. and Guth, J.A., 1976. Pestic. Sci., 7: 65—71.

Burkhardt, C.C. and Fairchild, M.L., 1967. J. Econ. Entomol., 60: 1602—1610.

Burns, R.G., 1972. Proc. Br. Weed Control Congr., 11th, Brighton, pp. 1203—1209.

Burns, I.G. and Hayes, M.H.B., 1974. Residue Rev., 52: 117—146.

Burns, I.G., Hayes, M.H.B. and Stacey, M., 1973a. Weed Res., 13: 67—78.

Burns, I.G., Hayes, M.H.B. and Stacey, M., 1973b. Weed Res., 79—90.

Burns, I.G., Hayes, M.H.B. and Stacey, M., 1973c. Pestic. Sci., 4: 201—209.

Burns, I.G., Hayes, M.H.B. and Stacey, M., 1973d. Pestic. Sci., 4: 629—641.

Burnside, O.C., Wicks, G.A. and Fenster, C.R., 1969. Weed Sci., 17: 241—245.

Calderbank, A., 1968. Adv. Pestic. Control Res., 8: 127—235.

Calderbank, A. and Tomlinson, T.E., 1969. PANS, 15: 466—472.

Carringer, R.D., Weber, J.B. and Monaco, T.J., 1975. J. Agric. Food Chem., 23: 568—572.

Castro, T.F. and Yoshida, T., 1974. Soil Sci. Plant Nutr., 20: 363—370.

Chiska, H. and Kearney, P.C., 1970. J. Agric. Food Chem., 18: 854—858.

Choi, J. and Aomine, S., 1972. Soil Sci. Plant Nutr., 18: 255—260.

Choi, J. and Aomine, S., 1974. Soil Sci. Plant Nutr., 20: 135—144.

Coffey, D.L. and Warren, G.F., 1969. Weed Sci., 17: 16—19.

Crosby, D.G., 1970. In: Pesticide in the Soil. Int. Symp. Pestic. Soil, Michigan State University, East Lansing, pp. 86—94.

Damanakis, M., Drennan, D.S.H., Fryer, J.D. and Holley, K., 1970. Weed Res., 10: 264—277.

Damielson, L.L., Gentner, W.A. and Jansen, L.L., 1961. Weeds, 9: 463—476.

Darding, R.L. and Freeman, J.F., 1968. Weed Sci., 16: 226—229.

Deli, J. and Warren, G.F., 1971. Weed Sci., 19: 67—69.

Deming, J.M., 1963. Weeds, 11: 91—96.

Doherty, P.J. and Warren, G.F., 1969. Weed Res., 9: 20—26.

Donaldson, T.W. and Foy, C.L., 1965. Weeds, 13: 195—202.

Dubey, H.D. and Freeman, J.F., 1965. Weeds, 13: 360—362.

Dunigan, E.P. and McIntosh, T.H., 1971. Weed Sci., 19: 279—282.

Edwards, C.A., Beck, S.D. and Lichtenstein, E.P., 1957. J. Econ. Entomol., 50: 622—626.
Fang, S.C., Theisen, P. and Freed, V.H., 1961. Weeds, 9: 569—574.
Farmer, W.J., Spencer, W.F., Shepherd, R.A. and Cliath, M.M., 1974. J. Environ. Quality, 3: 343—346.
Faust, S.D. and Zarins, A., 1969. Residue Rev., 29: 151—170.
Fleming, W.E., 1950. J. Econ. Entomol., 43: 87—89.
Fleming, W.E. and Maines, W.W., 1953. J. Econ. Entomol., 46: 445—449.
Fleming, W.E. and Maines, W.W., 1954. J. Econ. Entomol., 47: 165—169.
Gaillardon, P., 1975. Weed Res., 15: 393—399.
Gantz, R.L. and Slife, F.W., 1960. Weeds, 8: 599—606.
Gatterdam, P.E., Casida, J.E. and Stoutamire, D.W., 1959. J. Econ. Entomol., 52: 270—276.
Giles, C.H., MacEwan, T.H., Nakhwa, S.N. and Smith, D., 1960. J. Chem. Soc., pp. 3973—3993.
Gilmour, J.T. and Coleman, N.T., 1971. Soil Sci. Soc. Am. Proc., 35: 256—259.
Goring, C.A.I., Laskowski, D.A., Hamaker, J.W. and Meikle, R.W., 1975. In: R. Haque and V.H. Freed (Editors), Environmental Dynamics of Pesticides. Plenum Press, New York, N.Y., pp. 135—172.
Graetz, D.A., Chesters, G. and Lee, G.B., 1970. Agron. Abstr., 96.
Gray, R.A. and Weierich, A.J., 1968. Proc. 9th Br. Weed Contr. Conf., 1: 94—101.
Grice, R.E. and Hayes, M.H.B., 1970. Proc. 10th Br. Weed Contr. Conf., 3: 1089—1100.
Grice, R.E. and Hayes, M.H.B., 1972. Proc. 11th Br. Weed Contr. Conf., 2: 784—791.
Grice, R.E., Hayes, M.H.B. and Lundie, P.R., 1973a. Proc. 7th Br. Insect. Fungicide Conf., pp. 73—81.
Grice, R.E., Hayes, M.H.B., Lundie, P.R. and Cardew, M.H., 1973b. Chem. Ind. (Lond.), pp. 233—234.
Grover, R., 1968. Weed Res., 8: 226—232.
Hadzi, D., Klofutar, C. and Oblak, S., 1968. J. Chem. Soc., A, 905—908.
Hamaker, J.W. and Thompson, J.M., 1972. In: C.A.I. Goring and J.W. Hamaker (Editors), Organic Chemicals in the Soil Environment, 1. Dekker, New York, N.Y., pp. 49—143.
Hamaker, J.W., Goring, C.A.I. and Youngson, C.R., 1966. Advan. Chem. Ser., 60: 23—27.
Hance, R.J., 1965a. Weed Res., 5: 98—107.
Hance, R.J., 1965b. Weed Res., 5: 108—114.
Hance, R.J., 1967. Weed Res., 7: 29—36.
Hance, R.J., 1969a. Can. J. Soil Sci., 49: 357—364.
Hance, R.J., 1969b. Weed Res., 9: 108—113.
Hance, R.J., 1971. Weed Res., 11: 106—110.
Haque, R. and Sexton, R., 1968. J. Colloid Interface Sci., 27: 818—827.
Harris, C.I., 1967. J. Agric. Food Chem., 15: 157—162.
Harris, C.I. and Warren, G.F., 1964. Weeds, 12: 120—126.
Harris, C.I. and Sheets, T.J., 1965. Weeds, 13: 215—219.
Harris, C.R., 1966. J. Econ. Entomol., 59: 1221—1225.
Harris, C.R. and Sans, W.W., 1967. J. Agric. Food Chem., 15: 861—863.
Hayes, M.H.B., 1970. Residue Rev., 32: 131—174.
Hayes, M.H.B., Stacey, M. and Thompson, J.M., 1968. In: Isotopes and Radiation in Soil Organic-matter Studies. I.A.E.A., Vienna, pp. 75—90.
Hayes, M.H.B., Pick, M.E. and Toms, B.A., 1972. Science Tools, pp. 9—12.
Hayes, M.H.B., Pick, M.E. and Toms, B.A., 1975. Residue Rev., 57: 1—25.
Hermanson, H.P. and Forbes, C., 1966. Soil Sci. Soc. Am. Proc., 30: 748—752.
Herr, D.E., Stroube, E.W. and Ray, D.A., 1966. Weeds, 14: 248—250.
Hilton, H.W. and Yuen, Q.H., 1963. J. Agric. Food Chem., 11: 230—234.
Hollist, R.L. and Foy, C.L., 1971. Weed Sci., 19: 11—16.
Horowitz, M. and Blumenfeld, T., 1974. Phytoparasitica, 2: 19—24.

Hsu, T.S. and Bartha, R., 1974a. Soil Sci., 116: 444—452.
Hsu, T.S. and Bartha, R., 1974b. Soil Sci., 118: 213—220.
Hsu, T.S. and Bartha, R., 1976. J. Agric. Food Chem., 24: 118—122.
Hummel, J.P. and Dreyer, W.J., 1962. Biochem. Biophys. Acta, 63: 530—532.
Ivey, M.J. and Andrews, H., 1965. Proc. Southern Weed Conf., 18: 670—684.
Jordan, L.S. and Day, B.E., 1962. Weeds, 10: 212—215.
Kaufman, D.D., 1970. In: Pesticide in Soil. Int. Symp. Pestic. Soil, Mich. State Univ.,
    East Lansing, pp. 73—86.
Kay, B.D. and Elrick, D.E., 1967. Soil Sci., 104: 314—322.
Kemp, T.R., Stoltz, L.P., Herron, J.W. and Smith, W.T., 1969. Weed Sci., 17: 444—446.
Keys, C.H. and Friesen, H.A., 1968. Weed. Sci., 16: 341—343.
Khan, S.U., 1972. Environ. Letters, 3: 1—12.
Khan, S.U., 1973a. Environ. Letters, 4: 141—148.
Khan, S.U., 1973b. Can. J. Soil Sci., 53: 199—204.
Khan, S.U., 1973c. Can. J. Soil Sci., 53: 429—434.
Khan, S.U., 1973d. Can. J. Soil Sci., 24: 244—248.
Khan, S.U., 1974a. J. Environ. Quality, 3: 202—206.
Khan, S.U., 1974b. Residue Rev., 52: 1—26.
Khan, S.U., 1974c. Soil Sci., 118: 339—343.
Khan, S.U., 1977. Can. J. Soil Sci., 57: 9—13.
Khan, S.U. and Schnitzer, M., 1971. Can. J. Chem., 13: 2302—2309.
Khan, S.U. and Schnitzer, M., 1972. Geochim. Cosmochim. Acta, 36: 745—754.
Khan, S.U., Hamilton, H.A. and Hogue, E.J., 1976a. Pestic. Sci. in press.
Khan, S.U., Belanger, A., Hogue, E.J., Hamilton, H.A. and Mathur, S.P., 1976b. Can. J.
    Soil Sci., 56: 407—412.
Kirk, R.E. and Wilson, M.C., 1960. J. Econ. Entomol., 53: 771—774.
Knight, B.A. and Tomlinson, T.E., 1967. J. Soil Sci., 18: 233—243.
Kodama, H. and Schnitzer, M., 1971. Can. J. Soil Sci., 51: 509—512.
Koren, E.C., Foy, L. and Ashton, F.M., 1968. Weed Sci., 16: 172—175.
Koren, E.C., Foy, L. and Ashton, F.M., 1969. Weed Sci., 17: 148—153.
Lambert, S.M., 1967. J. Agr. Food Chem., 15: 572—576.
Lambert, S.M., 1968. J. Agr. Food Chem., 16: 340—343.
Lambert, S.M., Porter, P.E. and Schieferstein, R.H., 1965. Weeds, 13: 185—190.
Leenheer, J.A. and Ahlrichs, J.L., 1971. Soil Sci. Soc. Am. Proc., 35: 700—705.
Leopold, A.C., Van Schaik, P. and Neal, M., 1960. Weeds, 8: 48—54.
Li, G.C. and Felbeck, G.R., Jr., 1972a. Soil Sci., 113: 140—148.
Li, G.C. and Felbeck, G.T., Jr., 1972b. Soil Sci., 114: 201—209.
Lichtenstein, E.P., 1959. J. Agr. Food Chem., 7: 430—433.
Lichtenstein, E.P. and Schulz, K.R., 1959a. J. Econ. Entomol., 52: 118—124.
Lichtenstein, E.P. and Schulz, K.R., 1959b. J. Econ. Entomol., 52: 124—131.
Lichtenstein, E.P. and Schulz, K.R., 1960. J. Econ. Entomol., 53: 192—197.
Lichtenstein, E.P., Fuhremann, T.W. and Schulz, K.R., 1969. J. Agric. Food Chem., 16:
    348—355.
Licthenstein, E.P., Schulz, K.R. and Fuhremann, T.W., 1971. J. Econ. Entomol., 64:
    585—588.
Lichtenstein, E.P., DePew, L.J., Eshbaugh, E.L. and Sleesman, J.P., 1960. J. Econ.
    Entomol., 53: 136—142.
Lindstrom, F.T., Haque, R. and Coshow, W.R., 1970. J. Phys. Chem., 74: 495—502.
Lord, K.A., 1968. J. Chem. Soc. (Lond.), pp. 1657—1661.
Lundie, P.R., 1971. Adsorption of Organophorus Compounds by Soils and Soil Colloids.
    Ph.D. Thesis, University of Birmingham.
McGlamery, M.D. and Slife, F.W., 1966. Weeds, 14: 237—239.
Maslennikova, W.G. and Kruglow, J.W., 1975. Roczniki Gleboznawcze, 26: 25—29.

Massini, P., 1961. Weed Res., 1: 142—146.

Melnikov, N.N., 1971. Chemistry of Pesticides. Springer-Verlag, New York, N.Y., 480 pp.

Metcalf, R.L., 1971. In: R. White-Stevens (Editor). Pesticides in the Environment. Dekker, New York, N.Y., pp. 1—144.

Miller, R.M. and Faust, S.D., 1972. Environ. Letters, 2: 183—194.

Morita, H., 1976. Can. J. Soil Sci., 56: 105—109.

Mortland, M.M., 1968. J. Agric. Food Chem., 16: 706—707.

Nearpass, D.C., 1965. Weeds, 13: 314—316.

Nearpass, D.C., 1969. Soil Sci. Soc. Am. Proc., 33: 524—528.

Nearpass, D.C., 1971. Soil Sci. Soc. Am. Proc., 35: 64—68.

Nearpass, D.C., 1972. Soil Sci. Soc. Am. Proc., 36: 606—610.

Nearpass, D.C., 1976. Soil Sci., 121: 272—277.

Niemann, P. and Mass, G., 1972. Schriftenr. Ver. Wass. Boden-Lufthyg., Berlin-Dahlem, H37: 155—165.

O'Connor, G.A. and Anderson, J.U., 1974. Soil Sci. Soc. Am. Proc., 38: 433—436.

Ogle, R.E. and Warren, G.F., 1954. Weeds, 3: 257—273.

O'Toole, M.A., 1966. Irish Crop Protec. Conf. Proc., pp. 35—39.

Parkash, S., 1974. Carbon, 12: 483—491.

Peterson, J.R., Adams Jr., R.S. and Cutkomp, L.K., 1971. Soil Sci. Am. Proc., 35: 72—78.

Pierce Jr., R.H., Olney, C.E. and Felbeck Jr., G.T., 1971. Environ. Letters, 1: 157—172.

Pierce Jr., R.H., Olney, C.E. and Felbeck Jr., G.T., 1974. Geochim. Cosmochim. Acta, 38: 1061—1073.

Porter, L.K. and Beard, W.E., 1968. J. Agric. Food Chem., 16: 344—347.

Rahman, A., Burney, B., Whitham, J.M. and Manson, B.E., 1976. N.Z.J. Exp. Agric., 4: 79—84.

Rhodes, R.C., Belasco, I.J. and Pease, H.L., 1970. J. Agric. Food Chem., 18: 524—528.

Riley, D., Wilkinson, W. and Tucker, B.V., 1976. In: D.D. Kaufman, G.G. Still, G.D. Paulson and S.K. Bandal (Editors), Bound and Conjugated Pesticide Residues. ACS Symp. Ser., 29: 301—353.

Roberts, H.A. and Wilson, B.J., 1965. Weed Res., 5: 348—350.

Robinson, D.W., 1965. Weed Res., 5: 43—51.

Robocker, W.C. and Canode, C.L., 1965. Weeds, 13: 8—10.

Saltzman, S., Kliger, L. and Yaron, B., 1972. J. Agric. Food Chem., 20: 1224—1226.

Savage, K.E. and Wauchope, R.D., 1974. Weed Sci., 22: 106—110.

Schnitzer, M. and Khan, S.U., 1972. Humic Substances in the Environment. Dekker, New York, N.Y., pp. 327.

Schliebe, K.A., Burnside, O.C. and Lavy, T.L., 1965. Weeds, 13: 321—325.

Schwartz Jr., H.G., 1967. Environ. Sci. Technol., 1 : 332—337.

Scott, D.C. and Weber, J.B., 1967. Soil Sci., 104: 151—158.

Sheets, T.J., 1958. Weeds, 6: 413—424.

Sherbourne, H.R. and Freed, V.H., 1954. J. Agric. Food Chem., 2: 937—939.

Shin, Y.O., Chodan, J.J. and Wolcott, A.R., 1970. J. Agric. Food Chem., 18: 1129—1133.

Sigworth, E.A., 1965. J. Am. Water Works Assoc., 57: 1016—1022.

Spencer, W.F., Cliath, M.M., Farmer, W.J. and Shaperd, R.A., 1974. J. Environ. Quality, 3: 126—129.

Steelink, C. and Tollin, G., 1967. In: A.D. McLaren and G.H. Peterson (Editors), Soil Biochemistry. Dekker, New York, N.Y., pp. 147—172.

Stevenson, F.J., 1966. J. Am. Oil Chem. Soc., 43: 203—210.

Stevenson, F.J., 1972. J. Environ. Quality, 1: 333—343.

Stevenson, F.J., 1976. In: D.D. Kaufman, G.G. Still, G.D. Paulson and S.K. Bandal (Editors), Bound and Conjugated Pesticides Residues. ACS Symp. Ser., 29: 180—207.

Su, Y.H. and Lin, H.C., 1971. Chem. Abstr., 74: 301.

Sullivan Jr., J.D. and Felbeck Jr., G.T., 1968. Soil Sci., 106: 42—52.

Swoboda, A.R. and Thomas, G.W., 1968. J. Agric. Food Chem., 16: 923—927.

Talbert, R.E. and Fletchall, O.H., 1965. Weeds, 13: 46—52.

Tompkins, G.A., McIntosh, T.H. and Dunigan, E.P., 1968. Soil Sci. Soc. Am. Proc., 32: 373—377.

Tsunoda, H., 1965. J. Sci. Soil Manure, Jap., 36: 177—181.

Tucker, B.V., Pack, D.E. and Ospenson, J.N., 1967. J. Agric. Food Chem., 15: 1005—1008.

Tucker, B.V., Pack, D.E., Ospenson, J.N., Omid, A. and Thomas Jr., W.D., 1969. Weed Sci., 17: 448—451.

Upchurch, R.P. and Pierce, W.C., 1958. Weeds, 6: 24—33.

Upchurch, R.P. and Mason, D.D., 1962. Weeds, 10: 9—14.

Upchurch, R.P., Selman, F.D., Mason, D.D. and Kamprath, E.J., 1966. Weeds, 14: 42—49.

Walker, A. and Crawford, D.V., 1968. In: Isotopes and Radiation in Soil Organic-Matter Studies. I.A.E.A., Vienna, pp. 91—105.

Wang, W.G., 1968. Diss. Abstr., B29(3): 904B—905B.

Ward, T.M. and Upchurch, R.P., 1965. J. Agric. Food Chem., 13: 334—340.

Ward, T.M. and Holly, K., 1966. J. Colloid Interface Sci., 22: 221—230.

Warshaw, R.L., Burcar, P.J. and Goldberg, M.C., 1969. Environ. Sci. Technol., 3: 271—273.

Weber, J.B., 1966. Am. Miner., 51: 1657—1661.

Weber, J.B., 1970. Soil Sci. Soc. Am. Proc., 34: 401—404.

Weber, J.B., 1972. In: R.F. Gould (Editor), Fate of Organic Pesticides in the Aquatic Environment. Am. Chem. Soc., 111: 55—120.

Weber, J.B., Perry, P.W. and Upchurch, R.P., 1965. Soil Sci. Soc. Am. Proc., 29: 678—688.

Weber, J.B., Ward, T.M. and Weed, S.B., 1968. Soil Sci. Soc. Am. Proc., 32: 197—200.

Weber, J.B., Weed, S.B. and Ward, T.M., 1969. Weed Sci., 17: 417—421.

Weber, J.B., Weed, S.B. and Waldrep, T.W., 1974. Weed Sci., 22: 454—459.

Weber, W.J. and Gould, J.P., 1966. Adv. Chem. Ser., 60: 280—304.

Weed, S.B. and Weber, J.B., 1974. In: W.D. Guenzi (Editor), Pesticide in Soil and Water. Soil Sci. Soc. Am., Madison, Wisc., pp. 39—66.

Weil, L., Duré, G. and Quentin, K.E., 1973. Z. Wasser Abwasser Forsch., 6: 107—112.

Wolcott, A.R., 1970. In: Pesticide in the Soil. Int. Symp. Pestic. Soil, Mich. State Univ., East Lansing, pp. 128—138.

Yoshida, T. and Castro, T.F., 1970. Soil Sci. Soc. Am. Proc., 34: 440—442.

Yuen, Q.H. and Hilton, H.W., 1962. J. Agric. Food Chem., 10: 386—392.

Chapter 5

SOIL ORGANIC CARBON, NITROGEN AND FERTILITY

C.A. CAMPBELL

INTRODUCTION

Fertility and organic matter

Soil fertility is defined as the status of the soil in relation to the amount and availability to plants of elements necessary for plant production (Canada Department of Agriculture, 1972). A fertile arable soil has to meet the current and future needs of cultivated plants. All fertile soils have an adequate supply of organic matter. Soil organic matter refers to the organic fraction of the soil; it includes plant and animal residues at various stages of decomposition, cells (living and dead) and tissues of microbes, and substances synthesized by the soil population (Canada Department of Agriculture, 1972). In a fertile soil, the function of organic matter is both direct and indirect. Its direct role is concerned with the provision of plant nutrients via the processes of decomposition and mineralization; its indirect role is associated with its effect on the physicochemical properties of the soil. Although good crop yields are possible in systems devoid of organic matter (e.g., sand culture and hydroponics) the difficulties encountered in maintaining proper nutrient levels, pH and solution concentrations make them much less economical and attractive than a soil system with adequate organic matter.In the foreseeable future it is unlikely that such systems will be used on a large scale to meet the world's food needs. A soil with adequate soil organic matter is usually easily cultivated, has good tilth and aeration, and facilitates root and moisture penetration. It is therefore not difficult to envisage the continued concern of agriculturists as they keep a wary eye on any agronomic or environmental phenomenon which tends to result in diminished soil organic matter contents.

As stated earlier the direct role of soil organic matter is to provide nutrients for good plant growth. What are these nutrients? How much is required? What are the sources of plant nutrients? How much is supplied by soil organic matter? These are some of the questions we must answer.

Nutrients required by plants

The essential elements required by plants may be divided into macro- and micro-nutrients. The 9 macro-nutrients are O_2, H_2, C, N, P, K, S, Ca, and Mg.

There is some uncertainty as to the essentiality of some of the micro-
nutrients but the ones commonly regarded as essential are Fe, B, Mo, Zn, Cu,
Mn, and Cl_2. Not all of these are obtained from organic matter, although
they are all present in plant and animal tissue.

An indication of the amounts of N, P, K and S required by several crops
grown in temperate climates may be gleaned from the amount of each constituent used by these crops (Fig. 1). The crop does not take up all of the
nutrients available to it; some of the nutrients are often rendered unavailable
by leaching, fixation and gaseous means.

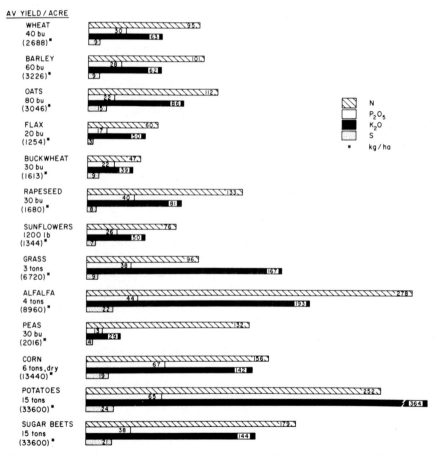

Fig. 1. Plant nutrients used by crops (compiled by Western Canada Fertilizer Assoc. from
research information obtained in western Canada).

Sources of plant nutrients

Under virgin conditions plants obtain all of their nutrients from soil, water, and air. C, H_2 and O_2 are readily obtained from air and water. The other nutrients are continuously being cycled by means of weathering and biological processes. Under virgin conditions the cycling may go on indefinitely with minimum losses from the system but, under crops some loss occurs regularly through livestock or human consumption. Under natural conditions plants absorb most of their N from the soil solution in mineral form; very small amounts are absorbed as organic substances. The source of this N may be NH_4 from rocks, N from air or N released upon decomposition of organic residues. Although soils often contain a third to a half of their P in organic forms this is almost completely unavailable to them unless it is mineralized into orthophosphate. K is present in soils in large quantities in primarily mineral form which is slowly released to plants by weathering and biological action. Ca and Mg are found in soils primarily as inorganic salts. S is present as inorganic salts and also in organic forms which decompose and is converted to salts in which form it is taken up by plants. Micro-nutrients are present in soils as relatively insoluble salts or as ions sorbed to clays or humus, or chelated by the latter. The traces of these used by plants are absorbed as ions from the exchange complex.

CARBON AND NITROGEN AND EFFECT OF SOIL FORMING FACTORS ON THEM

N required by the crop

According to Allison (1973), "N is the most important nutrient element in soil organic matter when considered from the economic standpoint. C is also equally essential but it is available in the atmosphere at no cost to the tiller of the soil." Crop yields are often directly proportional to the N released from organic matter; the other nutrients are also important but N is required in much larger amounts and is more likely to be deficient. Some idea of the amounts of N required by plants can be envisaged from the amounts of N present in the above-ground part of some crop and pasture plants and the vegetation of some forest communities (Table I). Agricultural crop plants have a higher annual N requirement than forest or natural plant communities and the annual demand for N under tropical conditions is higher than under temperate climates (Date, 1973).

Amount and distribution of N on earth

Of the $1972 \cdot 10^{20}$ g N present on earth, 97.82% is present in rocks in the lithosphere, 1.96% is in the atmosphere and only 0.02% in the biosphere

TABLE I

Nitrogen uptake for some selected crops, pastures and forest communities
(From Date, 1973)

	Yield of product (tops) (kg ha/yr.)	Nitrogen in (tops) (kg N/ha/ yr.)	References *
Grains			
Maize	10,800	160	Stanford, 1966
Wheat	5,600	190	Stanford, 1966
Vegetables			
Sugar beet	65,700	290	Viets, 1965
Cabbage	58,700	280	Viets, 1965
Forages			
Sorghum	31,800	380	Stanford, 1966
Pangola grass	39,400	750	Salette, 1965
Clover/grass	11,900	590	Melville and Sears, 1953
Lucerne	11,700	450	Melville and Sears, 1953
Native prairie	—	100—160	Dahlman et al., 1969
Tropical legume/ grass good	10,000— 20,000	300—500	Henzell, 1968
Tropical legume/ grass average	1,500— 7,500	40—190	Henzell, 1968
Forests			
Chaparral (13 yr.)	—	40	Zinke, 1969
Ponderosa Pine Assoc. (100 yr.)	—	7—10	Zinke, 1969
Redwood (alone) (1,000 yr.)	—	1—4	Zinke, 1969
Moist tropical forest (Ghana) (50 yr.)	7,800	41	Greenland and Kowal, 1960
Moist tropical forest (mean) (50 yr.)	10,340	427	Bazilevic and Rodin, 1966
Sub-tropical forest (mean) (50 yr.)	8,200	277	Bazilevic and Rodin, 1966
Sub-tropical "laurel" (Japan)	8,040	170	Bazilevic and Rodin, 1966
Dry savanna	540	81	Bazilevic and Rodin, 1966

* See Date (1973) for listing of these references.

(Porter, 1975). Of the 1913.17 · 10^{15} g N in the biosphere 47.04% is ocean bottom organic N, 39.72% is soil organic N, 7.32% soil inorganic N, 5.23% ocean inorganic N, 0.64% in plant and animals on land, and 0.05% in these organisms in the ocean (Porter, 1975). Thus, about 86.7% of the biosphere N is relatively inert and only slowly made available to plants by microbial degradation. The lithosphere N is of very low concentration and not available to plants.

The N contents of the surface soils of the U.S.A. vary from 0.01 to 1% or higher, the higher values being characteristic of organic soils (Shreiner and Brown, 1938). These scientists have also reported concentrations and amounts of N in several typical U.S.A. soils under virgin conditions (Table II). The chernozem and prairie soils which are well drained and developed on grassland have the highest amounts and concentration of N, the acid podzolic soils the lowest and the drier grassland chestnut and brown soils are intermediate. The grassland soils of the Great Plains of the U.S.A. contain a tremendous reservoir of N. Porter (1971) estimated the area to have about $2.39 \cdot 10^{12}$ kg N. Although we are not aware of any similar estimates for Canadian soils the amount of N to the bottom of the B horizon of some of the soils of the Canadian prairies (Campbell et al., 1976) is shown in Table III. The increasing depth of profile development as one proceeds from the brown to black soils in the chernozemic order is reflected in the total N in the soils. The proportion of the N which is located in the A horizon also increased from the brown to the black soil zone.

Effect of soil forming factors

According to Jenny (1930) the order of importance of the soil forming factors as they affect the N content of loams in the U.S.A. are climate > vegetation > topography = parent material > age. His concepts, although having certain shortcomings, have contributed considerably to our understanding of the factors affecting N content of soil and have facilitated our appreciation of the problems involved in maintaining the reserve of soil N on cultivated land.

Climate determines the plant species, the quantity of plant material produced, and the microbial activity and as such it is the main factor governing the organic matter levels in the soil. Under humid conditions podzolic-type soils are formed; semiarid conditions lead to the development of brunizems, chernozems and chestnut soils. Brunizems and chernozems have the highest N content of all well drained soils; desert, semidesert, and lateritic soils have the lowest, and chestnut and podzolic soils are usually intermediate. Soils formed under restricted drainage (gleysolic) do not follow a climatic pattern because poor aeration retards microbial decomposition of organic matter over a wide temperature range. N distribution in the profile of representative soils of various great groups of the U.S.A. are shown in Fig. 2 (Stevenson, 1965).

Parsons and Tinsley (1975) show profile distribution of C and N (Fig. 3) and C/N ratios (Table IV) for some great group soils all over the world. The organic C and N tends to accumulate near the surface of the cold tundra soil, in the surface litter of acid podzols, and to >1 m under the poor aeration of the peat (Fig. 3). Where organic matter was high the C/N ratio was wide (Table IV) reflecting the slow decay process. In soils with C ⩽ 5% and good

TABLE II

Average nitrogen content in various soil regions of the United States (Adapted from Schreiner and Brown, 1938)

Soil region	Approximate area of region (ha) (in millions)	Approximate nitrogen in surface 15 cm (%)	Average amount of nitrogen per hectare to depth of 15 cm (metric tons) *	Average nitrogen to depth of 100 cm (%)	Average amount of nitrogen per hectare to depth of 100 cm (metric tons)
Brown forest	72.8	0.05—0.20	2.8	0.05	7.5
Red and yellow	60.7	0.05—0.15	2.2	0.03	4.5
Prairie	45.7	0.10—0.25	3.9	0.12	17.9
Chernozem and chernozem-like	49.8	0.15—0.30	5.0	0.12	17.9
Chestnut	41.3	0.10—0.20	3.3	0.08	12.0
Brown	21.0	0.10—0.15	2.8	0.06	0.9

* 907 metric tons of soil per hectare.

TABLE III

Average weight of N in major horizons in soils of the Canadian prairies *
(From Campbell et al., 1976)

Soil order	Great group	Parent material	Cumulative N to bottom of horizon (kg/ha)			
			— — — — — horizon — — — — —			
			organic	A	transition	B
Chernozemic	brown	lacustrine	—	2,400	—	6,000
		till	—	2,600	4,500	7,200
	dark brown	lacustrine	—	3,200	6,900	7,900
		till	—	3,400	7,900	8,100
	black	lacustrine	—	5,800	9,400	11,000
		till	—	6,200	10,100	11,100
		alluvium	—	7,500	11,200	12,300
	dark gray	lacustrine	—	3,300	—	7,400
		till	1,500	4,800	7,500	8,500
Solonetzic	solonetz	lacustrine	—	3,000	4,200	7,300
		till	—	1,900	3,300	6,300
		alluvium	—	3,900	7,400	10,400
	solod	lacustrine	—	3,900	7,300	12,000
		till	—	3,400	5,700	8,500
Luvisolic	gray luvisol	lacustrine	1,300	—	2,600	5,300
		till	1,000	—	2,300	4,500
		alluvium	900	—	1,900	4,400
Brunisolic	eutric brunisol	various	1,100	1,400	1,800	2,600
	dystric brunisol	various	700	—	800	1,300
Regosolic	regosol	various	—	7,400	(to bottom of A or to 30 cm whichever was greatest)	
Gleysolic	humic gleysol	various	2,500	8,400	—	9,000
	eluviated gleysol	various	2,000	—	5,600	9,100

* Data from Can SIS files.

drainage C/N was 13 or less in the surface horizons.

In the forested podzols of the U.S.S.R. 50% of the organic matter of the profile was located in the top fifth of the profile (Kononova, 1966). In the chernozems the organic matter was more uniformly distributed throughout the profile, 24—32% being in the top fifth of the profile; in the soils of the drier regions (chestnuts and serozems) 43—45% was in the top fifth of the

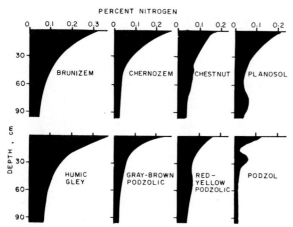

Fig. 2. Distribution of nitrogen in profiles representative of several great soil groups of the U.S.A. (Adapted from Stevenson, 1965.)

profile. The C/N ratios were >10 in the surface horizon of the chernozem and chestnut soils; in podzolics (except for the virgin and deep humic soils) it was between 10.5 and 9.7. Serozems which are rich in N had the narrowest C/N of about 8. Krasnozems of the humid subtropics contained large amounts of organic matter with $>50\%$ of it concentrated near the soil surface but, because the organic matter was low in N, the C/N ratio was very wide (18.9).

C and N data for typical Canadian soils are available in Can. SIS files (Dumanski et al., 1975) but very few of these data have been published in a systematic manner. Campbell et al. (1976) presented some data for the Canadian prairies. Within the chernozemic order percent N in soils was, as expected, higher in the black than in other great groups, and soils developed on till tended to have higher N concentrations than those developed on lacustrine material (Table V). However, in the solonetzic and luvisolic soils percent N was higher in soils developed on lacustrine than on till. The N content of alluvial soils was variable as might be expected from the variable nature of the materials deposited and the environment under which these soils were formed. The percent organic C of these soils followed the same general trends as percent N and generally C/N ratios of the A horizons were between 12 and 9 (Campbell et al., 1976). The C/N ratios in the B horizons were quite variable; they tended to be lower than in the A horizons in the chernozemic, solonetzic and humic gleysolics but the converse was true in the luvisolics and eluviated gleysol.

An inverse relationship exists between soil N content and mean annual temperature between $0°C$ in Canada to $20°C$ in southern U.S.A. (Jenny, 1930, 1941). Furthermore, Jenny found that the effect of temperature

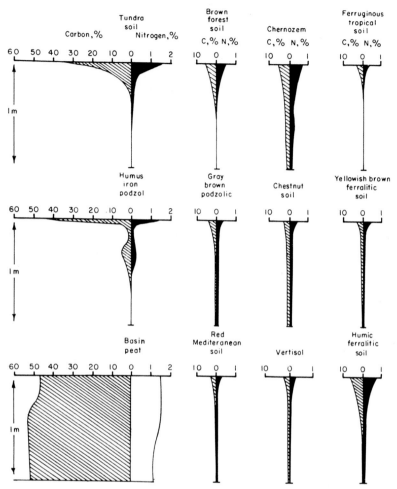

Fig. 3. Distribution of carbon and nitrogen in representative soil profiles. (From Parsons and Tinsley, 1975.)

obeyed Van't Hoff's temperature rule; thus the soil N content increased 2—3 times for each 10°C decrease in the mean annual temperature. The relationship is adequately defined by the expression:

N = a/(1 + C exp− kt)

where N is the total soil N content, t is the mean annual temperature in °C, exp is the base of the natural logarithm and a, C, and k are constants. Jenny (1930) also found that all other factors remaining equal, the soil N content increases with water supply. He expressed the combined effects of tempera-

TABLE IV

Carbon : nitrogen ratios of representative soils
(From Parsons and Tinsley, 1975)

Depth (cm)	Types of soil											
	1	2	3	4	5	6	7	8	9	10	11	12
0	25			12	13		11					
	20							11		16	11	
10			31	11	10	13						11
	31	29										
20									13		10	
	16				10	8	13	9				
30				11								9
		12										
			31		10						7	
40									8			
	12											
				11			12	9				
50		23				5						
					12					12	8	
60									10			
												8
70				32								
		6					12					
80						4						
					9			10			8	
90									9	12		
100												

1. Tundra soil, Alaska, Rieger (1966) *.
2. Humus iron podzol, England, Davies and Owen (1934).
3. Peat, Scotland, Glentworth and Muir (1963).
4. Brown forest soil (cultivated), England, Kay (1936).
5. Gray brown podzolic, U.S.A., Brown and Thorpe (1942).
6. Red Mediterranean, Spain, Guerra et al. (1966).
7. Chernozem (cultivated), U.S.S.R., Shuvalov (1964).
8. Chestnut soil (cultivated), Rumania, Cernescu (1964).
9. Vertisol (cultivated), Spain, Guerra et al. (1966).
10. Ferruginous tropical soil, Angola, D'Hoore (1964).
11. Yellowish brown ferralitic soil, Congo, D'Hoore (1964).
12. Humic ferralitic soil, Congo, D'Hoore (1964).

* See Parsons and Tinsley (1975) for individual references shown here.

TABLE V

Average N contents of major horizons in soils of the Canadian prairies [1]
(From Campbell et al., 1976)

Soil order	Great group	Parent material	A Horizon % N	A Horizon σ [3]	Transitional horizons [2] % N	Transitional horizons σ	B horizon % N	B horizon σ
Chernozemic	brown	lacustrine	0.17	0.06	—	—	0.12	0.06
		till	0.20	0.05	0.16	0.05	0.11	0.04
	dark brown	lacustrine	0.21	0.06	0.13	0.05	0.11	0.06
		till	0.26	0.11	0.15	0.08	0.12	0.06
	black	lacustrine	0.28	0.13	0.14	0.08	0.12	0.07
		till	0.32	0.15	0.13	0.04	0.11	0.05
		alluvium	0.26	0.12	0.12	0.08	0.10	0.06
	dark gray	lacustrine	0.22	0.10	—	—	0.05	0.02
		till	0.27	0.13	0.08	0.03	0.07	0.02
Solonetzic	solonetz	lacustrine	0.24	0.08	0.15	0.07	0.12	0.05
		till	0.21	0.04	0.12	0.04	0.11	0.04
		alluvium	0.31	0.16	0.23	0.07	0.13	0.04
	solod	lacustrine	0.25	0.08	0.16	0.09	0.10	0.03
		till	0.19	0.06	0.11	0.04	0.10	0.04
Luvisolic	gray luvisol	lacustrine	0.52	0.17	0.11	0.07	0.07	0.03
		till	0.35	0.29	0.05	0.03	0.06	0.04
		alluvium	0.34	0.30	—	—	0.05	0.03
Brunisolic	eutric brunisol	various	0.34	0.28	0.10	0.09	0.06	0.05
	dystric brunisol	various	—	—	0.06	0.05	0.03	0.03
Regosolic	regosol	various	0.28	0.11	0.08	0.02	—	—
Gleysolic	humic gleysol	various	0.40	0.21	0.10	0.08	0.08	0.07
	eluviated gleysol	various	0.38	0.26	0.11	0.05	0.07	0.04

[1] Data from Can SIS files.
[2] Ae, AB, etc.
[3] σ = standard deviation.

ture and water supply by the equation:

$$N = 0.55 \exp{-0.08t} \, (1 - \exp{-0.005H})$$

Here N, t and exp are as defined above and H is the "humidity" factor which is defined as the ratio of mean annual precipitation (in millimeters) to the absolute water vapor saturation-deficit of the air (in millimeters of Hg). Jenny et al. (1948, 1949) found similar results in Columbia, South America and in Costa Rica except that under similar temperature and moisture conditions Columbian soils had much higher N and organic matter contents than those of North America. In India Jenny and Raychaudhuri (1960) observed temperature and moisture effects similar to those found in North America, except that the organic matter levels were not as high as those in Central and equatorial South America.

Jenny (1950) attempted to explain the higher organic matter levels for tropical forest soils in terms of high litter fall and favorable climatic conditions plus high rates of N fixation by legumes. He also suggested that the decomposed forest litter was rapidly leached into the soil where it was fixed by soil minerals and decomposed slowly thereafter. Allison (1973) suggests that this is an inadequate explanation. Smith et al. (1951) gave a more plausible explanation. They suggest that the higher organic matter level in the tropics was due to no killing frosts plus conditions which favor year round luxuriant growth; dead plant material decomposes more rapidly than live material.

Walker and Brown (1936) have shown that there is a tendency for soils with finer texture to have higher contents of nitrogen (Table VI). Topography affects soil nitrogen content through its effect on the microclimate and runoff and through these its effect on vegetation. The types and numbers of organisms present in a soil affect the soil nitrogen but the significance of these variations cannot be quantified because microbial activities depend strongly on numerous other factors. Factors which restrict microbial activity and thus promote N accumulation are low temperatures (Jenny, 1930), restricted drainage, low pH, presence of toxic substances in the soil and the formation of metal—clay—organic complexes which protect proteins from attack (Stevenson, 1965).

Optimum level of soil organic matter

From the foregoing discussion it can be readily seen that soils vary considerably in their organic matter content. What is the optimum level of soil

TABLE VI

Average nitrogen content of upland soils differing in texture in northeastern Iowa (From Walker and Brown, 1936)

Soil texture	Nitrogen content of soil (%)
Sand	0.027
Fine sand	0.042
Sandy loam	0.100
Loam	0.188
Silt loam	0.230

organic matter? There is no single answer to this question even for soils of the same texture and from the same climatic zone. This is because the best organic matter level is a function of the crop to be grown — a level adequate for tobacco growing would be too low for general farming and much too low for market gardens. Furthermore, other factors such as drainage, type of clay, nutrient level and past management may make as significant a contribution to efficient production as the organic matter level.

EFFECTS OF MANAGEMENT ON SOIL ORGANIC MATTER

Several reviews have been written on this topic (Ensminger and Pearson, 1950; Stevenson, 1965; Allison, 1973). Very little reference was made to the significant Canadian experience in this area of research, most of the references being centred on American and British findings. Our aim is partly to rectify this oversight. Stevenson's coverage of the American experience is excellent. In general, he reports that in the U.S.A. the following was found: (a) that under average farming conditions about 25% of the N is lost in the first 20 years, about 10% in the second 20 years and about 7% in the third 20 years; (b) that steady-state conditions would probably be reached within 50 to 100 years after cultivation commenced; (c) that loss of N was greatest with intertilled crops, intermediate with cereal crops and smallest with legume and sod crops; and (d) that when soils increased in N it was usually because cropping systems were adopted which included legumes.

Effect of long term cropping

In temperate and humid regions of America and Australia where the profit motive has been strong, farmers have pursued cropping practices which have often hastened the destruction of once highly productive soils (Allison, 1973).

If the environment and management practices remain relatively static over many years the soil organic matter will equilibrate and gains and losses of N and C will be equal. If conditions are changed then the balance is disturbed and losses may exceed gains or vice versa. In Canada, as in the U.S.A. the effect of breaking virgin soil and cultivating it has led invariably to large losses of organic matter and N (Caldwell et al., 1939; Newton et al., 1945; Doughty et al., 1954; Hill, 1954). Losses during the early years of cultivation were also much more rapid than in later years (Fig. 4).

Newton et al. (1945) determined losses of C and N across the Canadian prairies by comparing paired virgin and cultivated soils. After 22 years of cropping the black, dark brown, and brown soils had lost on the average 20% of the C and 18% of the N from the top 15 cm of soil; losses of C and N were about 30% from the gray soils. Losses from the 15- to 30-cm segment

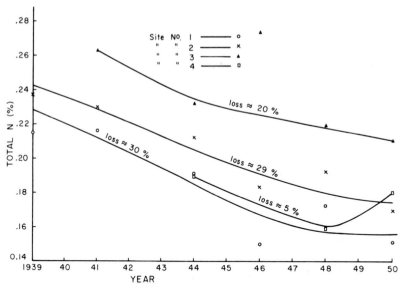

Fig. 4. Loss of soil N in top 30 cm after breaking. (From Campbell et al., 1976 — based on work of Doughty et al., 1954.)

were relatively small. Although the losses were significant in all cases, they were quite variable; at some locations they were as high as 50% while at others there were apparent gains. They also found that the C/N ratio narrowed with length of cultivation. Hill (1954) reported on the effect of several short- and long-term rotations on the total N and organic matter in the top 15 cm of a chestnut loam in southern Alberta. He found a consistent decrease in the soil N between 1910 and 1953; average loss for the six rotations was 24%. The smallest loss occurred in a 10-year rotation which included four years of alfalfa and the highest losses from a 9-year rotation which included three years of fallow and one intertilled crop, and from 2- and 3-year grain rotations. Jenny and Raychaudhuri (1960) have observed 60—70% loss of organic matter from dryland soils in India which have been cultivated for a long time.

Two separate studies were carried out in southern Saskatchewan to determine the loss of organic matter and N resulting from breaking native sod and cultivating to a wheat—fallow system. Doughty (1954) carried out the first study from 1939—50 and reported losses of as much as 30% of the initial N over a period of 11 years with the rate of loss being greater in the first few years. Doughty's work was based on sampling every 2 to 3 years and he took 3 replicates per site. A plot of his actual year-to-year data (Fig. 4) showed considerable variability in the data. A second study was carried out by Warder and Hinman (1953—1964) between 1953 and 1964. They sampled four soil types on land which was broken in 1953, taking 10 replicates (stepped off from a fixed point) each year. This study should be more precise than that of Doughty. Surprisingly, they found very little loss of total N (Fig. 5) or organic matter (data not shown). Losses ranged from a high of 8.5% for the Regina heavy clay to no change for the Hatton fine sandy loam after 11 years of cultivation. The reason for the difference between Doughty's and Warder's studies might very well lie in the inherent variability in the total N in a field. Campbell et al. (1976) calculated the coefficient of variability to be between 9 and 39% for Warder and Hinman's 1957 samples. Furthermore, the size of the difference that would be required in order that two means would be regarded as different at the 5% level of probability was calculated to be about 3½ times greater when 3 replicates were used as when 10 replicates were used. Thus when the number of replicates taken in this type of study is small, large errors in estimating the amount and rate of change of N or organic matter is quite probable. This point has also been emphasized by Kononova (1966), and Cameron et al. (1971). The latter concluded that even with as many as 30 samples per field there was considerable imprecision in values of N and P. This suggests that although there is little doubt that losses do occur due to cultivation, the magnitude of the losses is not at all certain.

It is possible for gains to exceed losses especially where legumes or grass— legume pastures are included as a major part of the cultural program as was the case in New Zealand (Jackman, 1964; Greenland, 1971). The two sandy

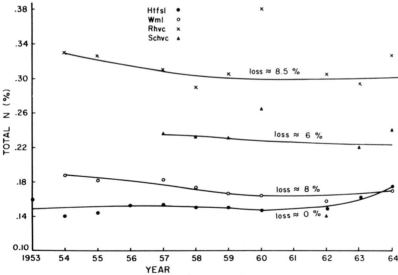

Fig. 5. Loss of soil N in top 30 cm after breaking. (From Campbell et al., 1976 — based on work of Warder and Hinman in Annual Reports 1953—1964, Soil Section, Research Station, Swift Current, Sask.)

soils (Fig. 6(a)) contained allophane and this clay seemed to stabilize organic matter in soils. The accumulation of N in the soil appeared to occur from the surface downwards (Fig. 6(b)). Greenland (1971) reported on a long-term permanent rotation trial at the Waite Institute in Australia which includes pasture—wheat rotations, some of which were started in 1926 and some in 1951 on plots of the same trial previously under more intensive cultivation (Fig. 7). The rotations which included the longer pasture periods attained higher levels of N. In the U.S.S.R. increase in organic matter after 2 years of leys was 20—50% for irrigated serozems, 10—25% for sod podzols, 5—10% for chernozems and 2—5% for chestnuts (Kononova, 1966). In a follow-up to a long term study carried out by Hobbs and Brown (1957) in Kansas, Hobbs and Thompson (1971) grew a fallow—wheat—sorghum rotation for a further 8 years. They found that the equilibrium level for the soil N was 0.103%; soils with N content >0.100% lost N and those with <0.100% gained.

Effect of cropping methods and rotations

Studies on the effects of rotations are difficult to design and to assess because of the numerous variables involved such as crops, crop sequence, fertility level and field arrangement.

The role of forages in the rotation

During the "dirty thirties" in western Canada when erosion wrecked havoc on the land the seeding of forages was encouraged to reduce soil drifting. At Swift Current, Saskatchewan, it was found that organic matter and N increased due to the growing of the forage (Campbell et al., 1976). Atkinson and Wright (1948) reported similar results in eastern Canada. In

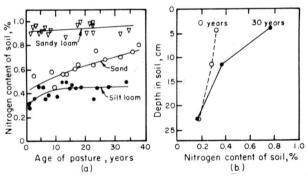

Fig. 6. Nitrogen in soils under grass-legume pasture vegetation in New Zealand. (a) Nitrogen content of the surface 7.6 cm of soil versus time after establishment of permanent pasture on three soils. (b) Vertical distribution of nitrogen in a soil profile at the time of pasture establishment and 30 years later. (From Jackman, 1964.)

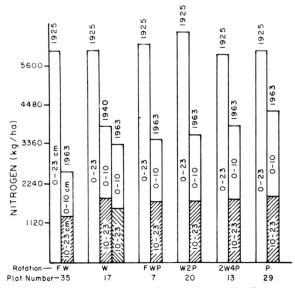

Fig. 7. Nitrogen contents of soils of the permanent rotation trial (Cl) at the Waite Institute, in 1925 when the experiment commenced and in 1963. (Adapted from Greenland, 1971.) *FW* = continuous fallow wheat since 1925; *W* = continuous wheat since 1925; *FWP* = fallow-wheat-1 year ryegrass pasture 1925—1945, 1 year ryegrass-clover 1946 onwards; *W2P* = peas-wheat-oats 1925 to 1949, then wheat-2 years ryegrass clover pasture; *2W4P* = fallow—wheat—peas 1925 to 1948, then from 1949 2 years wheat and 4 years *Phalaris*—ryegrass—clover pasture. *P* = wheat—peas 1925 to 1949, then continuous pasture, composed predominantly of annual grasses, but including some lucerne.

one of the more complete studies in western Canada nine perennial grasses differing in rooting habit were seeded at 15- and 30-cm spacings and with and without alfalfa (1954 Annual Report, Soils Section, Agriculture Canada, Research Station, Swift Current, Saskatchewan). Differences between spacings, and with and without alfalfa were small, therefore, the results from these four treatments on each grass were averaged (Table VII). Over 5 years these grasses did not change the N or organic matter of the surface 15 cm of soil but there was a small increase in the 15- to 30-cm segment. There were no differences between the grasses. They were highly beneficial in conserving the fertility of the soil since under cereal cultivation a large decrease in N and organic matter might have occurred.

In the U.S.A. Gupta and Reuszer (1967) grew continuous alfalfa, bromegrass and corn for 7 years. The soil N and organic matter content were greater under alfalfa than under grass which was greater than under corn. Shorter rotations had no effect on soil organic matter or N. Dubetz and Hill (1964) in southern Alberta compared short cash crop rotations versus longer rotations which contained alfalfa. They found that it required at least 3

TABLE VII

Effect of various grasses on soil nitrogen and organic matter
(1954 Annual Report, Soil Section, C.D.A. Research Station, Swift Current, Saskatchewan)

Grass	Organic carbon (%)				Nitrogen (%)			
	0—15 cm		15—30 cm		0—15 cm		15—30 cm	
	1947	1952	1947	1952	1947	1952	1947	1952
Brome	1.38	1.44	0.82	1.00	0.15	0.14	0.10	0.11
Green stipa	1.49	1.37	0.67	0.93	0.15	0.14	0.10	0.10
Streambank	1.36	1.31	0.65	0.90	0.14	0.14	0.10	0.10
Crested wheat	1.29	1.21	0.72	0.94	0.13	0.14	0.10	0.11
Virginia wild rye	1.34	1.23	0.67	0.86	0.13	0.13	0.10	0.10
Mandan wild rye	1.31	1.27	0.64	0.89	0.14	0.14	0.10	0.10
Tall wheat	1.29	1.28	0.72	0.93	0.15	0.14	0.10	0.11
Intermediate wheat	1.36	1.35	0.69	0.99	0.14	0.14	0.10	0.11
Russian wild rye	1.27	1.23	0.79	0.83	0.14	0.15	0.11	0.11
Average	1.34	1.30	0.71	0.91	0.14	0.14	0.10	0.11

years of alfalfa or brome-alfalfa in the rotation to maintain the soil organic matter and N.

In Australia (Greenland, 1971; Russell, 1966) legume-based pastures are used in rotation with cereals as an effective method of improving the soil fertility. When these leys or "resting crops" are ploughed up they might not increase the total humus content but they do increase the "light fraction" of the organic matter which is more active in nutrient supply than is the rest of the organic matter (Greenland, 1971).

The role of the forages in the rotation cannot be overlooked. When they contain legumes these fix N from the air and their deep roots bring up nutrients from below the zone of penetration of cereal roots. Grass roots provide tremendous amounts of dry matter for slow steady decomposition throughout the body of the soil and also will tend to improve the structure and permeability of fine textured soils. Growing legumes and non-legumes together results in transfer of N to the non-legumes due to sloughing off of nodules from legume roots; under certain conditions, e.g., severe drought, the legumes may excrete N from their roots (Allison, 1973).

Row crop — small grains — summerfallow

In Manitoba, Ridley and Hedlin (1968) found little difference in soil organic matter between the close-growing cereals: wheat, oats and barley after 37 years of continuous cropping but the organic matter was lower when an intertilled row crop such as corn was grown. The effect on total N though similar in trend was not as conclusive. They also found that the organic matter and N content of a black lacustrine soil was maintained at a high level by intensive cropping to cereals; frequent summerfallowing resulted in the greatest decline in organic matter and total N. Similarly Ferguson and Gorby (1971) reported a 24% loss of total N from a coarse textured chernozemic black soil which was continuously fallowed for 12 years.

Effect of trees

Salisbury and Delong (1940) studied the organic matter changes in the Appalachian upland podzols of Quebec which had been deforested and either cultivated or pastured for 75 years after deforestation. They found that the cultivated and pastured soils tended to have slightly lower organic matter than the forested soil. It is not surprising that the difference was small because although this soil is high in organic matter the latter is reported to be very inert. Some workers in the U.S.A. have reported that there appears to be greater mineralization of N under conifers (Stone and Fisher, 1969) and mesquite trees (Tiedemann and Klemmedson, 1973) than on adjacent open areas. Consequently, herbaceous vegetation and grasses were more luxurious under the trees. No reasonable explanation has been advanced to account for this observation.

Effect of manures and residues

Farmyard manure

Feeds contain certain nutrients, some of which are recovered in the manure; how much is recovered depends on the quality and amount of feed, the age and kind of animal (Allison, 1973). The recovery of N—P—K in the urine is nearly as great, or even greater than in feces. Manure supplies numerous micro-nutrients and is an excellent source of humus. Unlike green manures, farmyard manure and composts decompose slowly because they are usually already well rotted. Farm manure is of most value under systems of intensive cropping such as gardening and floriculture. Barnyard cow manure usually contains 5 kg N, 2.5 kg P_2O_5 and 5 kg K per ton; poultry and sheep manure are higher in N and P (Allison, 1973). Farmyard manure even if handled carefully loses a large proportion of the nutrients it contains by volatilization as NH_3 and by leaching of NO_3.

Newton et al. (1945) reported on work carried out in western Canada over a period of 25—30 years in which farmyard manure reduced the rate of loss of soil C and N compared to losses resulting from grain—summerfallow rotations. In Manitoba, Ridley and Hedlin (1968) found that an application of barnyard manure at 8.9 metric tons/ha to wheat maintained organic matter and N at slightly higher levels than in unmanured plots. However, manure was not enough to maintain the organic matter of the frequently fallowed soils (4.1%) at the same level as the non-manured continuously cropped soils

(7.2%). In studies carried out in southern Alberta on a dark brown chernozemic soil, Dubetz et al. (1975) applied 27 tons/ha/4 yr. for 16 years to an irrigated 4-crop rotation and found that the manure increased soil organic matter, N—P—K, and yields of beets and sweet corn but not of spring wheat; the benefits of manure increased with time. Sommerfeldt et al. (1974) found that when feedlot manure was applied at 67 metric tons/ha/yr. for 40 years, total and plant available N increased significantly but not excessively to a depth of 3 m. Scientists in eastern Canada (Horton, 1942; Wright et al., 1950; Cordukes et al., 1954; Bishop et al., 1962) have reported results similar to those found in western Canada. In growing flue-cured tobacco, the level of organic matter cannot be too high because coarse dark green, late maturing leaves will result; it cannot be too low because the sandy soils will erode rapidly. Horton (1942) working in southern Ontario was able to increase tobacco yields without changing the level of soil organic matter by applying 11 metric tons manure/ha in the spring prior to ploughing.

In the U.S.A. Anderson and Peterson (1973) grew continuous corn on irrigated Typic Haplustoll soil (Soil Survey Staff, 1960) in western Nebraska from 1912—1972. Corn yields declined rapidly for the first 10 years when no manure or fertilizer N was applied. When manure was applied to half the plots at 27 metric tons/ha beginning in 1942, total N (Fig. 8) and yields increased till 1953 and then remained constant. By 1972 the manure had returned the soil N to 90% of its original value. In a 5-year study, also with continuous corn, carried out on a Typic Ochraqualf in Vermont, McIntosh and Varney (1973) applied barnyard manure at rates from 0—66 tons/ha.

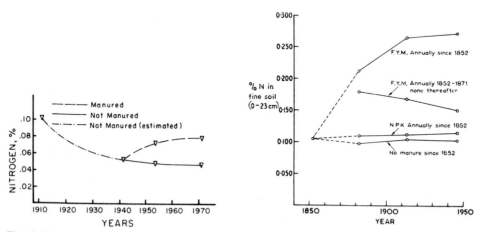

Fig. 8. Total soil N content of the surface 30 cm from 1910 to the present as related to time and manuring. (From Anderson and Peterson, 1973.)

Fig. 9. Nitrogen content of Hoosfield soil. (From Jenkinson, 1966a.)

They found that cultivation and cropping of unmanured plots reduced organic N and C by 8.7 and 17.7%, respectively; an annual application of 44 tons/ha of fresh dairy manure was required to maintain the soil organic matter.

At Rothamsted in Britain, Warren's results (1956) for continuous barley experiments showed that where no manure was applied the N content of soil remained constant for 94 years (Fig. 9). While the larger root and stubble additions from N—P—K produced a 10% increase in N, an annual dressing of 34 metric tons of farmyard manure per hectare trebled the total N in 94 years. Evidence that manure breaks down slowly can be seen from the fact that about 1/3 of the N added in manure between 1851—1871 still remained in the soil in 1946 (Jenkinson, 1966a).

In the U.S.S.R. systematic application of farmyard manure increased soil organic matter primarily in the top 20 cm with smaller increases in the 20—40 cm depth (Kononova, 1966). Only 1/3 to 1/4 of the manure remained in the soil, the rest being decomposed and increasing the soil's fertility.

Allison (1973) states that on most farms the cost of handling manure balances the nutrient gains, and that all things considered the benefits do not justify the costs under modern farming conditions when fertilizers are available. We feel that in today's pollution-conscious society agrologists and agronomists have few degrees of freedom; they must encourage the agronomic use of manures even at the cost of a small loss in income to the farmer. If we do not use these wastes and if we continue to abuse their handling they shall in the long run return to haunt our children!

Green manure

Allison (1973) gives an excellent review of this subject. The question of priming will be dealt with separately later.

Green manures were used for manuring rice in China for as long as 3,000 years ago; they were also used in Greece and Rome before the birth of Christ (Allison, 1973). Today they are seldom grown just for maintaining soil organic matter although they are grown to increase the active fraction of the organic matter. Allison (1973) states that contrary to past prevailing opinion, although green manure is rapidly oxidised by micro-organisms, this C is no less valuable to humus formation per unit of C added than is C from more carbonaceous crop residues.

Green manure is usually very beneficial, and its occasional harmful effects can be circumvented by good management. Its effects can be listed as follows: it is involved in N fixation, humus formation, improving soil physical condition, conservation of nutrients, supplies CO_2 as a nutrient, controls erosion, controls plant diseases, and may cause harmful chemical and biological effects. (For a complete discussion of these factors the reader may wish to consult Allison, 1973, chapter 22).

In western Canada, Poyser et al. (1957) reported on a comprehensive study started in 1919 and carried out on the Fort Garry and Red River clay in Manitoba. It consisted of a rotation of green crop—fallow, wheat, corn, wheat, laid out on four blocks, one for each year of the rotation. Each block was divided into eleven 0.01-ha plots permitting three check plots, seven green manure treatments, and one barnyard manure treatment. Green crops used were: weeds, buckwheat, spring rye, corn, peas, sweet clover and red clover. Check plots were kept free of weeds. Rotted farmyard manure was applied at 22.4 metric tons/ha and ploughed down at the same time that the green crops were. Over the period 1930—1955 there was an overall average decrease of 28% of the organic matter and 16% of the N. Although there was a decrease in the N level under all treatments (Fig. 10) the rate of loss was retarded where farm manure and legumes were used. The rate of loss was not influenced by the green manure treatment. Only sweet clover and farm manure maintained the organic matter (data not shown) and N at a significantly higher level than the check. Similar results were obtained by Ridley and Hedlin (1968), and Ferguson and Gorby (1971). In eastern Canada, Sowden and Atkinson (1968) made annual additions of green rye, straw, alfalfa, leaves and other manures to an Uplands sand and a Rideau clay. After 20 years they observed a loss of organic matter on the untreated and rye plots, with most of the loss occurring in the first 10 years. The other materials maintained the organic C level on the clay. On the sand, the control plot lost organic matter but all other treatments increased it. Halstead and Sowden (1968) also showed that these green manure treatments had positive effects on the yield of oats.

According to Allison (1973), "A careful evaluation of all the information has led to the conclusion that it is usually most profitable and satisfactory to buy the N in the bag rather than grow it. Purchased N can be applied at any rate and time desired, and the crop yields are likely to be higher than when

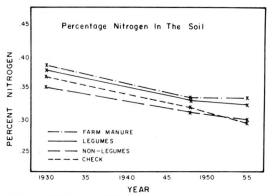

Fig. 10. Effect of different types of green manure crops on soil N. (From Poyser et al., 1957.)

legume is the sole source of N." We agree with Allison to a certain extent. Unlike the case of farmyard manures, the pollution hazard is not great. However, green manures would still be recommended to reduce the incidence of diseases and toxins; to prevent erosion (e.g., in tobacco growing and on sandy soils); to improve the structure of heavy clays; and in Australia and Africa to be used as resting crops or leys. There is also a possibility that it might prove beneficial in reducing the incidence of salinity in western Canada.

Crop residues

The term crop residues is being used here arbitrarily to refer only to dry plant tops and roots as opposed to green tops which are turned under at a succulent stage of growth. Although the tie-up of inorganic N in organic form by micro-organisms which decompose residues (immobilization) will be mentioned in this section, immobilization per se will be discussed fully in a later section.

The value of plant residues to the fertility of the soil will depend upon the amount, its N content, its relative availability and so on. Newton et al. (1939) found that while the roots and stubble of hay plants contained about 71% of the dry matter and 68% of the nitrogen, the corresponding value for wheat was 19% of the dry matter and 9% of the N. There are differences between species of grasses regarding the amount of N that is available for decomposition. Haas (1958) showed that the N in grass roots of several species ranged from 56 to 150 kg N/ha — 60 cm. Vallis and Jones (1973) found that although leaves and litter of legumes D. intortum cv Greenleaf and P. atropurpureus cv Siratro had similar N and lignin content, N mineralized from the former was less than that from P. atropurpureus. They suggested that the fact that D. intortum had a much higher polyphenol content in its leaves was partly responsible for this difference in mineralization. Although there is considerable N present in grassland soils only a small proportion of the N present is released each year unless the sod is ploughed up (Porter, 1971).

The method of addition of plant residues to soil affects the rate of decomposition and buildup of organic matter reserves. When left on the surface as a mulch it often becomes desiccated and decomposes more slowly than if incorporated (Parker, 1962; Brown and Dickey, 1970; Shields and Paul, 1973). The rate of decomposition varies with depth of placement; this is particularly true since depth affects temperature, aeration, and moisture conditions of decomposition. Generally, organic residues decompose more rapidly at shallow depths and leave less humus than at lower depths. Burial of residues under wet, cold conditions (Kononova, 1966) or very dry conditions tends to preserve organic residues (Shields and Paul, 1973).

In western Canada, Hedlin et al. (1957) studied the effect of alfalfa, grass, and wheat crop residues on wheat yields over a 5-year period. Wheat residues reduced yields after summerfallow but reduced them even more after one crop (wheat on stubble). The depressing effects were partly alleviated by applying N—P fertilizer. Yields were greater after one crop if the stubble was burnt or if clover residues were ploughed under during the fallow year. Ferguson and Gorby (1964) report that when straw was applied to fallow fields nitrate release from the soil was reduced markedly; the same was true when soil samples with straw were incubated in the laboratory. This was no doubt due to immobilization caused by the wide C/N ratio of straw. In a field study carried out in Manitoba, Ferguson (1967) found that after 8 years of straw application with continuous cropping, the ninth and tenth crops showed a residual benefit from the heaviest rate of straw (8,967 kg/ha). The soil NO_3—N analyses seemed to suggest that this yield increase was due to the residual mineralization of the N from these large straw applications. In the U.S.A., Larson et al. (1972) carried out a field experiment on Marshall silty clay loam (Typic Hapludoll) in which residues of alfalfa, cornstalks, saw dust, oat straw, and bromegrass were applied at rates from 0 to 16 tons/ha/yr. for 11 years. They found that average increases over the check for C, N, S and P were 47, 37, 45 and 14%, respectively, for the 16 ton/ha rate. Residues added at 8 tons/ha did not affect soil C or P (Fig. 11), and N and S were lower with saw dust than with the other residues. The amount of cornstalk residue required to prevent net soil organic C loss was estimated to be 6 tons/ha/yr. The C/N, C/P, N/P, and N/S ratios increased with increasing rate of residue addition; the S/P ratio remained relatively constant.

In general it may be concluded that the well established recommendation that all crop residues be returned to soil to reduce the rate of loss of humus

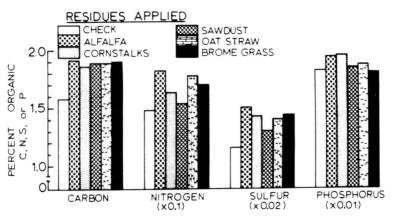

Fig. 11. Distribution of soil organic C, N, S, and P for the check and five types of residues applied at 8 tons/ha/yr. for 11 years. (From Larson et al., 1972.)

is still a sound one; however, more emphasis should be put on conserving root residues and less on top residues, at least in regions of adequate rainfall.

Effect of burning

Burning is often used to clear land of unwanted debris in forests and grasslands. Quite often the burning in forests eliminates accumulation of pine needles and facilitates quick regrowth of grasses and legumes. Daubenmire (1968) has written a good review of this subject. After a fire all shoots stand stiffly erect when they dry; they therefore decompose slowly. The reduced shade facilitates drying of the soil and reduces microbial activity. However, these effects are transient (Daubenmire, 1968). Some changes induced by burning grasslands tend to reduce soil moisture while others favor an increase. In the combustion of plant tissue, N and S may be volatilized but other nutrients form simple water soluble salts which are available immediately. In forests when a holocaust burns the ash serves as fertilizer of considerable significance but in grasslands the ash is so scanty as to be of negligible importance. In Ghana the nutrients released from burning the herb layer of a savanna was much less than that released by the burning of a forest (Nye, 1959). For example, the P, K, Ca, and Mg released per hectare of savanna burnt was 8.4, 49.2, 37.2 and 27.6 kg, respectively. In a forest there was 16 times the P, 18 times the K, 73 times the Ca and 13 times the Mg released. Some scientists report small reductions in total N of the upper 3.8 cm of soil but in the grasslands of central Kansas no reduction of N was detected after 6 years of annual burning (Daubenmire, 1968). Increases in total N after burning has also been reported (perhaps due to increase in legume growth after the fires). Burning raises the pH of soil due to the abundance of Ca, Mg and K in the ash; however, in grasslands the degree of change is not sufficient to significantly influence microbial or subsequent plant growth; in any event the change persists for only a year or two.

Birch and Friend (1956) suggest that burning crop residues may result in desiccation of the surface soil and expedite the release of bound organic matter which may then decompose more rapidly when the soil is rewetted. But Daubenmire (1968) points out that the temperature gradient in the soil during a burn is so steep from the surface downwards that direct alteration of soil structure or chemistry is inconsequential in grasslands. In Brazil, Baldanzi (1960) states that burning is a common practice because if they do not burn off the great mass of grass residues (10—15 ton/ha) they have to plough it in. The latter is laborious; furthermore the residues will depress yields unless N fertilizer is applied, but this is expensive. He found no effect on the organic C and N after 3 successive years of burning. In Ghana, Greenland and Nye (1960) applied straw at rates of 22.4 metric tons/ha twice per year. They found that incorporation and burning had little influence on N mineralization but mulching increased it. Christensen (1973) examined soils under burned and unburned chaparral in California, U.S.A., and found higher levels of NO_3—N under burned conditions. He credited this to NH_4 in the ash and to the fact that mineralization was inhibited under unburned chaparral due to the toxic effects of foliar leachates. In Canada, Smith (1970) found that large amounts of nutrients from the L—H horizon and 0—2 cm depth of a sandy podzol were either redistributed at depth or leached from the profile in 15 months after severe fire (>1,000°C at soil surface) in Jack pine barren lands in northern Ontario. There was a large decrease of the organic matter (Fig. 12) and a small decrease in exchange capacity of L—H horizon, and an increase in pH and solubility of nutrients. In another study carried out in southern Ontario, Smith and Bowes (1974) found that in a series of 11 low temperature (200°—375°C) surface burns in old fields changes in soil chemistry were restricted to the surface litter and were transitory. As much as 30% of the nutrient loss was recovered in downwind deposits of fly-ash adjacent to the burned area.

Is burning a good or bad practice? Although at first glance one is tempted to condemn burning as bad for soil fertility, its net effect will depend on the severity and frequency of burning. Fires do cause a rapid release of nutrients at the soil surface instead of gradually; nutrients can therefore be leached and are subject to loss by erosion. Fires also may cause the mellow, permeable soil surface to become compact; this varies with soil texture, amount of litter and severity of burn (Allison, 1973). On the other hand fires tend to decrease soil acidity making the environment more suitable for plant and

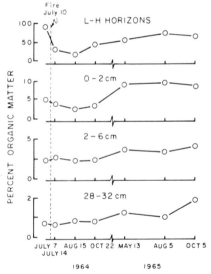

Fig. 12. Percent organic matter at four soil depths before and after burning. Differences between dates significant at $P < 0.01$. (Adapted from Smith, 1970.)

microbial activity and growth. In forested areas burning is generally less harmful than one might surmise and might even be beneficial. In grasslands especially prairie-grassland the fires are likely to be of such low intensity as to be of little consequence. In Saskatchewan when heavy crops are obtained cereal straw is often removed by fall burning. This is sometimes done even when the field is going to be summerfallowed. This practice is undesirable because under prairie conditions moisture is usually the most limiting factor influencing yield and the straw should be left on the field over the winter so as to trap snow and to reduce erosion of valuable top soil. After 18 years of burning straw, scientists at the Agriculture Canada Research Station at Melfort, Saskatchewan, have found that neither fall nor spring burning have influenced the yield of spring wheat (K. Bowren, Agric. Canada, Res. Station, Melfort, Sask., personal communication, 1976). Instead of burning cereal straw it is preferable to operate the combine with a chopper on it; this will spread the straw uniformly over the field and reduce problems in seeding which are usually encountered due to bunching of the straw. If burning must be done, it is preferable to do it in early spring. It should be noted that the soil at Melfort is a black chernozem with high organic matter content (ca. 10%); it might therefore take a considerable number of years of burning to produce any visible effect on the fertility of such a soil. It is quite possible also that there could be a loss of organic matter over the years without any apparent effect to date on the yield; this has not been determined.

Effect of tillage and mulching

Tillage

Clean fallowing by cultivation invariably promotes good aeration, rapid decomposition and loss of native organic C, mineralization of organic N, P, and S at parallel rates resulting in narrower ratios of C/N, C/P and C/S (Kononova, 1966; Swaby, 1966; Allison, 1973). Furthermore, ploughing kills the plants present and dead plants decompose rapidly; quite often the nutrients are lost before the crop can use them. The most intensive decomposition occurs under irrigated conditions. In the U.S.S.R. irrigated serozems supporting a row crop of cotton lost 53% of their total organic matter in 3—5 years; in contrast losses in the drier chernozems and chestnuts were only 1%/year (Kononova, 1966). Ploughing which buries surface residues or weeds will cause more rapid decomposition and release of available N. However, the tool used to turn under residues is important. The plough often buries the residues in a furrow in layers where they decompose more slowly than when they are stirred into soil by discing (Swaby, 1966). Furthermore, layers of residues rich in carbohydrates may produce toxic butyric acid while those rich in proteins may produce amines especially in heavy wet soils which are poorly aerated. The losses in organic matter due to ploughing seem to come mostly from the active organic matter (Kononova, 1966). Rovira and Greacen (1957) suggest that the reason for the accelerated rate of organic matter decomposition after ploughing is because the soil aggregates are disrupted and the microsites where organic matter were previously physically inaccessible to microbes then become available for attack. Robinson (1967), however, was unable to show that disruption of aggregates had this effect in East African soils.

Dowdell and Cannell (1975) found that mineralization of N was 2—5 times greater in a clay soil after ploughing as compared to direct drilling; this difference was transitory and only lasted for a few months. Similar results were obtained by Bakermans and De Wit (1970) and Free (1970); however, Tomlinson (1973) found no difference between the two methods. It has been reported that direct drilling causes more rapid drainage because the channels in soil are more continuous (Baeumer, 1970), thus, leaching losses of N might be enhanced.

Brown and Ferguson (1956), McCalla et al. (1962), Dev et al. (1970), and Anderson (1971) compared chemical versus conventional tillage and found that mineralization of organic matter was enhanced by cultivation. In the studies by Brown and Ferguson (1956), and Anderson (1971) there was evidence that the herbicides had a detrimental effect on mineralization and/ or nitrification (Campbell et al., 1976). In a field study in southwestern Saskatchewan, Chandra (1964) found that spraying of some herbicides at normal rates decreased NH_4 and NO_3 production and microbial populations for two months. However, the decreases were temporary and repeated annual application of the herbicides had much less residual effects. Unger (1968) examined four methods of tillage on a wheat—summerfallow and a continuous wheat system. After 24 years he found that in the wheat—fallow system soil organic matter remained highest where stubble mulch was delayed; this was because the residues were maintained on the soil surface for the major part of the fallow period (Table VIII). The greatest differences occurred in the top 7.6 cm. On the continuous wheat plots there was no difference between the tillage treatments. Similar results to those obtained for organic matter were obtained for N.

Mulching

Throughout the section on management effects (p. 185) mulching has been mentioned from time to time. Also, there is an excellent review of this subject in Allison's book (1973). Consequently, this topic will only be mentioned briefly here.

Mulches are used for increasing water infiltration, reducing soil drifting, reducing evaporation, modifying soil temperature, controlling weeds, and increasing yields. They also increase biological activity, modify the level of available nutrients and maintain or increase the organic matter level. Under minimum tillage they may have a favorable effect on the soil physical condition.

TABLE VIII

Soil organic matter content after 24 years of tillage and cropping practices *
(From Unger, 1968)

Tillage practice	Depth of sampling (cm)			
	0.0—7.6	7.6—15.2	15.2—22.9	22.9—30.5
	— — — — — — Percent organic matter — — — — — — —			
	Wheat—Fallow			
Oneway	1.61	1.58	1.38	1.24
Stubble-mulch	1.72	1.61	1.46	1.27
Field cultivator	1.79	1.69	1.45	1.31
Delayed stubble-mulch	2.26	1.80	1.46	1.34
LSD (5%) for tillage =	0.13			
for depth =	0.10			
	Continuous wheat			
Oneway	1.79	1.73	1.46	1.29
Stubble-much	1.93	1.72	1.44	1.25
Field cultivator	1.97	1.76	1.53	1.40
LSD (5%) for depth =	0.06			

* Soil from plot area in 1941 contained 2.44% organic matter in the 0.0—15.2 cm depth increment.

On the prairies of western Canada and the U.S.A. where soil moisture is often the limiting factor affecting growth, and where one in every two or three years the field may be summerfallowed, stubble mulching is practised. Its purpose is twofold: the 20 cm stubble helps in trapping snow over the winter months, thus it protects the soil from dry-freezing and wind erosion (Bisal and Nielsen, 1964; Anderson, 1968; Anderson and Bisal, 1969); secondly, it results in increased water storage from the snow-melt in the spring. In southern Ontario, Stevenson and Chase (1953) compared the effect of mulch, sod, and clean cultivation on the microbial activity of the soil in a peach orchard. Over a 4-year period they found that the microbes were maintained at a higher level and mineralization was greater under the mulch than under sod or clean cultivation.

NITROGEN TRANSFORMATIONS IN SOIL

"In their interaction the N-transforming processes build up a pattern of N pools in and outside the soil connected by pathways along which N is transported. The functional pattern of these pathways and pools is commonly referred to as the N Cycle" (Jansson, 1971). For simplification Jansson

(1971) discussed the N-cycle in terms of three subcycles, viz., the elemental, the autotrophic, and the heterotrophic subcycles (Fig. 13). Atmospheric N forms the primary source of the elemental pool of N, and N-fixation the primary pathway of the N-subcycle. The second pool involves the death, decomposition and mineralization of the organic biomass. At the same time some of this biomass goes into humus formation or organic N stabilization and joins the humus pool. Ammonium formed during ammonification may be immobilized by heterotrophic microflora, or be nitrified, fixed chemically, or taken up by plants. Immobilization results in inorganic N being cycled into soil biomass and then being remineralized or passed into the humus pool. The NO_3–N may be taken up by plants, lost by leaching or by denitrification, or re-enter the biomass via heterotrophic immobilization. The pool of humus N may undergo slow mineralization.

This is a telegraphic overview of the processes involved in N transformation in soil. In this section we shall deal in some detail with some of the processes mentioned above, e.g., decomposition, mineralization, immobilization and so on. In soil, several of these processes may proceed simulta-

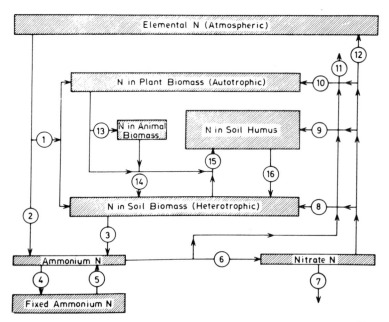

Fig. 13. Pathways and pools in the cycling of soil nitrogen. *1* = Nitrogen fixation, biochemical; *2* = nitrogen fixation, industrial; *3* = mineralization; *4* = chemical NH_4 fixation; *5* = NH_4 defixation; *6* = nitrification; *7* = leaching; *8* = immobilization; *9* = chemical fixation of NH_3 and oxidized forms of nitrogen; *10* = plant uptake; *11* = NH_3 evaporation; *12* = denitrification; *13* = animal consumption; *14* = microbial consumption; *15* = humus formation; *16* = humus decomposition. (From Jansson, 1971.)

neously and they interact with each other which makes the system complicated to study.

Decomposition of organic matter

Aerobic conditions

We shall arbitrarily define decomposition to refer to breakdown of organic matter to simple organic compounds. The decomposition of organic matter depends on the activities of numerous different populations that reside in the soil. The meso- and micro-fauna and the microflora play the dominant role. "Although the animal population, especially the earthworms, is often very important in the incorporation of the plant residues into the soil, and in the mixing process, it is the bacteria and fungi that play the major role. The decomposition of debris can proceed in the complete absence of all animals, but humus formation in the absence of microflora does not occur" (Allison, 1973).

Residues added to soil are first broken down to their basic organic components by the extracellular enzymes produced by heterotrophs. In order to derive energy other organisms oxidize these smaller units by means of intracellular enzymes. Irrespective of dissimilarities in the initial chemical nature of the residues the metabolic sequences within the microbial cell follow similar pathways. The organisms are continuously excreting into the soil a whole range of organic compounds, and also building their own body tissue by immobilizing some of the elements just released.

The number and type of the flora involved in the decomposition process depends on the type, quantity, and availability of the organic matter. Each individual organism has its own complex enzymes which allow it to decompose certain chemical compounds. The first group of heterotrophs (the primary flora) attack the basic components of the added carbonaceous substances; these are succeeded by the secondary flora which thrive on the cells and by-products of the primary flora. Addition of simple sugars cause proliferation of bacteria, starch stimulates the actinomycetes, cellulose benefits the fungi, and proteins and amino acids encourage spore-forming bacilli (Alexander, 1961). The water soluble material is first to be decomposed followed by cellulose and hemicellulose at equal rates; lignins are the most resistant and tend to accumulate in the soil (Alexander, 1961).

Numerous scientists have used tracers to follow the decomposition of organic matter in the field and in the laboratory. Jenkinson (1971) in a review paper reported that excluding very acid soils, the proportion of the added residue carbon remaining in field soils after one and five years was about one-third and one-fifth, respectively. He noted that this proportion was not affected by climatic condition, or plant material; even fresh green manure behaved in this way, contrary to the widespread opinion that such residues decompose rapidly and completely in soil. Similar results have been obtained in the laboratory (Jenkinson, 1971; Marumoto et al., 1972).

Whether the material is added as whole residues, microbial cells, plant protein, or carbohydrates they rapidly appear in all chemical fractions of the soil even before humification could possibly have occurred (Jenkinson, 1971). Jenkinson therefore concluded that there appears to be no satisfactory chemical fractionation method available which will separate soil organic matter into meaningful entities. (This is not completely true, as will be seen from the discussion of Stanford's work in the nitrogen-availability section, p. 248.) In the field there is a rapid flush of decomposition which accompanies the addition of residues, this flush subsides in a few months but several years later the added organic C (labelled material) will still be decomposing at a more rapid rate than the native soil humus (Jenkinson, 1971). In soil there appear to be at least five organic fractions of differing biological stability: (a) fresh residues; (b) lignin from previous additions; (c) soil biomass including microbial cells and by-products of microbial synthesis; (d) material sorbed to soil colloids; and (e) old humus (Jenkinson, 1971).

Jenkinson (1971) estimated that of the added carbon about 10% was present as living biomass after one year and about 4% after four years. Jansson (1971) points out that the living biomass makes up only a minor part of the soil organic N; the main part of the soil N is in the dead stabilized substances which decompose slowly. Parsons and Tinsley (1975) reported that there was about 250 kg N/ha in the total biomass (i.e., living and dead microbes and dead roots) of the top 15 cm of a Swiss meadow soil. This compared with a total N content of 7,000 kg N/ha in the top 15 cm segment of the soil. In a virgin grassland clay in western Canada, McGill et al. (1974) found 128 kg N/ha tied up in the fungi and bacteria alone, and about 200 kg N/ha in the biomass excluding roots. The N in green vegetation was only 22 kg/ha and was thus much less than that tied up in the biomass or potentially very active fraction of soil organic matter. Greenland (1971) reports that Hénin and Dupuis were two of the first to suggest that non-humified organic materials should constitute the active fraction of soil organic matter; this led to a separation of these materials (light fraction) by flotation on heavy liquids (Hénin et al., 1959), and subsequently an improved separation by use of ultrasonic vibration in heavy liquids (Greenland and Ford, 1964; Ford et al., 1969). Incubation tests showed that the N contained in this "light fraction" comprised only 7—23% of the total N in the soil but accounted for 25—60% of the N mineralized (Greenland, 1971). This fraction which includes much of the biomass therefore provides an estimate of the active or readily decomposable soil organic matter.

Simonart and Mayoudon (1961) were first to point out that when readily decomposable materials were decomposed in soil the residual C was distributed throughout the soil organic matter in a pattern similar to that of soil organic N. Marumoto et al. (1972) observed that of the C remaining in soil after decomposition part went into amino acids and part into amino sugars. But these constituents were not the same as that added in the residue and were probably of microbial origin. In a review Black (1968) came to a similar conclusion. It is not known to what extent this nitrogen is further changed. Proteinaceous materials certainly exist in soils (Biederbeck and Paul, 1973). The surprising and perplexing thing is that generally only about 25—50% of the total N in surface soils has been identified (20—40% as amino acid—N and 5—10% as amino sugar—N). Keeney and Bremner (1964) fractionated 10 pairs of virgin and cultivated soils (Table IX) and determined the changes in the organic N fractions in relation to their relative susceptibility to decomposition. They found little difference in susceptibility; there was a tendency for the chemical pattern of organic N in both systems to be similar. Porter et al. (1964) used a different fractionation procedure to make the same type of comparison and obtained comparable results. Campbell et al. (1976) have reviewed the Canadian literature on the effect of management on the different chemical components of soil organic matter; therefore nothing further will be said about this facet here.

The reasons for the difficulty in identifying the soil organic matter components are partly analytical and partly due to the stability of some of the components. Campbell et al. (1967) carbon-dated soil organic matter fractions and found the mean residence time of the humic C to be >1,000 years while that of the humic hydrolysates was <100 years. Allison (1973) summarized the reasons advanced to explain this stability as: ligno-protein complexes formed; oxidized lignin reacts with NH_3; carbohydrates or their derivatives react with amino compounds; nitrites react with organic matter; the non-hydrolyzable N might be in the form of heterocyclic compounds; organic compounds are protected by clays; organic matter reacts with metals

TABLE IX

Nitrogen fractions in ten pairs of virgin and cultivated soils and losses due to cultivation (From Keeney and Bremner, 1964)

Nitrogen fraction	Average content as percentage of total nitrogen		Average loss due to cultivation as percentage of total nitrogen in virgin soils
	virgin soils	cultivated soils	
Nonhydrolyzable	25	24	39
Hydrolyzable:			
ammonium	22	25	29
hexosamine	5	5	28
amino acid	27	23	43
unidentified	21	23	35
Nonexchangeable ammonium	5	7	0

to form complexes; tanins form complexes with proteins; phenolic compounds or phenols react with amino acids. Biederbeck and Paul (1973) suggest that humo-proteins were originally reversibly combined with humic polyphenols by hydrogen bonding. Rowell (1974) found evidence that enzymes were complexed to humic substances by means of covalent cross-linking reactions between the quinone and protein molecules with the result that these complexes were much more stable than the free enzymes. Bremner (1965) suggests that several of these mechanisms are involved in the stability process. Allison (1973) disagrees; he suggests that organic matter stability is primarily a function of its chemical and physical composition, and secondarily due to steric hindrance (i.e., the microbial cells may be too large to get to the humus in some instances while the enzymes may become sorbed before they can penetrate into micropores).

Anaerobic decomposition

Anaerobic decomposition is dependent on anaerobic bacteria; if NO_3—N is present facultative anaerobes may use O_2 from this source as electron acceptor in their respiration. They operate at a much lower energy level and are less efficient than aerobic organisms. Thus metabolic processes of decomposition and synthesis are slowed. This is one reason why there is greater accumulation of plant residues in organic soils.

Some workers have reported that the lignin in oak leaves did not decompose under anaerobic conditions while hemicellulose decomposed very slowly, and even alfalfa with a high N content did not decompose rapidly (Tusneem and Patrick, 1971). Accumulation of proteins was greater under anaerobic than under aerobic conditions in all instances perhaps due to more economic use of N, smaller loss of NH_3, or lesser decomposition of synthesized protein in the anaerobic system.

In an aerobic soil the end products of decomposition are CO_2, NO_3, SO_4, H_2O and resistant residues, but in anaerobic soils they are CO_2, CH_4, H_2, R-COOH, NH_3, RNH_2, RSH, H_2S and resistant residues (Alexander, 1961). In the aerated soil the degradation of carbohydrates goes through pyruvic acid, then the Krebs cycle with the production of CO_2 and H_2O in the presence of molecular O_2 through the mediation of terminal oxidases. Under anaerobic conditions the system is the same up to the formation of pyruvic acid; but there is no free O_2 thus the terminal oxidation is suppressed, and pyruvic acid and reduced nicotinamide—adenine dinucleotide accumulates. The pyruvic acid then undergoes various transformations not specific to any organism with the formation of various organic acids (e.g., formic, acetic, and butyric). When protein decomposes under anaerobic soil conditions the final products are NH_3, R-COOH, RNH_2, RSH, and H_2S. Under aerobic and anaerobic conditions the process is probably the same until amino acids are formed. Under aerobic conditions deamination then occurs and R-COOH is oxidized via the Krebs cycle. The NH_3 is nitrified. Under anaerobic conditions the products of deamination and decarboxylation may accumulate or are transformed to gases.

Mineralization—immobilization (turnover)

In the previous subsection we arbitrarily stopped the decomposition process at the stage where organic substances were converted to small, simple organic molecules. In nature the conversion process involving breakdown from organic to mineral form and resynthesis to organic form is in continual motion. We shall now discuss the conversion of simple organic compounds to NH_4 with the release of CO_2 via the heterotrophic micro-organisms in soil (i.e., N mineralization or ammonification). We shall also discuss the resynthesis of inorganic constituents (e.g., NH_4 and NO_3) into organic compounds by heterotrophs (N immobilization). Mineralization and immobilization proceed simultaneously and constantly in opposition in soils; we shall refer to this mini-cycle as the turnover cycle. Two other terms will be used which will require definition: "reversion" will be used to describe the case where fertilizer N is immobilized into microbial tissue or by-products; and "availability ratio" (which is a measure of the change in relative availability of applied fertilizer N with time) is defined as the fraction of plant N that is ^{15}N, divided by the fraction of soil N that is initially ^{15}N for each cropping period (Broadbent and Nakashima, 1967). The discussion which follows will attempt to update the voluminous literature on this subject. Thorough reviews have been presented by Harmsen and Van Schreven (1955), Jansson (1958), Bartholomew (1965), Allison (1966), and others.

Aerobic turnover

As defined above, turnover refers to the heterotrophic N cycle of soils, i.e., the mineralization—immobilization relationships. The latter processes work in opposite directions, are always related to each other, and are constantly at work; it is therefore necessary to refer to net mineralization or net immobilization. It is possible for these two processes to balance each other such that there is no apparent change in mineral N and yet the processes are at work. Thus the best method of assessing the scope of the processes can be obtained by measuring the energy dissipated from the soil as revealed by CO_2 produced via heterotrophic respiration. The amount of turnover will be directly proportional to the CO_2 production. The magnitude and direction of the net effect will be a function of the amount and availability of the organic matter being decomposed and turned over; specifically it will depend on the ratio of the relative quantity of available energy to the available N in the system. Black (1968) used Fig. 14 to elaborate this process. Fig. 14(a) shows the case where soil is summerfallowed, with no additions of organic residues; it is assumed that environmental conditions are optimal, and that there are no losses due to leaching or denitrification. Here there is net mineralization

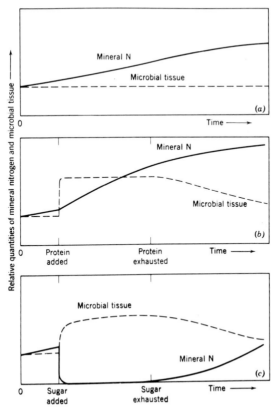

Fig. 14. Schematic representation of changes in content of mineral nitrogen and microbial tissue with time in fallow soil. (a) No organic matter added; (b) protein added; (c) sugar added. (From Black, 1968.)

of the active soil organic matter as the microbial population stays relatively constant. (There would, of course, be small flushes due to rainfall.) The addition of organic material containing excess N relative to C, Fig. 14(b), causes the microbial population to increase rapidly to a limiting value which depends on the environmental conditions; the mineral N increases beyond the time when the organic substrate is completely exhausted because of mineralization of native soil N and because of a decrease in the microbial population; eventually the microbial population will approach their initial value and the mineral N in the soil will equilibrate at a higher value than in the case of Fig. 14(a). Fig. 14(c) demonstrates the situation where the organic substrate added is much higher in C relative to N. The microbial population increases but at the same time uses up any N present or mineralized in the soil. Thus the content of mineral N remains low until the substrate is exhausted and the microbial population decreases in size; in time the mineral N content will surpass the amount initially present in the soil.

The conditions determining N turnover, its scope and effects on N availability to plants are very complex, fraught with pitfalls for the researcher and still requires considerable research effort. An excellent attempt at such research was carried out by Jansson (1958) who spawned the term "continuous internal cycle" to describe biological turnover of N. He postulated that mineralization—immobilization repeats itself repeatedly until any added material is exhausted. Jansson (1958) found evidence in support of his hypothesis by adding inorganic ^{15}N to decomposing plant residues in quantities in excess of the needs for decomposition. He sampled at intervals, determined ^{15}N in the inorganic and organic phases of the system and found that the tracer content of the two phases approached equivalence as a result of turnover. He reasoned that this behavior would require that the plant residue N be mostly mineralized such that the resulting NH_4 joins the inorganic pool of N. The microbes would then obtain their N for cell growth from this active pool.

Where is the equivalence point when N mineralization equals N immobilization? Iritani and Arnold (1960) found this point to be a C/N ratio of about 22 and a N percentage of about 2. Smaller ratios (higher N percentage) are associated with net mineralization, and larger ratios with net immobilization. Harmsen and Van Schreven (1955) reported that the generally accepted values for equilibrium are C/N = 20 to 25, and % N = 1.5 to 2.0%; Black (1968) gives 1.2 to 2.6%. These values are obviously not constant and depend on several factors, e.g., temperature and time allowed for turnover, the supply and kind of mineral N in the soil, and the amount and composition of the organic substrates.

Goring and Clark (1948) have shown, using nontracer studies, that less apparent net mineralization occurs in cropped soils than in summerfallowed soils even after allowances have been made for N in the crop. This effect has usually been attributed to possible increased immobilization of N by rhizosphere population due to root excretions and due to the mass of root debris that accumulates with increased age of plants. Allison (1973) discusses this phenomenon and cites other possible explanations, none of which seem completely adequate.

Immobilization takes place more rapidly upon addition of soluble carbohydrates than on addition of ordinary plant residues (Bartholomew, 1965; Agarwal et al., 1972); the amount of N immobilized is greater for the more soluble carbohydrates (Agarwal et al., 1972; Ahmad et al., 1969). There is evidence that for a given C/N ratio N mineralization decreases with increasing content of lignin in the residue and with decreasing content of water soluble N (Black, 1968). Bartholomew (1965) reports that the remineralization of immobilized N takes place more rapidly upon addition of soluble carbohydrates than on addition of residues; however, Ahmad et al. (1969) found no difference between the two sources. Turnover after the

addition of residues, takes place very rapidly during the first month, thereafter, the change in organic N is slow even though mineral N is present for
immobilization and the C/N ratio is in the range in which mineralization
would be expected in fresh residues (Black, 1968; Shields et al., 1973).
There is some question as to whether the heterotrophs prefer NH_4- to NO_3-
N; the concensus seems to be that they prefer NH_4-N (Jansson, 1958;
Ahmad et al., 1972). Ahmad et al. (1972) also found that the amount of C
required to immobilize a unit of N was greater for NO_3- than for NH_4-N,
and that the fraction of the immobilized N which was remineralized was
greater for NO_3- than for NH_4-N. Kai et al. (1969) found that irrespective
of the temperature ($10°-40°C$) maximum immobilization of N occurred
very rapidly; that temperature had little effect on the amount of net maximum immobilization; but net remineralization of N was directly proportional
to temperature. Craswell and Waring (1972) found that N mineralization
increased several fold when clay soils were ground; also that the increase was
greater where the clay content was montmorillonitic compared to where it
was kaolinitic. They suggested that clays play an important role in protecting
organic matter from decomposition, thus finer textured soils tend to
accumulate more organic matter and lose organic matter less readily than
coarse textured soils.

The use of ^{15}N in studies of N transformations in soil has provided considerable insight into the turnover process. In general after immobilization
there is a gradual decline in the organically bound ^{15}N. In this respect the
work of Broadbent and his coworkers, and Legg, Stanford and coworkers are
outstanding. We will cite two examples: Legg et al. (1971) found that the
reversion of ^{15}N tagged fertilizer into organic forms was accelerated by
growing a series of oat crops in the greenhouse. The availability ratios
declined from 10.4 for the Chester surface soil to approximately 2 after
eleven crops. Using an incubation method of assessing the availability
principle they obtained similar results; mineralization tests of soil samples
taken after two, five and eight cropping periods showed that availability
ratios levelled out at about 2. This was interpreted by the authors to mean
that after the fertilizer N was incorporated into relatively stable forms in the
soil it was either still twice as available as the native soil N or that this ^{15}N
tagged organic N had the same availability as one-half of the total soil N. The
latter would indicate that the native soil N was relatively inactive biologically.
The findings of Broadbent and Nakashima (1967) had been similar. They
found that when straw was added to soils which were incubated in the greenhouse, as much as two-thirds of the fertilizer N remained in the soil after
nearly 1.5 years of continuous cropping to sudangrass. Almost all of the
labelled fertilizer N was immobilized in the first 10 days thus the mineral N
they obtained in subsequent leachings was regarded as remineralized N; the
latter was very small.

Some of the mechanisms advanced to explain this stability were listed in the subsection dealing with decomposition. We will elaborate briefly on three of these at this time. Jansson (1958) suggests that this stability of immobilized N may be the result of greater susceptibility to decomposition of the most recent additions of organic N relative to native forms. Broadbent and Tyler (1962) and Broadbent and Nakashima (1967) suggest that the increased stability of the immobilized N with time is due to reversion of the recently incorporated N to more difficultly mineralizable forms. Another theory advanced by Broadbent and Nakashima (1967) is that there is a nonbiological reaction involving NH_3 and resulting in the formation of N-containing compounds which are much more stable than microbial N and which are independent of the energy status of the soil. Allison (1973) suggests that this stability is no doubt responsible for the difficulty scientists have often experienced in explaining the frequent failure of crops to benefit to the extent expected from residual soil N from a preceding cash crop or a green manure crop even though leaching and denitrification losses were known to be negligible.

The chemical fractions of the organic matter which are mainly involved in the turnover process appear to be the amino acids and amino sugars (Cheng and Kurtz, 1963; Chu and Knowles, 1966; Freney and Simpson, 1969; McGill et al., 1973). Cheng and Kurtz (1963) found that after adding labelled mineral N along with C to soil, immobilization occurred mainly in the amino acid. Chu and Knowles (1966) found that 50—68% of the NH_4—N was immobilized in this same fraction, while Stewart et al. (1963a) found it in an equivalent fraction, i.e., the nondistillable acid-soluble fraction. McGill et al. (1973) added ^{14}C-acetate and ^{15}N-ammonium sulfate to soil and incubated. N was rapidly immobilized in 3—4 days followed by net remineralization over a long period. The immobilization products were carbohydrates, amino acids and amino sugars; these were degraded more slowly than the original acetate. The turnover of C was highly correlated to that of N. Stewart et al. (1963b), and Freney and Simpson (1969) found that remineralization of immobilized N came primarily from the nondistillable acid-soluble fraction. Thus these scientists found good evidence that there existed a fraction of the organic matter which could be regarded as more active than the rest.

Anaerobic turnover

Under anaerobic conditions both mineralization and immobilization rates are considerably retarded. The characteristic features of anaerobic turnover are: (a) there is incomplete decomposition as explained earlier; (b) low energy of fermentation which results in the synthesis of fewer microbial cells per unit of C degraded: e.g., only 2—5% of the substrate C is assimilated by anaerobic bacteria compared with 30—40% by fungi in aerobic systems (Alexander, 1961); (c) low N requirement of the anaerobic metabolism leading to a more rapid release of NH_4 ions than would normally be expected because of the wide C/N ratio of the residues and the much slower rate of anaerobic decomposition. Thus NH_4 release is higher under anaerobic than under aerobic conditions where C/N is large despite the more rapid decomposition under aerobic conditions. It is also noteworthy that NH_4 accumulates in anaerobic soils (in contrast to NO_3 in aerobic soils) because the nitrifiers are completely suppressed.

Tusneem and Patrick (1971) found that the mineralization rate of organic N of added residues was initially greater under waterlogged than under optimum moisture conditions (Fig. 15). The difference between the two moisture regimes was more pronounced at higher C/N ratios (21/1 and 35/1) and persisted for about two months. Thereafter there was a gradual decline in NH_4 accumulation under waterlogged conditions in contrast with a rapid increase in inorganic N under optimum moisture. Immobilization increased with added straw irrespective of aeration and while net immobilization was much more rapid under aerobic conditions net mineralization tended to dominate under anaerobic conditions. Remineralization of immobilized N was slow under aerobic and slower under anaerobic conditions. Irrespective of aeration the N immobilized was found primarily in the amino acid, amino

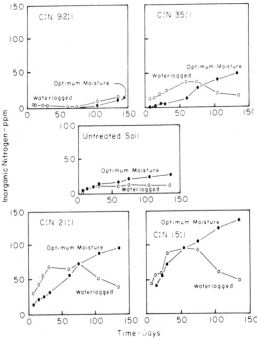

Fig. 15. Mineralization of organic nitrogen at various C : N ratios of added plant material under waterlogged and optimum moisture conditions. (From Tusneem and Patrick, 1971.)

sugar and hydrolyzed ammonium fractions with 2—3 times as much N entering the amino acid fraction as entered the other fractions.

The turnover process is of practical importance since the addition of residues will often mean that N fertilizer must be applied to overcome the immobilization of soil N caused by large C/N ratios. The amount of N to be applied is a product of the dry weight of the residues times the difference between the equilibrium value (lets assume 1.7% N) and the N content of the residue. For straw of say 0.7% N the fertilizer N requirement is 20 lb. or 9.1 kg per ton of dry residue. When should the N be applied? Allison (1973) states categorically that it is best to fertilize the crop and not the residue or green manure. Crop residues deficient in N decompose just as rapidly with or without added N, in decomposing they will immobilize any soil N being formed. If there is no crop growing they will be performing the good task of saving N from possible loss by leaching or erosion. If a crop is present there is a competition between the crop and the microflora for any available N and the microbes are more efficient competitors.

Priming effect

Jenkinson (1966b), Jansson (1971) and others have written excellent reviews on this topic. In this section we will attempt to update the literature with respect to: (1) priming due to the addition of carbonaceous material; and (2) apparent priming due to the addition of soluble salts and fertilizers.

Priming due to C additions

The use of radioactive isotopes in mineralization studies resulted in the discovery of the priming effect. This effect is due to the fact that newly added organic residues may cause positive (stimulate) or negative (depress) mineralization of the native organic matter. In Fig. 16, taken from Jenkinson (1971) two samples of soil removed from the same plot were incubated with

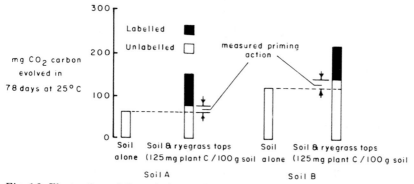

Fig. 16. Illustration of the priming action. (From Jenkinson, 1971.)

uniformly labelled ryegrass. Soil A was taken after one year of fallow and soil B six weeks after the addition of 34 metric tons of farmyard manure per hectare plus the stubble from the preceding wheat crop was ploughed in. The labelled ryegrass increased the rate of decomposition of the native organic matter. Mathematically Jenkinson defines the amount of organic C lost by priming action in a given time as $X-Y-Z$, where X is the total CO_2 lost from the soil plus organic matter system, Y is that part of X which comes from the labelled added plant material, and Z is the CO_2 evolved by an untreated soil which is incubated under the same conditions at the same time. Jenkinson (1971) stresses that what appears to be a priming action might just as easily be the result of isotopic exchange, or the result of errors in calculating Y from the isotopic data or due to poor statistical techniques in designing the experiment. Jansson (1971) stresses that one must be careful, both in planning and interpreting the experimental work, to differentiate between priming and turnover because, in the early stages of incubation experiments, the turnover process normally gives results that can be erroneously interpreted as priming. He also suggests that priming action is a phenomenon that should be expected because all soils contain a native heterotrophic biomass which will react upon additions of energy rich materials. This reaction may include variations in the rate of mineralization of biomass C and N (priming action). For example, a change in the microbial species may occur, the defeated species die and their biomass becomes liable for mineralization. Possible examples of this are presented later when we discuss the effects of wetting and drying and freezing and thawing (following section).

Some workers have reported positive, some negative and some both types of priming in a single study. Sauerbeck (1966) investigated the effect of ^{14}C labelled young rape and their possible priming action on a calcareous soil with high organic matter content (3.28% C), and on a low organic matter soil (0.31% C). He found that there was no difference between the green manure and straw residue in their effects; they both caused positive priming of the native C of the low organic matter soil throughout the experiment and the priming was proportional to the amount of C added. This initial positive effect lasted for only a few days and then green manure had a negative priming effect. Smith (1966) applied ^{14}C labelled plant tops and roots to a prairie soil with high organic matter (3.5% C) and found that the organic matter frequently had a negative (i.e., a protective) effect upon the native organic matter. The latter effect was not related to the maturity or N content of the plant tissue. He speculated that this strong suppression of decomposition by root material of corn and wheat was due to the formation of a toxic compound. In the discussion which followed this symposium paper Sauerbeck suggested that Smith's results might be due to: (a) the high C content of the soil used; and (b) the high rate of residues applied. Bartholomew on the other hand suggested that percent C was not always correlated with degree of stimulation; furthermore, soil type seems to play some role in the results obtained. Sørenson (1976) applied unlabelled glucose to a loam and sandy soil which had been incubated with ^{14}C labelled straw for 3 and 12 years, respectively. He determined the biomass and the evolution of labelled and native CO_2. The addition of 50, and 200 mg glucose-C/100 g soil, each resulted in priming. The priming was largest after the small addition of glucose C if the calculation was expressed per unit of glucose-C added. A single addition of glucose primed the labelled biomass but after repeated additions there was no difference between the glucose-treated and the control soil.

These are but a few of the more recent studies on this subject, and the findings are similar to numerous others reported in the literature. It can be readily seen that with regard to the occurrence and implications of the priming action there is need for further research and clarification. Sauerbeck (1966) stresses that the distinction of the two CO_2 sources (i.e., the native organic matter versus the green manure) by means of the radioactivity found in them is questionable, because after short incubation times the soil organic matter is known to become radioactive itself. This means that part of the labelled material added turns into soil organic matter quite soon, thus not all of the radioactivity of the CO_2 evolved can be actually attributed to the green manure. He concludes that as long as the total C level after manurial treatments does not decrease to a greater extent than without manure, the so-called priming action may be only considered as a turnover within the soil, which might even be desirable. Jenkinson (1971) concludes that although priming action may well occur in the field one would be rash to postulate, based on the present laboratory evidence, that the annual loss of soil organic matter by priming is greater than that left in the soil by the manure residues. Furthermore, even the effects observed in laboratory studies are transient and small compared to the organic matter in the soil and their practical importance is minimal.

Apparent priming due to addition of soluble salts and fertilizers
It is well known that high salt concentrations will injure plants. The autotrophic bacteria such as nitrifiers are more sensitive to unfavorable conditions in soil than are other soil microbes thus they will be more easily killed by high concentrations of salts and fertilizers, especially if the latter are high in soluble salts (Allison, 1973). It should not be surprising either if higher concentrations of salts were found to be extremely deleterious to the activity of all soil organisms. Thus it is somewhat surprising to find that the addition of salts and fertilizers cause positive apparent priming of soil organic matter. Several workers have found that additions of labelled fertilizer N cause increases in the nontagged inorganic N in agricultural soils (Legg and Stanford, 1967; Broadbent and Nakashima, 1971; Westerman and Tucker, 1974) and forest soils (Overrein, 1971). This has resulted in an

increase in plant uptake of native N as shown by Sapozhnikov et al. (1968) and Westerman and Kurtz (1973).

Laura (1974) in a review states that the addition of neutral salts to soils has given varied results with respect to C and N mineralization. Some workers have observed increases and others decreases in ammonification with increased salinity. The mechanisms advanced to explain this apparent priming effect will be discussed later in this subsection; first we shall review the more recent findings on this subject. Broadbent (1970) discussed the priming effect of fertilizers by examining published research findings in terms of Fried and Dean's A-value concept. A-value is defined as the quantity of soil N which is equivalent in availability to the fertilizer N added as determined by tracer method. Broadbent showed that several scientists found increases, some decreases, and others no change in A-value resulting from different levels of N fertilizer application. He further found that A-values were usually greater when NH_4-fertilizer was applied compared to when NO_3-fertilizer was applied. Broadbent and Nakashima (1971) found that $(NH_4)_2SO_4$ and NH_4Cl increased mineralization of soil N; KCl, $CaCl_2$ and $AlCl_3$ had a similar influence in one soil but their effect was not consistent in two other soils. On the other hand $CuSO_4$ depressed N mineralization in all soils. The magnitude of the effect was a function of the concentration of salt and nature of the soil. Westerman and Tucker (1974) used 0 to 1.0 M concentrations of Na, Cu and Ca salts with and without $^{15}NH_4Cl$ to investigate mineralization and immobilization over a 49-day period. They found that dilute concentrations of salts stimulated mineralization of soil N, while immobilization of $^{15}NH_4$—N was decreased by high concentrations of salt. In the U.S.A., Heilman (1975) determined the effect of various salts and concentrations of salts (0—0.2 M) on NH_4— plus NO_3—N changes in 10 forest soils of the Pacific Northwest (PNW) in a laboratory incubation study. Like Westerman and Tucker (1974) they found that mineralization increased with time and concentration; the order of effectiveness of the salts in releasing N was $AlCl_3 > CaCl_2 > KCl > K_2SO_4 > K_2CO_3 > KHPO_4$. El-Shakweer et al. (1976) reported that when clover straw was added to soil, CO_2 evolution increased in the presence of Na_2CO_3 and $CaCO_3$ and was depressed by NaCl, $CaCl_2$, $MgCl_2$, Na_2SO_4 and $MgSO_4$ at 25 me./100 g soil, and $CaSO_4$ and $MgCO_3$ at 250 me./100 g soil. Thus in general the rate of decomposition was inversely related to the concentration of salt.

Laura (1974) in reviewing previous research findings on this topic observed that in most experiments volatilization losses of N were not measured. Furthermore, results were usually expressed in terms of quantity of salt added rather than as a function of the saturation of the exchange complex of the soil (Laura, 1976). The latter method would have allowed more meaningful comparison among different soils. Soil biological processes are affected by the pH or exchangeable sodium percentage (ESP) rather than the quantity of an alkali salt added to the soil because different soils have different exchange capacities. Laura (1974) summarized the findings of previous researchers as follows: in low base-saturated soils, addition of Cl caused very little increase in mineralization; addition of NO_3 and SO_4 increased N mineralization; and addition of salts of transition metals generally decreased N mineralization.

Laura's recent work on this topic is worth detailed examination. In a 6-month incubation experiment Laura (1974) found that CO_2 evolution and total C mineralization decreased with increasing concentrations of salt mixture (NaCl plus $CaCl_2$) from 0.1 to 5.1% (Table X) and that the process of nitrification was completely inhibited between 0.6 and 0.9% (or greater) salt concentration (Fig. 17). The amount of N lost was about 10.4% from the control up to 0.6% salt, and about 17.8% from 0.9 to 5.1% (Table X). In contrast to C, mineralization of N was equal at all levels of salt (Table X). Laura (1974) reasoned that since N was produced and N mineralization was equivalent at all levels of salt, and it was not progressively inhibited with increased salt concentration as was C mineralization, and since the heterotrophs involved in mineralization of C and N are the same, then the results indicate that N mineralization was not of biological origin but likely

TABLE X

Total mineralization and losses of nitrogen and carbon in soils after six months of incubation *
(From Laura, 1974).

Salt added (%)	Mg of mineralized N in 100 g of soil			Nitrogen loss (%)	Mg of C lost as CO_2 and/or volatile organics from 100 g of soil	Carbon loss (%)
	present in soil	unaccounted nitrogen	total			
0.0 (control)	16.0	13.0	29.0	9.9	450	42.1
0.1	15.8	13.0	28.8	9.9	450	42.1
0.2	14.9	14.0	28.9	10.8	430	40.2
0.3	14.5	14.0	28.5	10.8	420	39.3
0.6	13.9	14.0	27.9	10.8	410	38.3
0.9	4.5	22.0	26.5	16.8	350	32.7
1.5	4.5	22.0	26.5	16.8	320	30.0
2.1	4.5	24.0	28.5	18.3	310	29.0
2.7	4.5	23.0	27.5	17.6	290	27.1
3.3	4.4	24.0	28.4	18.3	270	25.2
3.9	4.4	24.0	28.4	18.3	250	23.4
4.5	4.4	23.0	27.4	17.6	240	22.4
5.1	4.4	22.0	26.4	16.8	220	20.6

* After addition of 2% gulmohur leaves the initial N and C percentages of soil under all treatments were 0.131 and 1.07, respectively.

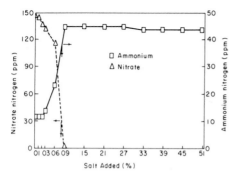

Fig. 17. Nitrogen in different treatments after six months of incubation. (Arrows on a given curve point towards its abscissa and ordinate.) (From Laura, 1974.)

due to some physicochemical process. Laura (1976) found further evidence in support of the latter hypothesis when he studied the mineralization as a function of ESP. After 6 months of incubation he found that CO_2 evolution (data not shown here) and total mineralization of C (Table XI) increased with increased ESP while nitrification was completely inhibited between 70 and 92 ESP. Mineralized N (Table XI) and N losses (data not shown here) were equal to the control up to 70 ESP but were considerably higher at 92 ESP. Thus while C mineralization increased with increasing ESP, N mineralization remained constant up to 48.6 ESP, increased slightly at 70 ESP and increased considerably at 92 ESP (Table XI). The argument used earlier is therefore again applicable. Laura therefore suggests that the exceptionally high N mineralization at 92 ESP compared to C mineralization could be due to some chemical ammonification of organic N without benefit of soil micro-organisms.

Let us now consider some of the mechanisms which have been advanced in an attempt to explain this apparent priming by salts. Jansson (1958) suggested that the phenomenon, with respect to additions of fertilizer N, was

TABLE XI

Total mineralized carbon and nitrogen and the C/N ratio of the mineralized material of soils after six months of incubation
(From Laura, 1976)

ESP level	Mineralized N (mg)	Mineralized C (mg)	C/N ratio of the mineralized material
2.2	18.8	172	8.6
23.4	18.4	216	11.7
48.6	18.3	254	13.9
70.0	22.1	269	12.2
92.0	32.1	306	9.5

related to microbial activity of soil since it did not occur in the absence of biological activity. He suggested that it was a temporary phenomenon which occurs when the inorganic N pool is very large because of added fertilizer; the extent to which the mineral pool is diluted depends on the amount of fertilizer added. Thus this apparent priming is merely turnover, not stimulation of mineralization. Broadbent (1970) disagrees with Jansson's explanation and suggests that mineralization of native organic matter is definitely stimulated by fertilizer N additions. The causes he thinks are at least partly physical or chemical in nature, e.g., salt effects, osmotic, pH changes, etc. Legg and Stanford (1967) advanced the theory that the increased mineralization of native N is related to the rhizosphere organisms. Jansson (1971) agrees that in the presence of a crop, rhizosphere effect might be a factor but points out that since the phenomenon can occur in noncropped soils in the laboratory, rhizosphere can only be a partial explanation. Agarwal et al. (1971b) suggested that cationic and anionic phenomena are involved. Aleksic et al. (1968) like others found that N application greatly increased N uptake. However, when they calculated A values they found these to be independent of the rate of addition of fertilizer. This indicated that fertilizer had therefore not affected the amount of available native soil N, i.e., no priming had occurred. They explained their results by suggesting that the increased soil N uptake resulted from more rapid development of roots and shoots caused by the fertilizer application.

Laura (1974, 1976) dismisses most of the above explanations as being inadequate. In his work salts did not stimulate mineralization from 0.1 to 5.1% salt (Laura, 1974). Furthermore as he quite rightly points out salts are poor sources of energy for any organism. He ruled out "salt" or "osmotic" effects on the basis that the effects of salts were not dependent on concentration, and there was no increase in mineralization of N or C with increased salinity (Laura, 1974). This latter criticism is not completely valid since Laura used a single soil type and Broadbent and Nakashima (1971) indicated that the salt effect depended on the nature of the soil. Both Laura and Broadbent seem to agree that a nonbiological type of mechanism is at least partly involved in the explanation. Laura suggests that the apparent priming is due to "protolytic action".

As evidence in support of the protolytic theory Laura (1974) cites literature which shows that NH_4 can be formed through purely chemical reactions from organic nitrogenous compounds adsorbed to soil clays when they were heated. Furthermore, he states that protolytic action of water plays a major role in these transformations. He states that "any factor which enhances the degree of dissociation of water increases the rate of formation of NH_4. Since salts affect the degree of dissociation of water (which is amphoteric) it is quite possible that the effects of added salts in soils on mineralization of N and C might be partly chemical." He uses the same theory to explain the results of his 1976 study. For example, he states that factors which increase the proton supply of water in soil should increase mineralization. Since water is amphoteric it will act as a base in acid medium and accept protons from other proton donors, while in alkaline medium it will donate protons to other bases. Thus the increase in pH (ESP) will tend to induce proton donation from water. Consequently, the rates of reactions catalyzed by the proton from water will increase with increase in pH of the medium. This would explain the observed increase in mineralization with increased ESP (Table XI). This theory seems quite plausible but is still only a theory. In our opinion its main weakness is the fact that although NH_4 formation has been shown to accrue from heating of organic nitrogenous compounds in soil, Laura's and all such experiments are carried out at room temperature.

In conclusion let us consider what are some practical implications of the fertilizer and salt effect on apparent priming. Where fertilizer is applied in the field its priming effect will only occur within about 3 cm of the fertilizer band where the solution emanating from it is highly concentrated. It may therefore influence the young seedling but only temporarily since the effect is transient as the solution becomes rapidly diluted. Heilman (1975) points out that the salt levels which showed significant release of soil N in soils studied by Singh et al. (1969) and by others are comparable both to those found following burning (Grier and Cole, 1971) and to those likely to occur in the surface horizons of forest soils when they are fertilized with urea at the commonly recommended rate of 242 kg N/ha. Thus the release of native soil N may be a significant factor to be considered with regard to both burning and fertilizer application. The fact that NH_4 fertilizer was more effective than NO_3 fertilizer in its priming effect might also be of significance in choosing which to use. It might not be desirable to increase the mining of the soil N and possibly increase pollution. In areas such as western Canada where salinity is becoming an ever increasing problem the early effects of increased salt concentrations in soil will be to increase the mineralization of organic N and cause its loss by leaching and possibly ruin the soil's physical structure. It would appear that this subject is in need of increased investigation since unlike true priming its influence could be more dramatic and of greater practical significance.

ENVIRONMENTAL FACTORS AFFECTING MINERALIZATION

When a soil is kept under fairly constant conditions its biological activity and the stabilization of its organic substances seem to operate efficiently and the net mineralization declines to a low level. However, if the current state is interrupted by unfavorable conditions, for example drought or frost, and thereafter re-established, a stimulation of biological activity is often found. Consequently, such factors as drying and wetting, freezing and thawing, ploughing, partial sterilization (e.g., by chemical sprays) and so on, are likely to cause a subsequent flush; the magnitude of the flush will be a function of the amount of disruption caused by the disruptive process.

Moisture

The effect of soil moisture on changes in mineral N in soil is a function of chemical, physical, and biological processes. Moisture affects ammonification, nitrification, denitrification, movement of nitrates, and immobilization. We will mainly be concerned with ammonification here.

It has been established (Robinson, 1957; Miller and Johnson, 1964; Stanford and Epstein, 1974) that ammonification increases with moisture content between 15 and about 0.1 to 0.5 bar (i.e., wilting point to field capacity) and that above and below these limits the rate of ammonification decreases (Miller and Johnson, 1964; Reichman et al., 1966; Stanford and Epstein, 1974). The results of Stanford and Epstein (1974) were based on nine soils of the U.S.A. which differed widely in properties (Fig. 18). Using the results for the 0.1—15 bar region they normalized the N-mineralized (Y) and water content (X) such that the maximum rate of mineralization was 100 and the water level associated with this mineralization rate was 100. They found by regression that $Y = X$, and that the regression did not differ among the soils. They discussed how this principle could be used to estimate the soil N mineralization under fluctuating moisture conditions.

While ammonification can continue at suctions well below 15 bar, nitrification ceases at about 15 (Robinson, 1957) to 50 bar (Wetselaar, 1968). Perhaps the most important effect of water occurs when there are cycles of wetting and drying (Calder, 1957; Campbell et al., 1975); this facet will be discussed separately later.

Most of the foregoing discussion is based on laboratory findings when soil moisture was kept at constant levels. In the field the moisture changes often and sometimes rapidly. Its effect is a function of temperature, wind, drainage, soil, and so on. It is, therefore, difficult to assess the contribution of moisture per se to the mineralization of N under natural conditions. There needs to be more research carried out on this facet.

Temperature

Mineralization is very slow near the freezing point because of restricted microbial activity. Over a considerable range of temperatures above 35°C ammonification continues but nitrification ceases at 45°C (Harmsen and Kolenbrander, 1965). Justice and Smith (1962) and Alexander (1965) state that the optimum constant temperature for ammonification and nitrification is between 25° and 35°C.

As mentioned earlier, Jenny (1930) found that organic matter content of

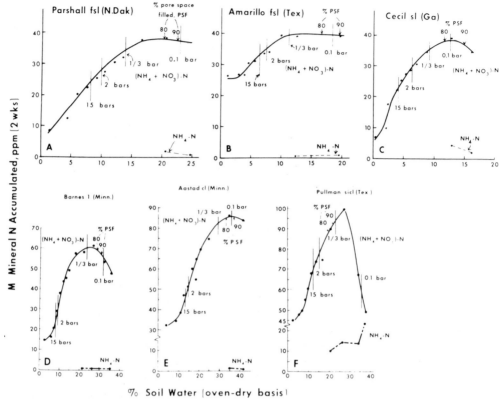

Fig. 18. Mineral N accumulated in relation to soil water content, matric suction, and percent of water-filled pore space (PSF). (Adapted from Stanford and Epstein, 1974.)

soils in the U.S.A. decreased 2 to 3 times per 10°C increase in the mean annual temperature. This is in agreement with the findings of Stanford et al. (1973a) that in the temperature range of 5°—35°C the temperature coefficient (Q_{10}) for mineralization was 2 for several diverse soils of the U.S.A. (Fig. 19). Stanford et al. (1973b) also found close correspondence between soil N A-values and amounts of N mineralized under fluctuating greenhouse temperatures as calculated using Q_{10} derived from constant temperature incubations. Furthermore, Stanford et al. (1975) showed that different sequences of temperature fluctuations between 5° and 35°C imposed on 3 soils during incubation had no effect upon the amount of N mineralized. Studies by Sabey et al. (1956) and Campbell et al. (1971) had shown that nitrification rate under fluctuating low temperatures differed from that under the corresponding mean temperatures.

Since present concepts of N mineralization—temperature relations in soils are based primarily on laboratory incubation studies conducted at constant temperatures while under field conditions marked diurnal fluctuations are

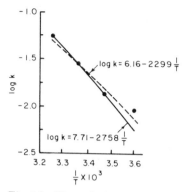

Fig. 19. The relation between log k and the reciprocal of absolute temperature. The regression lines obtained by pooling the regressions of log k on $1/T$ for 11 soils are identified as follows. The broken line is fitted to all points; the solid line is fitted to 3 points, representing 35°, 25° and 15°C. (k is the mineralization rate constant and T is the absolute temperature). (From Stanford et al., 1973.)

commonplace, the results of Stanford et al. (1973, 1975) are quite encouraging and should prove useful in quantifying the contributions of temperature to N mineralization under field conditions. The fact that fluctuating low temperatures tend to retard nitrification (at least temporarily) should be of little consequence in practice since plants can use NH_4–N as well as NO_3–N.

Drying, and wetting and drying

When soils are dried NH_4 may increase, perhaps because nitrification ceases at a higher moisture content than does ammonification. Thus, under conditions of extended drought there is usually a buildup of NH_4. The rate of drying is important. For example, rapid drying at 37°C is used in several western Canadian laboratories to arrest mineralization so that soils can be stored for subsequent mineral analysis. On the other hand air drying calcareous soils can lead to decreases in NH_4 due to volatilization of NH_3 at high pH. In soils high in expanding lattice clays some NH_4 can be fixed when drying is carried out at 105°C. Agarwal et al. (1971) found that generally air drying caused greater N mineralization upon incubation than when soil was heated to 60°C before incubation; but the converse was true in an organic-rich soil. Birch (1960), Birch and Friend (1956), and others have shown that when soil is incubated under constant moisture conditions the rate of mineralization is steady but slow while if the soil is first dried and then remoistened mineralization is rapid for 1 to 2 weeks, then falls to a steady state. Air drying can cause the release of simple organic N compounds (Van Schreven, 1967) and larger quantities of these compounds are released if drying is carried out at 105°C.

Cycles of drying and wetting cause flushes in C, N and other nutrients; each successive cycle causes a slightly smaller flush (Birch, 1960). The size of the flush is positively related to the humus content, the dryness of the soil, and the length of time the soil has remained dry. If the soil was kept dry too long or dried at too high a temperature it has to be reinoculated with a little fresh soil. Heating at high temperatures enhanced the effect of subsequent wetting and drying (Broadbent et al., 1964; Agarwal et al., 1971a). Mineralization of N and C due to wetting and drying was highly correlated (Agarwal et al., 1971a). Van Schreven (1968) found that while drying stimulated the mineralization of C and N of the humus it retarded the mineralization of C and N of the fresh plant materials.

The mechanisms by which the flush in NH_4 production is brought about is still a matter of conjecture. There are at least four good theories all of which may be involved to some extent.

(a) It has been suggested by some workers (Black, 1968) that some of the active microbial cells in soil are probably killed by the drying treatments and their tissue may then undergo autolysis and decompose. In support of this theory are the findings of some workers (Stevenson, 1956; Takai and Harada, 1959) that drying a soil increased the free amino acid concentration in water extracts.

(b) Waksman and Starkey (1923) concluded that drying a soil brings about a chemical change in the organic matter making it more available as a source of energy for micro-organisms; the rapid increase in the number of microbes after remoistening is at the expense of this newly released organic matter.

(c) A third theory (Russell, 1966) is that wetting and drying causes swelling and shrinking of the soil which physically disrupts some of the otherwise stable organic matter causing exposure of new surfaces for microbial attack. This is much like the theory put forward by Rovira and Greacen (1957) with respect to the effect of ploughing. Some supporting evidence for this theory is provided by the findings of Soulides and Allison (1961) that wetting and drying decreased the amount of water-stable aggregates in soil.

(d) A fourth hypothesis was recently advanced (Laura, 1974, 1975, 1976) and may be referred to as the protolytic theory. (Already discussed under "priming".) He suggests that "the decomposition of organic matter in soil is affected by the availability of the protons in the soil environment. Generally the factors which increase the supply of protons in soil increase the mineralization of soil organic matter." The dissociation of water increases as water content decreases. Since a dry soil contains residual water he suggests that it is quite possible that the supply of protons from this water might cause some chemical changes in humus during drying which would result in the formation of NH_4 and CO_2 on wetting of the soil and thus a flush of decomposition would be observed. He further suggests that while drying of soil increases the proton supply and thus mineralization, drying fresh plant material would save the organic colloids from protons and thus decrease their subsequent mineralization in soil. Although this theory has not yet been proven it does sound as plausible as the other three.

All four mechanisms may be involved in this flushing phenomenon to some extent but as to which is the more important will require further research. Some practical implications of this process are discussed by Russell (1966).

Freezing, and freezing and thawing

Gasser (1958) reported that freezing soil at $-22°C$ for 32 days did not affect mineralization on subsequent incubation; however, frequent freeze—

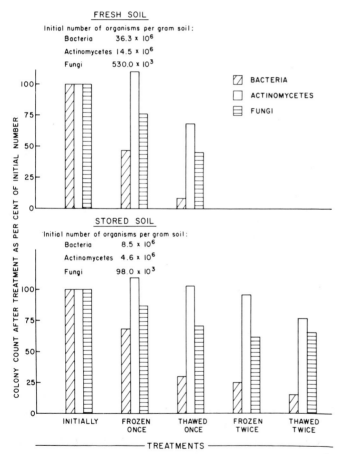

Fig. 20. Effect of slow freezing followed by thawing at 14° to 3°C on viability and composition of microflora in fresh and stored soil. (From Biederbeck and Campbell, 1971.)

thaw cycles did increase it. In general, the results of Soulides and Allison (1961) and Mack (1962, 1963) indicate that freezing results in less but similar physical, chemical and biological changes as those caused by drying.

Biederbeck and Campbell in a series of experiments (Biederbeck and Campbell, 1971, 1973; Campbell and Biederbeck, 1972; Campbell et al., 1970, 1971) determined the effect of simulated western Canadian spring conditions (i.e., freezing soil followed by incubation at diurnal temperatures of $14/3°C$) and also simulated autumn conditions ($14/3°C$ without prior freezing) on microbial and N changes in soil. They found that: (a) freezing caused a small reduction in the active microflora (Fig. 20) and no change in the mineral N; (b) simulated spring conditions were extremely lethal to the microflora, especially the bacteria (Fig. 20), and while it had no effect on net mineralization it reduced nitrification considerably (Table XII); (c) simulated autumn conditions had an effect similar to spring's but more subdued; (d) addition of NH_4— or peptone—N reduced the detrimental effect on

TABLE XII

Effect of simulated spring conditions on N transformations in a fresh loam soil

(From Campbell and Biederbeck, 1972)

	Control	NO_3—added	NH_3—added
	14—3°C	14—3°C	14—3°C
Mineralization [1]:			
after freezing	71	119	112
after 16 days incubation	81	130	128
net mineralization	+10	+11	+16
Nitrification:			
NO_3 after freezing	65	114	70
NO_3 after 16 days incubation	67	115	103
Change in NO_3	+2	+1	+33

[1] NO_3 + exch. NH_4—N.

the microflora considerably, consequently, nitrification was unaffected by simulated conditions; (e) field (Table XIII) and laboratory results were presented in support of the hypothesis that immediately after transient cold spells vegetative microbial cells are killed and their protoplasm then serves as a source of readily available substrate for the surviving and adapting micro-

TABLE XIII

Effects of sudden decreases in air temperature on changes in soil NO_3-N in the field
(From Biederbeck and Campbell, 1973)

Date 1968	Temp. (°C) max	min	NO_3-N (kg/ha)	Date 1969	Temp. (°C) max	min	NO_3-N (kg/ha)	Date 1970	Temp. (°C) max	min	NO_3-N (kg/ha)	Date 1971	Temp. (°C) max	min	NO_3-N (kg/ha)
25.9 [1]	19	4	63	5.6 [1]	28	11	87	2.9 [1]	25	12	26	7.9 [1]	20	6	85
26.9	16	4		6.6	24	11		3.9	28	10		8.9	27	9	
27.9	17	5		7.6	16	9		4.9	25	11		9.9	22	11	
28.9	19	3		8.6	22	4		5.9	22	10		10.9	26	4	
29.9	22	8		9.6	26	9		6.9	23	11		11.9	31	14	
30.9	16	8		10.6	15	2		7.9 [1]	18	9	28	12.9	22	5	
1.10 [1]	6	3	40	11.6 [1]	15	5	86	ΔNO_3-N			+2	13.9 [1]	16	5	71
ΔNO_3-N [2]			−23	ΔNO_3-N			−1					ΔNO_3-N			−14
2.10	8	−2		12.6	14	−4		8.9	13	5		14.9	12	4	
3.10	12	−4		13.6	17	−2		9.9	15	1		15.9	7	0	
4.10	21	1		14.6	20	3		10.9	10	2		16.9	8	4	
5.10	14	3		15.6	21	7		11.9	2	0		17.9	8	1	
6.10	13	0		16.6	25	6		12.9	2	−2		18.9	18	0	
7.10	9	−1		17.6	25	9		13.9 [1]	4	−3	68	19.9	14	4	
8.10 [1]	6	−2	77	18.6 [1]	27	14	120	ΔNO_3-N			+40	20.9 [1]	6	−1	126
ΔNO_3-N			+37	ΔNO_3-N			+34					ΔNO_3-N			+55

[1] Designates dates NO_3-N samples (0—60 cm depth).
[2] ΔNO_3 = change in NO_3-N between sampling dates.

flora. The nitrifiers are relatively harder hit than the more numerous heterotrophs, consequently while mineralization is hardly affected nitrification is curtailed; however, the effect is transient only lasting for 1 to 2 weeks. This is only one hypothesis; it is similar to explanation (a) in the wetting and drying section. Explanations (a), (b) and (c) from the wetting and drying section have been used as possible explanations for freeze—thaw effects; no doubt the more recently advanced protolytic theory will soon be put forward as well. In essence, these are all still theories and more research is required before the true facts will be known.

The drying and rewetting, and freezing and thawing effects may influence the N economy in agricultural areas where vegetation periods alternate with dry seasons or cold winters. The flush will sometimes coincide with rapid nutrient uptake and growth of the young crops except where the freezing occurs as a late frost. In the latter case the flush of NO_3 may lead to high NO_3—N in damaged plant tissue such as forage crops and could result in injury to livestock. Another possible although hypothetical benefit of the effect of spring and autumn conditions on N may be to retard nitrification and preserve mineral N in the less leachable NH_4 form. This would be especially beneficial in early spring when spring melt and runoff NO_3—N losses may prevail.

DYNAMICS OF ORGANIC MATTER TRANSFORMATIONS

At least some fractions of soil organic matter must be very resistant to decomposition, since peat bogs which accumulate a few cm/100 years, (Durno, 1961) are sometimes several meters deep, and thousands of years old as determined by pollen analysis (Durno, 1961) and carbon dating (Libby, 1955). Furthermore, mineral soils which usually develop under conditions less favorable for organic matter accumulation than peats, often have A horizons which are 30 cm deep with an organic matter content of 10—15% (Mitchell et al., 1944).

The N content of soil is seldom static under natural conditions, and it can change dramatically under the influence of man's activities. Some soils presently contain less than half the N originally present in them. This section will deal with quantitative data demonstrating how various factors affect the rate of soil N change under man's influence. It will also deal with quantitative estimates of organic matter turnover in soil, touch briefly on the use of the carbon dating technique in studying organic matter dynamics, and show how these methods may be used to estimate and predict the state of the organic matter in soil. Numerous review articles have been written on this subject (Jenkinson, 1966a; Stevenson, 1965; and Jansson, 1958).

Using long term data

Soil organic matter is an equilibrium system. Consequently, some workers have used long term field experiments (Bartholomew and Kirkham, 1960; Jenkinson, 1966a; Russell, 1975) and others have used isotopes (Jansson, 1963; McGill et al., 1974) in simulated field experiments to make mathematical evaluation of the dynamics of the system. These studies have led invariably to the proposal of various mathematical expressions to describe this system. Salter and Green (1933) proposed the following equation:

$$N_t = N_0 K^t \tag{1}$$

where N_t = amount of organic N per unit mass of soil remaining in soil after t years; N_0 = amount of organic N per unit mass of soil at the beginning of the study; and K = the fraction of the original organic nitrogen per unit mass of soil remaining after a year. This expression is based on the assumption that the amount of N lost is proportional to the total amount of N in the soil.

Eq. 1 can be expressed as:

$$N = N_0 (1 - r)^t \tag{2}$$

where r is the annual rate of N loss.

From eq. 2 the following differential equation can be obtained:

$$dN/dt = -rN \tag{3}$$

which by integration becomes:

$$N = N_0 \exp -rt \tag{4}$$

According to eqs. 2, 3 and 4, the soil N should decline until an absolute minimum is reached, e.g., until the N content of the soil reaches zero. In agronomic practice, this situation never exists, for when soils are placed under cultivation the N level assumes a new equilibrium value. Usually, but not always, the new level is lower than the original.

Eq. 4 did not allow for the accumulating effect of humus, which results from the addition of organic materials. Jenny (1941) therefore proposed the following equation:

$$dN/dt = A - rN \tag{5}$$

Here he assumed that the losses were proportional to the amount of organic matter per unit mass of soil (N), present at any time (t); that the rate of addition of new N per unit mass of soil (A) was constant; and that the annual rate of decomposition of N per unit mass of soil (r) was a constant for all forms of organic matter. Bartholomew and Kirkham (1960) used the integrated form of Jenny's equation:

$$N = N_0(\exp -rt) + A/r(1 - \exp -rt) \tag{6}$$

and graphical methods to obtain the constants A and r and used these to calculate the probable equilibrium level (A/r) for long term rotation plots in the U.S.A. (Stevenson, 1965).

Woodruff (1949) had modified eq. 6 by recognizing that different components of soil organic matter decomposed at different rates. Thus eq. 6 became:

$$N = N_1 (\exp -r_1 t) + A_1/r_1 (1 - \exp -r_1 t) + N_2 (\exp -r_2 t)$$
$$+ A_2/r_2 (1 - \exp -r_2 t) + ... \text{ etc.} \tag{7}$$

Where the subscripts 1, 2, etc., refer to humus fraction with different decomposition rates. Woodruff (1949) using data from Sanborn plots found the average rate of decomposition of the organic matter for cultivated crops, small grain, and nonleguminous meadow crops to be 2%, 1% and 0.75%, respectively.

Only a few results suitable for testing eq. 6 are available, and these come from long term experiments. When such data are used in this manner Jenkinson (1966a) points out that two assumptions additional to those already introduced in deriving the equations must be made. Since it is easier to measure N than C almost all data available are for N. But only in the later stages of decomposition are losses of soil N proportional to losses in organic C. A second difficulty is that most published data are expressed in percent N at different sampling dates, all samples being taken to the same depth. Changes in the organic content of a soil are usually associated with changes in bulk density and if the amount of soil organic matter is not uniform throughout the sampling depth then soil samples taken at different times to the same depth are not comparable. For this reason data for arable soils are preferable to those from grassland or forest soils since the arable soils will be relatively uniform down to plough depth.

Jenkinson (1966a) cites an example of how data from the Rothamsted experiment on continuous barley can be used to test eq. 6. The constants in the quation were calculated from the percentages of N in the soil 0, 30, and 61 years after the experiment had started. The curve (Fig. 21) was then drawn using these constants ($r = 0.0251$, corresponding to a turnover time of 40 years; $A = 0.0078\%$ N per year). There were 2,900 metric tons of fine soil

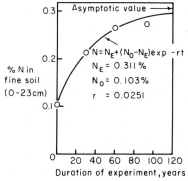

Fig. 21. Calculated and measured nitrogen content of soil under continuous barley (Hoosfield plot 7—2). (From Jenkinson, 1966a.)

per hectare in the top 23 cm of the experimental field so that the calculated annual addition was 226 kg N/ha. The actual annual addition of farmyard manure N was found to correspond closely with the amount of N entering the soil organic matter per year (A) as calculated by eq. 6.

Greenland (1971) used eq. 5 to examine organic N dynamics under various rotations which included pasture in Australia. Under pasture, A will normally exceed rN in eq. 5 so that the N level will tend to increase while during cropping A will be small, kN will exceed A and the N level will fall. After several cycles of pasture and crop, the N level will tend to fluctuate about a mean, N_m, dependent on the relative lengths of the crop and pasture periods and the values of k and A for the two phases of the rotation (Fig. 7). The mean, N_m, should be of such a magnitude that the mineral N released, kN_m, is sufficient to produce a crop of desired size. For example, a crop of 25 bushels of wheat per acre requires about 25 lb. N available to it in the soil. If k is 5% then the minimum level of N_m must be $25 \times 100/5 = 500$ lb./acre, corresponding to about $500/2 \cdot 10^6$ or 0.025% of N in the top 6 inches (2 million poinds) of soil. Since some N will be lost by leaching, etc., a higher level of N than this will be required. Greenland (1971) gives another useful example of how a mathematical model can be used to estimate what lengths of crop and pasture are required to maintain N_m. If the lengths of the cropping and pasture phases are t_c and t_p, respectively, the loss of N during cropping is $(dN/dt)_c t_c$, and the gain under pasture is $(dN/dt)_p t_p$. The N contents during the crop and pasture phases will be approximately equal to N_m. If equilibrium is to be maintained, losses and gains must be equal, so that:

$$(-k_c N_m + A_c)t_c + (-k_p N_m + A_p)t_p = 0$$

and:

$$t_c/t_p = (A_p - k_p N_m)/(k_c N_m - A_c) \tag{8}$$

If values can be assigned to the decomposition and addition constants under crop and pasture, the value of N_m which will be attained under any relative lengths of crop and pasture can be readily calculated.

For instance, suppose the addition of N to the soil—plant system during the pasture phase (A_p) is 80 lb. per acre per annum and the decomposition constant (k_p) 1% per annum, and the corresponding values during cropping are A_c = 20 lb. per acre per annum and k_c = 4%. If a total N level in the top soil of 2,500 lb. per acre is required, by substituting into eq. 8 the relative lengths of crop and pasture must be roughly in the ratio 2 to 3.

Greenland (1971) further used eq. 5 together with certain reasonable assumptions (Fig. 22) to estimate the N content of the top 10 cm of the soil under a long term rotation at Waite Institute.

Based on eq. 5 it is possible to deduce the half life (the period for half the change from the original organic matter level to the new equilibrium level to take place). Half life $(t1/2)$ = 0.693/r. If r = 1%, $t1/2$ = 69 yr. (Russell, 1962). Some representative data for soils under pasture in temperate regions (Table XIV) have been presented by Russell (1962). Campbell et al. (1976) presented similar data for Canadian prairie soils and discussed these in terms of the effect of length of cultivation, rotation, soil zone, texture, slope and crop residues. Table XV shows the effect of length of cultivation on the rate of N loss from chernozemic soils under western Canadian conditions (Campbell et al., 1976). It confirms that the rate of N depletion is more rapid during the first 5—10 years after breaking land and then gradually declines.

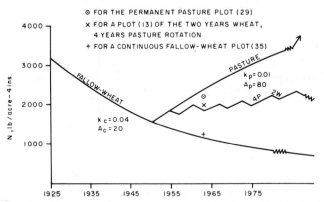

Fig. 22. Changes in the nitrogen content of the 0—4 inch soil horizon under continuous fallow—wheat, and under pasture and pasture—wheat rotations, assuming the changes to follow the equation $dN/dt = -kN + A$. Under fallow—wheat it has been assumed that k_c = 0.04 and A_c = 20 lb. per acre, and under pasture k_p = 0.01 and A_p = 80 lb. per acre. Years and nitrogen levels have been chosen to correspond to plots in the permanent rotation trial $(C1)$ at the Waite Institute. Actual levels of nitrogen are shown in 1963. (From Greenland, 1971.)

TABLE XIV

Half-lifes and calculated (r) values based on assumption that observed data can be fitted by first order-kinetics
(Adapted from Russell, 1962)

Locality	Depth of soil		Period studied	Calculated r value	Half-life	Source *
	inches	cm	years	(%/yr.)	years	
Rothamsted, U.K.	0—8	0—20	200	2.8	25	Richardson (1938)
Mypolonga, South Aust.	0—9	0—23	29	2.2	31	Russell and Harvey (1959)
Kybybolite, South Aust.	0—2	0—5	39	2.9	24	Russell (1960a)

* See Russell (1962) for these references.

TABLE XV

Effect of time under cultuvation on rate of N loss from chernozemic soils of the Canadian prairies
(From Campbell et al., 1976)

Great group	Texture and original soil N (%)	Years of culti-vation	Rotation and segment	Avg. rate of loss, k (y^{-1})	$T_{1/2}$ (yr.)	Reference *
Brown	loam (0.19)	2—14 14—32 32—45	Various (0—15 cm)	0.013 0.005 0.002	53 139 347	Hill, 1954
	loam (0.19)	2—14 14—32 32—45	W-F (0—15 cm)	0.010 0.007 0.003	69 99 231	Hill, 1954
	loam (0.20)	0—14 14—35	W-F (0—30 cm)	0.025 0.011	28 63	Campbell et al., 1975
Dark brown	loam (0.17)	0—14 14—35	W-F (0—30 cm)	0.009 0.019	77 36	Campbell et al., 1975
Black	loam (0.45)	0—15 15—60	Various (0—10 cm)	0.019 0.005	36 139	Martel and Paul, 1974
Black	loam (?)	0—12 12—24 24—36 36—60	Various (0—15 cm)	0.019 0.014 0.008 0.007	36 50 87 99	Paul, 1971

* See Campbell et al., 1976, for these references.

Minderman (1968), used long term data from forest soils in which the annual addition of the litter components and the amount of organic matter in the soil had been measured and, calculated an overall degradation rate from knowledge of the degradation rate of the individual components. By summation of the decomposition rates of the individual organic components of the leaf litter, he showed that the calculated decomposition curve corresponded in shape with that actually determined over a period of 15 years. The actual curve under field conditions showed a lower degradation rate in later years than did the calculated values.

Some workers (Jenkinson, 1966a; Paul, 1970), recognizing that organic matter is composed of components which decompose at different first order rates, have used a technique commonly employed in radiochemistry to graphically separate the decomposition curve they obtain into its component parts (Fig. 23). This treatment is analogous to the reduction of the decay curve for a mixture of radioisotopes into its component first-order decay reactions, (Friedlander and Kennedy, 1960, p. 128). It is questionable whether this treatment is correct because the reactions involved are not independent and are equilibrium reactions rather than ones that go to completion as does the radioisotope decay reaction. Reid and Miller (1963) have cited a relationship for reversible exchange into more than one compartment which can be generalized to the following:

$$C = C_{eq} + a_1(\exp - b_1 t) + a_2(\exp - b_2 t) + \dots a_n(\exp - b_n t) \qquad (9)$$

where C = concentration of exchanging ion in solution at time = t; C_{eq} =

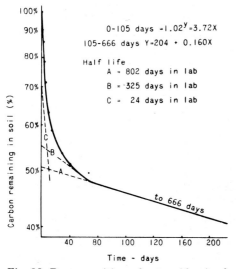

Fig. 23. Decomposition of oat residue in chernozem. (From Paul, 1970.)

concentration of exchanging ion in solution at equilibrium; a_1, a_2, a_n = concentration constants; and b_1, b_2, b_n = rate constants for the respective exchange reactions.

In eq. 7 the addition and decomposition parameters were regarded as constants. Russell (1962) points out that from the point of view of long term changes a more comprehensive equation is:

$$dN/dt = -PN + Q \tag{10}$$

where P and Q are decomposition and addition parameters, respectively, and both are functions of time (t). Thus eq. 5 is a special case of eq. 10 in which P and Q are constants.

Russell (1975) recently tested eq. 10 satisfactorily. We shall reproduce some of this important paper here. First let us change symbols so as to conform to Russell's report. We start by rewriting eq. 5 as follows:

$$dN/dt = -K_1 \cdot N + K_2 \tag{11}$$

where $K_1 = r$ and $K_2 = A$ in eq. 5. Then using the principle expressed in eq. 10 Russell proposes that:

$$dN/dt = K_1(t) \cdot N + K_2 + K_3(t) \cdot Y(t) \tag{12}$$

This equation, called model 1, recognizes that the decomposition coefficient, $K_1(t)$, may change with time. This occurs with agricultural land used for different crops in rotation, and in a fallow—wheat—pasture system $K_1(t)$ changes each year in a 3-year cycle. Eq. 10 also considers the addition term as comprising two components. The term K_2 represents the addition to soil organic matter from noncrop sources, e.g., manure. The term $K_3(t) \cdot Y(t)$ represents the addition term due to plant residues. $Y(t)$ refers to the crop yield and $K_3(t)$ is a coefficient related to the specific crop at time t. In a rotation it is likely that different crops or land treatments (e.g., corn, wheat) have different K_3 coefficients, and $K_3(t)$ will vary with the crop in the rotation at a certain time t. The main advantage of this model is that it allows the effects of crop yield within a rotation on soil organic matter levels to be examined. In the analysis of some crop sequential systems model 1 involves estimation of a large number of parameters. It is desirable to test whether the use of all these parameters can be justified by the experimental data available.

Russell proposed three additional restricted models. In the first of these (model 2) K_2 was assumed to be equal to zero. The equation for this model is then:

$$dN/dt = K_1(t) \cdot N + K_3(t) \cdot Y(t) \tag{13}$$

where the parameters have the same meaning as in model 1. In the second restricted model (model 3) it was proposed that in a sequential crop system the relationship between decomposition and addition coefficients is the same for all crops in the system, i.e.:

$$K_3/K_1 = a \qquad\qquad (14)$$

where 1 is the crop and a is a constant. The equation for model 3 is then:

$$dN/dt = -K_1(t) \cdot N + K_2 + a \cdot K_1(t) \cdot Y(t) \qquad\qquad (15)$$

The third restricted model (model 4) includes both the restrictions in models 2 and 3. Its equation is then:

$$dN/dt = -K_1(t) \cdot N + a \cdot K_1(t) \cdot Y(t) \qquad\qquad (16)$$

Using these models it is possible to estimate the effects of crop yield on soil N equilibrium levels. At a given equilibrium level of soil N content and plant yield and where $K_1(t)$ and $K_3(t)$ do not vary, i.e., with continuous treatments, then it can be assumed that

$$dN/dt = 0 \qquad\qquad (17)$$

At this point the equilibrium soil level is N_E and the plant yield level is Y_E, thus for model 1:

$$-K_1 \cdot N_E + K_2 + K_3 \cdot Y_E = 0 \qquad\qquad (18)$$

and:

$$N_E = (K_2 + K_3 \cdot Y_E)/K_1 \qquad\qquad (19)$$

From the estimated coefficients it is possible to plot the effect of crop yield on equilibrium soil N level. The intercept on the Y axis when Y_E is zero is K_2/K_1 and the slope of the line relating plant yield and soil equilibrium level is K_3/K_1.

For models 2, 3, and 4 the equivalent equations relating equilibrium soil N and plant yields are, respectively:

$$N_E = K_3 \cdot Y_E/K_1 \qquad\qquad (20)$$

$$N_E = K_2/K_1 + a \cdot Y_E \qquad\qquad (21)$$

and:

$$N_E = a \cdot Y_E \qquad\qquad (22)$$

Models 2 and 4 are not realistic since they imply that if plant yields are zero then soil N levels will also be zero. Russell tested the four models by fitting them to plant yield and soil N data from the Morrow plots (Urbana, Illinois) and the Sanborn plots (Columbia, Missouri). He used a minimization technique to provide a least square fit and thus estimated the parameters for each model. These estimates were done by computer analysis. Using the estimated parameter values (Table XVI) and estimated initial values (Table XVII) obtained from the Morrow and Sanborn plots together with the recorded plant yields Russell calculated the trend in soil N for each plot. The results for the Morrow plots are shown in Fig. 24. It can be seen that the fit of the data is excellent.

One advantage of Russell's method is that it allows the interaction between crop sequential practice, crop yield and soil organic matter levels to be assessed. This excellent work should prove valuable to scientists who have access to long term crop rotation data and computer facilities since it will allow them to re-examine their data more meaningfully.

Mineralization—immobilization (turnover)

Quantification of N transfer from organic substrates through soil micro-organisms and into organic components or the mineral form is required to define the soil system with respect to N turnover and the N-supplying power of soils.

A tracer study was undertaken by McGill et al. (1974) as a part of the International Biological Program at Matador, Saskatchewan, to trace N transfer through soil micro-organisms and their metabolites into more resistant forms in soil, and to elucidate the relationship between recently synthesized microbial metabolites and soil constituents. The mineralization rate of immobilized N and of N contained in plant debris was determined by field experiments. In the laboratory, a nondegradative fractionation system was developed to separate microbial constituents in soil. This information, together with measurements of microbial populations was applied to soils incubated under laboratory and field conditions to interpret results of studies of N mineralization and the contribution of the microbial component to N turnover. The results of these excellent studies are briefly summarized here.

Two models were developed based on the results. The first model was a mathematical description of the relationships between C and N turnover through soil micro-organisms and N transfer into a large labile pool of microbial products (Fig. 25). The concepts applied were as follows: (a) two biochemically separate populations developed sequentially; (b) the primary population (mostly fungi) was the sink for acetate—C, but it also assimilated more complex substrates; (c) the secondary population (mainly bacteria and actinomycetes) utilized microbial metabolites and soil organic matter, but

TABLE XVI

Estimated parameter values for Morrow and Sanborn
(From Russell, 1975)

Experiment	Coefficient	Crop	Symbol	Units	Model 1	2	3	4
Morrow 1904–1953	decomposition	corn	K_{1c}	%/yr.	0.84	1.21	1.12	1.48
		oats	K_{1o}	%/yr.	1.35	1.78	0.38	0.46
		clover	K_{1p}	%/yr.	0.49	1.29	0.47	1.92
	addition	no manure	K_2	kgN/ha/yr.	0.0			
		with manure	K_{2m}	kgN/ha/yr.	15.8		7.2	
		corn	K_{3c}	kgN/1000 kg grain/yr.	$4.0 \cdot 10^{-5}$	9.8	29.0	
		oats	K_{3o}	kgN/1000 kg grain/yr.	27.7	38.5		
		clover	K_{3p}	kgN/1000 kg dry matter/yr.	0.59	12.8		
			a	kgN/kg grain or dry matter			$1.21 \cdot 10^{-5}$	$9.69 \cdot 10^{-1}$
Morrow 1904–1967	decomposition	corn	K_{1c}	%/yr.	1.19	0.96		
	addition	no manure	K_2	kgN/ha/yr.	15.3			
		with manure	K_{3m}	kgN/ha/yr.	30.3			
		corn	K_{2c}	kgN/1000 kg grain/yr.	$4.4 \cdot 10^{-4}$	5.53		
Sanborn 1889–1938	decomposition	corn	K_{1c}	%/yr.	5.71	5.01	3.92	4.73
		oats	K_{1o}	%/yr.	3.97	3.02	2.59	0.24
		wheat	K_{1w}	%/yr.	3.62	3.01	3.11	0.82
		timothy	K_{1p}	%/yr.	2.21	1.53	2.11	0.49
	addition	no manure	K_2	kgN/ha/yr.	58.6		60.5	
		with manure	K_{2m}	kgN/ha/yr.	97.7		107.4	
		corn	K_{3c}	kgN/1000 kg grain/yr.	30.8	6.8		
		oats	K_{3o}	kgN/1000 kg grain/yr.	47.9	30.2		
		wheat	K_{3w}	kgN/1000 kg grain/yr.	17.9	7.8		
		timothy	K_{3p}	kgN/1000 kg dry matter/yr.	1.1	12.9		
			a	kgN/kg grain or dry matter			$7.96 \cdot 10^{-6}$	1.37

TABLE XVII

Estimated initial values of the equations expressed as kg soil N/ha
(From Russell, 1975)

Experiment	Rotation	Model			
		1	2	3	4
Morrow 1904—1953	continuous corn	4058	4010	4117	4068
(estimated for 1904)	continuous corn + manure	4379	4463	4336	4434
	corn—oats	4549	4522	4467	4423
	corn—oats + manure	4659	4713	4678	4806
	corn—oats—clover	5065	5070	5027	5025
	corn—oats—clover + manure	5330	5414	5266	5361
Morrow 1904—1967	continuous corn	4015	3946		
(estimated for 1904)	continuous corn + manure	4385	4492		
Sanborn 1889—1938	all plots				
(estimated for 1889)		3491	3675	3204	3329

not acetate—C; (d) the quantity of C and N entering or removed from a given
population was a function of the population size (i.e., C content); (e) N turnover was strictly dependent upon C turnover; and (f) the relative proportions
of soil-C and labelled-C (or N) in a population could be determined using the
specific activity and quantity of evolved CO_2 as the experimental input to
the model.

This first model defined the interaction of C and N in soil and N transfer
into a large labile organic pool of microbial debris. We will not give the
mathematics of the first model since space does not permit.

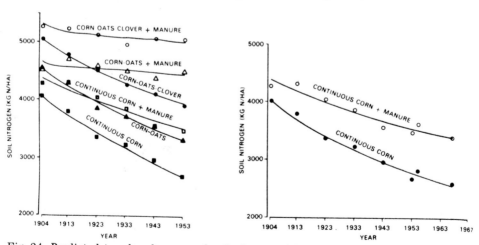

Fig. 24. Predicted trend and measured soil nitrogen values of model 1 for Morrow plots.
(Adapted from Russell, 1975.)

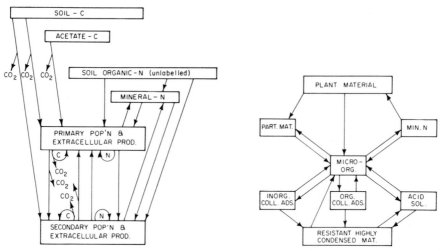

Fig. 25. Conceptual model of C and N transfer through various microbial and soil components. (From McGill et al., 1974.)

Fig. 26. Conceptual model of soil-N and transformations. (From McGill et al., 1974.)

The objective of the second model (Fig. 26) was to define some of the components of the labile organic pool or microbial debris with respect to their importance in N turnover. The model was described with first order reaction constants for the transfer of N from straw into the total organic pool and its subsequent remineralization. The various fractions were then separated out, and first order constants were determined for each.

N present as recognizable straw particles rapidly declined in the two soils studied. The degradation rate of straw-N was calculated from these values. The second model, was simplified to:

Since first order kinetics and a nonreversible reaction were assumed, N immobilization was omitted; hence the degradation constants obtained were net values. Calculations were performed such that N was transferred from "A" to "B" at a rate defined by the decay constant of the straw particles (Table XVIII). This N was mineralized and transferred to box "C" at a rate by the decay constant of the labelled organic moiety (Table XVIII). The calculations were performed using various decay rates and the final results obtained were compared with those obtained from field measurements. The calculated and measured apparent losses compared favorably in both soils. Thus, it was concluded that the constants chosen and the conditions of the

TABLE XVIII

Decay constants of straw-N and labelled organic-N in cultivated Sceptre and Waitville soils (From McGill et al., 1974)

Soil	Interval (yr.)	K	
		straw-N	labelled organic-N
Sceptre	0.1—0.4	1.15	1.15
	0.5—2.0	2.31	0.53
Waitville	0.1—0.7	2.77	1.15
	0.8—1.0	2.77	0.69
	1.1—2.0	2.77	0.43

$K = 1/dt \cdot \ln C_1/C_2 = 2.303/T_1 - T_2 \log C_1/C_2$

where: dt = time in years;
C_1 = conc. of substrate at time t_1;
C_2 = conc. of substrate at time t_2;
$T_{1/2}$ (half-life) = 0.693/K.

model adequately approximated the real soil system. The predicted labelled-N content of the organic moiety also closely approximated that measured in soil samples taken at various times (Fig. 27).

After N transfer through the total organic moiety had been defined, an effort was made to measure the quantity of N entering each fraction as a result of straw decomposition and to determine the contribution of each fraction to the mineral-N pool.

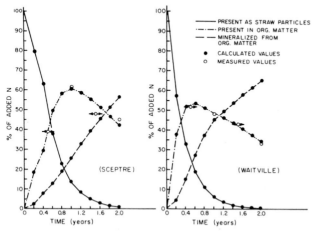

Fig. 27. Calculated and measured distribution of added straw-N in Sceptre and Waitville soils under field conditions. (From McGill et al., 1974.)

The decay constants for each fraction could be determined from the quantities of N in them at various times. This, however, yielded only net values due to continued additions from labelled straw decomposition. Similarly, the decay constant calculated for the total organic moiety (from measured values) was also a net value. The actual values, derived from the model, are reported in Table XVIII. Actual values for the various fractions were obtained by multiplying the net value (from experimental results) by the ratio: actual K (total organic moiety)/net K (total organic moiety).

The portion of decomposing straw entering each fraction was approximated by the following argument. Let: $KA = K$ of straw particles (fraction A); $KB = K$ of given fraction B (actual); $NA = N$ in straw at start; $NB = N$ in fraction B at start; $NR = N$ remaining in fraction B at end; $A =$ portion of decomposing straw entering fraction B.

McGill et al. (1974) proceeded as shown below. They assumed that: amount remaining in X after T units of time is $X_0 \exp -KT$, and amount removed from $X = X_0 - (X_0 \exp -KT) = X_0(1 - \exp -KT)$.

Thus: total amount removed from A is: $NA \cdot (1 - \exp -KT)$
transferred to B is: $A \cdot NA \cdot (1 - \exp -KA \cdot T)$
total in B is: $NB + A \cdot NA \cdot (1 - \exp -KA \cdot T)$
$NR = N$ remaining in B after T units of time
$NR = [NB + A \cdot NA (1 - \exp -KA \cdot T)] \cdot [\exp -KB \cdot T]$
$NR = NB \cdot (\exp -KB \cdot T) + A \cdot NA \cdot (1 - \exp -KA \cdot T) \cdot (\exp -KB \cdot T)$
$NR = [NB(\exp -KB \cdot T)] = A \cdot NA \cdot (1 - \exp -KA \cdot T)(\exp -KB \cdot T)$
Therefore:

$$A = \frac{NR - [NB \cdot (\exp -KB \cdot T)]}{NA \cdot (1 - \exp -KA \cdot T) \cdot (\exp -KB \cdot T)} \qquad (23)$$

Eq. 23 was used to calculate "A" for each soil fraction. This was then used to calculate the N transferred from straw to each organic fraction and its subsequent conversion to mineral N. Total N in the organic and mineral forms calculated by summing results from each individual fraction were found to correspond favorably to that predicted in the calculations based on the total organic moiety.

Another approach to the solution of the turnover cycle was provided by Jansson (1958). Space will not permit us to reproduce the mathematical development of this approach here. In brief, Jansson illustrated the organic matter turnover cycle as follows:

Organic N	i	Inorganic N
u = gross	(immobilization per unit time)	x = gross
v = tagged	m	y = tagged
	(mineralization per unit time)	

The definitions of the various symbols used in the mathematical development were as follows:

t = time

u, x, a = mass, per unit mass of soil of all organic, inorganic, and organic plus inorganic N atoms, respectively (tagged plus nontagged).

v, y, b = mass, per unit mass of soil of all organic, inorganic and organic plus inorganic tagged N atoms, respectively.

u_0, v_0, x_0, y_0 are values of u, v, x, y, when t = 0 (at the start of the experiment).

m, and i are mineralization and immobilization rates, respectively (mass of organic N, per unit mass of soil, including tagged and nontagged atoms, which are transformed per unit time to inorganic form).

After making various reasonable assumptions Jansson obtained the following:

$$\frac{dy}{dx} = \frac{b-y}{a-x} \cdot \frac{m}{m-i} - \frac{i}{m-i} \cdot \frac{y}{x} \qquad (24)$$

In calculating the mineralization or immobilization rates or the absolute magnitude of the N transformations (mt and it, respectively), eq. 24 is integrated between the limits (x_0, x) and (y_0, y). Jansson (1958) has discussed the limitations and given examples of how this turnover approach can be used.

Another type of calculation which might be required in considering organic matter turnover processes is that of estimating the biomass. For example, Jansson (1971) determined experimentally the net mineralization outflow (m) and the tagged part (y). Since the amount of tagged N in the organic state (B) is known, then the amount of soil N (A) equally mineralizable as tagged organic N is:

$$A = Bm(1-y)/my = B(1-y)/y \qquad (25)$$

Jansson (1971) emphasizes that this "active phase" calculation must be carefully scrutinized since after some time the calculations will give increasing amounts of soil N in the active phase. The latter is an expression of the stabilization of the tagged N, that is, the passing of active biomass N into passive conditions.

In the first part of this section we showed how knowledge of the long term history of soil N changes can be used to predict rates of change of N, equilibrium levels, and what effects various rotations might have on organic N, and so on. There is another approach to solving this problem which does not require as long a wait to obtain the data to be used in the computations. N exists in soil in forms having various degrees of stability. If we can determine experimentally the decomposition and addition rates of these components we should be able to obtain the same information as that from the long term studies. This was the approach used by Campbell et al. (1976); we shall briefly outline how this was done.

Based on knowledge obtained on N cycling at the International Biological Program project in southwestern Saskatchewan, together with data available on residue decomposition rates and N cycling rates in chernozemic cultivated soils, a N model was developed (Fig. 28). The source of the constants is primarily McGill et al. (1974); also see Campbell et al. (1976).

The model was used to calculate expected N contents of soil during 60 years of cultivation under two rotations. It is probable that this model has underestimated the biomass N; Jansson (1971) suggests that this fraction makes up about 10—15% of the total N. In our opinion compartment "C" has also probably been overestimated. Martel and Paul (1974) have shown that turnover rates must be based on the acid resistant-C and the C/N ratios. They should not be based on hydrolysable-N. The data in this model, therefore, may have underestimated the amount of resistant N and overestimated the amount of labile N and the turnover time of the labile N. Nonetheless the results (Fig. 29) are a useful first approximation. They indicate that the constants in the model provide a reasonable representation of the soil system. There is an initial rapid N loss followed by a gradually decreasing rate of loss. The wheat—fallow rotation (W—F) was generally less efficient in conserving soil N than was the W—W—W—F system but the differences were not very apparent until after about 45 years.

The model can be modified to represent different cropping systems, management practices, and native soil conditions by changing the turnover rate for the labile pool, the proportion of N in dead plant residues, and the amount of plant residue N reincorporated into the soil annually. The model was used to assess cereal rotation effects on organic N, to calculate when to fertilize and how much to apply, to determine the fate of soil N and to calculate the equilibrium level and persistence of N (Campbell et al., 1976). For example, if we consider a 4-year wheat—wheat—wheat—fallow rotation

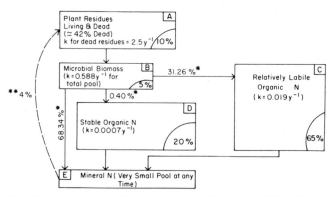

Fig. 28. Model for predicting long term N status of Canadian prairie soils. (From Campbell et al., 1976.)

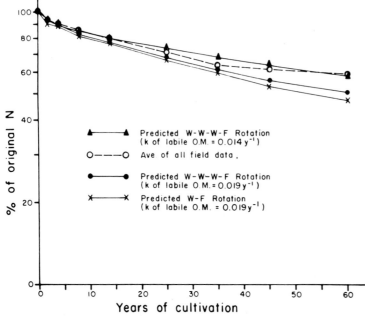

Fig. 29. Average N content of soils of the Canadian prairies during 60 years of cropping compared to the predicted N content using the model. In running the model it was assumed that 34 kg N/ha were returned with crop residues, fixed N and N absorbed from the atmosphere. It was assumed that no N entered the soil during fallow years. The original soil was assumed to contain 4,500 kg N/ha, i.e., a brown or dark brown chernozem of medium texture. (From Campbell et al., 1976.)

on a soil originally having 4,500 kg N/ha, and if we assume the yield of wheat to be 1,350 kg/ha/yr., also that 0.2 kg N was required to produce 1 kg of wheat then from the model it was calculated that the equilibrium N level would be 25 to 28% of the original N and this level would be reached in about 260 years after the land was broken. In a 2-year wheat—fallow rotation the equilibrium level would be about 22% of the original N and this level would be reached in 430 years; in the latter rotation the total N would reach 25% of the original in about 230 years.

Use of carbon dating

Only brief treatment of this subject will be presented here. For a review of this topic the reader may wish to read Paul (1970), and Jenkinson (1966a).

Bombardment of the atmosphere by cosmic radiation produces high energy secondary neutrons which react with N to produce ^{14}C. The ^{14}C is oxidized to $^{14}CO_2$. This is mixed by winds with inert $^{12}CO_2$ and $^{13}CO_2$. Of the total CO_2 photosynthesized, 1 part in 10^{10} is $^{14}CO_2$, and 1.1 part in 100 is $^{13}CO_2$ (Libby, 1955).

Libby (1955) showed by measuring the radiocarbon activity of a sample and that of a modern reference sample that it is possible to estimate the sample's age from the following equation:

$$\text{Age (years before present: B.P.)} = 18,500 \log_{10}(A_0/A) \qquad (26)$$

Where A_0, the activity of the modern ^{14}C standard was based on 95% of the activity of the National Bureau of Standards (NBS), oxalic acid, and A was the activity of the unknown sample. B.P. is taken as before the year 1950.

Factors which affect carbon dating are: (a) isotopic discrimination (Craig, 1954); (b) Suess effect (Suess, 1955), which results from the dilution of radiocarbon in the atmosphere by the nonradioactive carbon of fossil fuels; (c) bomb carbon effect resulting from atmospheric testing of hydrogen bombs with the production of fairly large concentrations of ^{14}C (Broecker and Olson, 1960); and (d) there are variations in the natural ^{14}C content in different parts of the world (De Vries, 1958). Campbell et al. (1967) have discussed these factors with respect to their effect on the mean residence time of soil organic C. These authors used the term mean residence time (mrt) to describe the soil "age" in recognition that the soil organic matter is composed of materials ranging from the very old to the very young. Thus the mrt is a measure of the length of time a hypothetical average component has resided in the soil or fraction of organic matter.

Campbell et al. (1967) and Martel and Paul (1974) have used this technique together with chemical fractionation methods to study the dynamics of the soil humus. For example, Campbell (1965) fractionated soil from the Ap horizon of a Melfort orthic black chernozem into several humic constituents, determined the distribution of the soil C and N in and carbon dated these fractions (Table XIX). By making a few pertinent assumptions he estimated that about 84 kg N/ha is released from the soil humus each year with 80% of this coming from the humic acid hydrolysate fraction. The non-hydrolysable humin and non-hydrolysable humic acids although comprising 57% of the C, accounted for only a small proportion of the soil metabolism. These fractions are therefore regarded as being physically and chemically active affecting soil aggregation and exchange reactions but are biologically inactive. Martel and Paul (1974) concluded that the high degree of hydrolysability of N is not necessarily related to its turnover rate and the internal cycling of N complicates the modelling of the flow of this nutrient.

From the foregoing it can be seen that considerable progress is being made in determining the dynamics of soil organic matter. In this respect the work of Russell, McGill and coworkers, and Jansson stands out. Use of long term data analysis as well as the method of using isotopes to determine turnover of various fractions is proving very useful in helping us to estimate and predict the effect that different management and natural processes have had and will have on soil organic matter.

TABLE XIX

Estimate of carbon and nitrogen released per annum from the humus of the melfort orthic black chernozem (From Campbell, 1965)

	% of total carbon	Carbon (kg/ha)	Mean rate of decomp. ($r = yr^{-1}$)	Carbon released per yr (kg/ha)	% total nitrogen	Nitrogen (kg/ha)	Estimated nitrogen released/yr. (kg/ha)
Humic acids-hydrolysate	7	8,820	0.04	346	16	1,690	67
Acid extract	14	17,640	0.003	54	20	2,170	7
Humin hydrolysate + fulvic acids III	22	27,720	0.002	60	37	4,050	8
Non-hydrolysable humic acids	33	41,580	0.0007	30	15	1,590	1
Non-hydrolysable humin	24	30,240	0.008	50	12	1,305	1
Total	100	126,000		540	100	10,805	84

NITROGEN AVAILABILITY AND ITS ESTIMATION

We have discussed the amounts of soil organic matter and N in soils, the factors which affect the amounts present in soil, and the rates at which transformations occur. In considering organic matter—soil fertility interactions another aspect which must be considered is concerned with the availability of organic N to the crop. We need to know how availability can be related to crop yield and uptake and how this affects the amount of fertilizer to be applied. Several excellent reviews have been written on this subject (Harmsen and Van Schreven, 1955; Bremner, 1965).

Nitrogen availability

Although there are several thousand kilograms per hectare of N in most soils of temperate climates (Introduction of this chapter) only a very small proportion is made available to plants each year (perhaps 1—2%). However, this small amount does not accrue from the entire soil organic N. As already mentioned the soil organic matter is composed of several components of varying stability. There is some difference of opinion as to how many components (from a stability view point) there are; however, the concensus is that there are at least three, and probably four components. In order of ease of decomposition these are fresh residues, biomass material, microbial metabolites and cell wall constituents adsorbed to colloids, and the old very stable humus. It is primarily the fresh residues and the biomass which give rise to most of the N released to plants.

Available N is defined as N in a chemical form that can be readily absorbed by plant roots; it is understood that this N is within the root zone. The most important forms of available N are NH_4, and NO_3 forms; nitrites and certain simple organic compounds containing free amide or amino groups are a minor source. The simple organic compounds mineralize so rapidly in soil that they are not important. There is some controversy as to whether the NH_4- or NO_3-N is more available to plants. It is known that rice prefers the NH_4 form at all growth stages (Scarsbrook, 1965) and that some plants such as cotton prefer NH_4-N at an early growth stage, but other crops are either believed to have no preference or prefer NO_3-N.

Apart from soil organic matter, there are organic materials which farmers and gardeners apply to soil as fertilizers. Allison (1973, p. 477), gives a list of many of these compounds together with their organic N content and the availability of the N. These materials have been used by gardeners who wish to avoid high salt concentrations. Although the high grade organic fertilizers are often advertised as being slow, gradual release materials, Allison (1973) points out that this is a myth since they release N almost as rapidly as does urea, i.e., much faster than required by young seedlings. On the other hand, the residues of the high grade materials are so slowly available that they release very little N during the first growing season after application. Low grade materials such as sewage sludge, bone meal and tankage are very slow in releasing their N.

Availability indexes

There has long been much interest in methods of testing soils for their ability to supply N to crops. Field and greenhouse experiments give the most accurate values, but vegetative tests are too expensive and time consuming and do not meet the needs (Allison, 1973). A quick chemical test would be excellent if it met the requirements, but most quick tests tend to over-emphasize the N that is immediately available rather than give a good idea of the mineralization potential for the entire growing season. According to Black (1968), "Indexes of N availability obtained in any manner on single samples of soil are empirical in nature and cannot be expected to provide information on the many uncontrolled variations in conditions from one place to another in the field and from time to time at a given location. Some of these variations, but not all, can be taken into account in interpreting the laboratory measurements if suitable supplementary information is available."

A test of availability is a function of the usefulness of the nutrient to the plant, thus laboratory results must be calibrated against actual crop yield data obtained under field conditions. Without such a standardization the test results may serve to separate the soils into a meaningful order but the results alone will not allow us to: (a) predict the amount of available N produced from soil organic matter; and (b) serve as a basis for recommendations for specific amounts of N fertilizer. Usually correlations of available N are better with nutrient uptake than with yield. This is because in some soils there are other factors besides nutrient supply which limit the yield and response to added nutrients but which have less effect on nutrient uptake.

Although available P and K tests are standard procedures in most soil testing laboratories, routine chemical or biological tests for soil N are seldom employed. This is partly because except for fairly recent work by Stanford and his coworkers (see later discussion) most methods are empirical and not founded on rational principles. The nature of the readily available N in soils has not been completely established although progress has been made.

In soil test models which have been used to predict the quantity of N available to a crop in a growing season the implied relationships described by the models are that N in fertilizer = N required minus N in soil, where N in the soil is derived by one of the following models:

$$N_{soil} = K \times N_{mineralized} + N_{initial \ (mineral)} \tag{27}$$

$$\text{or } N_{soil} = K \times N_{mineralized} \tag{28}$$

$$\text{or } N_{soil} = K \times N_{initial \ (mineral)} \tag{29}$$

The appropriate model is a function of the generally prevailing environmental factors. For example eq. 29 applies to the Canadian prairies where initial mineral N and mineralizable N are correlated (Nyborg et al., 1976); eq. 28 is used in areas where leaching affects the effectiveness of residual NO_3.

Various tests and indexes

Many of the various tests developed up to 1965 were reviewed and discussed of Bremner (1965), Keeney and Bremner (1966a). Thus we shall only summarize these briefly here and later elaborate on some of the more promising new indexes.

(a) *Aerobic incubation*. This test is perhaps the most widely used; it was perfected by workers at Iowa Experiment Station. In most procedures soil is leached free of soluble N, then incubated under constant temperature for 2 to 6 weeks. The relative amount of NO_3—N produced is used as a measure of the N-supplying power of the soil. It has the advantage of providing automatically an integration of the effects of amount and composition of substrate that affect the natural process in the field. The mineral N present at the time of sampling may or may not be taken into account.

(b) *Anaerobic incubation*. This is Waring and Bremner's (1964) method. It involves incubating soil at 40°C for 7 days and measuring the NH_4—N increase. This is the best N index for rice (Sims et al., 1967).

(c) *Plants*. Utilization of higher plants as extracting agents in field plots is the most accurate means of determining available N in soil. This method is used mostly as a standard of measure for other methods more adapted for routine determinations.

These three methods together with a few others such as microbial assays are commonly referred to as biological indexes. The following 8 methods which are by no means all the remainder are some of the commonly used chemical methods. Generally, chemical methods give disappointing results (Black, 1968).

(d) *Total N or C*. This is probably the second most used index. There is a general relationship between total soil N and the amount of available N released during a single growing season. However, there is some controversy regarding the usefulness of this method because of the many factors that affect the release of available N from organic matter (Vlassak, 1970; Allison, 1973). Usually organic C is measured instead of total N because the two are related and C is easier to measure.

(e) *Truog's (1954) method* for available nitrogen involves the distillation of NH_3 from soil for 5 min in the presence of $KMnO_4$ and Na_2CO_3.

(f) *Boiling water method*. This is the best of the chemical methods and is perhaps as good as the incubation methods. It was developed by Livens (1959) and later modified by Keeney and Bremner (1966b). It involves the determination of the total N in boiling water containing 10% K_2SO_4.

(g) The latter method was further modified by Stanford (1968a) who used

0.01 M CaCl$_2$ instead of water and 0.5 N K$_2$SO$_4$ instead of 10% K$_2$SO$_4$. Advantages of adding CaCl$_2$ are that the pH tends to remain constant during incubation and the soil does not become dispersed.

(h) *Reducing sugars extracted by Ba(OH)$_2$.* This is Jenkinson's (1968) method. Soil and Ba(OH)$_2$ are shaken for 30 min, the suspension filtered and the reducing sugars in the filtrate estimated colorimetrically with anthrone reagent.

(i) *Cornfields method (1960).* This is an alkaline hydrolysis method.

(j) *Stanford and DeMar's (1970) method.* Soil is autoclaved for 16 h at 121°C and the hydrolysed NH$_4$ measured by the Conway micro diffusion technique.

(k) *Nitrate method.* This is the method used on the Canadian prairies. It involves extraction of NO$_3$ by shaking soil with K$_2$SO$_4$ or NaHCO$_3$ and measuring the extracted NO$_3$ colorimetrically.

Use of some indexes

In the prairie provinces of Canada cereals are grown either on summer-fallowed land, in which case there is usually sufficient NO$_3$–N to produce the entire crop, or it is grown on stubble land and here fertilizer N is usually added. The rationale behind the "NO$_3$-test" can be gleaned from Fig. 30, which gives a sketch of mineralization of soil N throughout a year for a typical soil cropped to barley in north-central Alberta. It also indicates times of soil sampling, the usual time of fertilization with N, and length of time of crop uptake of N. In western Canada losses of mineral-N from soils by leaching and denitrification are usually small. Thus the mineral N found in the soil will be mostly available to the crop. The amount of mineral N released from the soil during the nine months when the soil has no crop on it

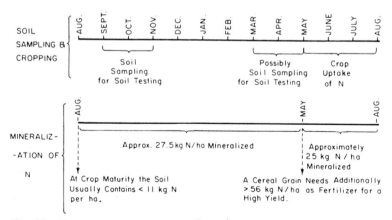

Fig. 30. The mineral N in a typical non-fallowed soil in north-central Alberta during a year in which it is cropped to barley. (Adapted from Nyborg et al., 1976.)

(from about August 15 to May 15) may equal the amount of mineral N released from the soil during the three months when the soil is growing a crop (Nyborg et al., 1976). Thus a crop grown on stubble land will obtain about half of its N from mineral N present in the soil prior to seeding. For a crop grown on a soil low in mineral-N an addition of fertilizer N is needed for a high yield. The "NO$_3$-test", under western Canadian conditions, measures half or more of the soil N that will be available to the crop.

There are several factors which can affect the usefulness of the "NO$_3$-test". It is assumed that most of the mineral-N in soils occurs as NO$_3$—N as opposed to NH$_4$—N, which is true for most soils under most conditions. The depth of soil sampling may have to go to 90—120 cm on some soils to obtain most of the NO$_3$—N in the profile, while on other soils depths of 15 cm or 30 cm may be sufficient. Although soils are normally sampled in the fall, sampling at time of seeding in the spring may give a more accurate measure of crop-available N. This test is suitable for annual cropping as practised in western Canada; however, with continuous cropping (e.g., forage crop) it is of little use because there is little mineral-N in the soil at any time of year. If the soil is broken out of forage, and especially a legume, the rate of mineralization of soil N will be increased greatly compared to mineralization experienced under the usual annual cropping.

Results from Manitoba (Soper et al., 1971) showed a very close correlation ($r^2 = 0.84$) between barley uptake of N and the amount of NO$_3$—N to a depth of 61 cm in soil sampled at time of seeding.

Where rainfall is high one or several of the other indexes of N availability has been used singly or in combination with the initial mineral N in multiple regression type equations. In Kentucky, U.S.A., Ryan et al. (1971) tested several chemical and biological indexes as to their ability to predict N uptake and yield of sorghum. They found that initial mineral N was a poor index; aerobic and anaerobic incubation were the best and their correlation with N uptake was improved when the soils were amended with P, K and Ca. Geist et al. (1970) developed multiple regression models which accounted for fertilizer N added plus mineral N initially present in soil, and organic N mineralized during the growing season as estimated by one of five availability indexes. Although all the indexes gave good estimates of available N, the best index was one which used NaOH + Devarda distillation. Geist et al. (1970) did not test the Keeney and Bremner boiling water index. Since this appears to be one of the most promising indexes it is reasonable to assume that they might have obtained even higher correlations than the 0.49 to 0.73 achieved.

In Britain, Gasser and Kalembasa (1976) compared several indexes (Table XX) and found the best to be aerobic and anaerobic incubation and the boiling water methods; these methods were also the most highly predictive of N uptake in the greenhouse. Lathwell et al. (1972) obtained similar results in testing soils in Puerto Rico. In other parts of the West Indies, Cornforth and Walmsley (1971) carried out an extensive testing of 155 soils

TABLE XX

Values of the correlation coefficients, r[1] between measurements of available-N in soils by various methods, between these measurements and the total-N or organic-C of soils, and between these soil properties (From Gasser and Kalembasa, 1976)

| | Method | | | | | Total-N | Organic-C | |
	A aerobic incubation	B anaerobic incubation	C N by boiling water	D sugars by Ba(OH)$_2$	E sugars by K$_2$SO$_4$		Walkley-Black	Tinsley
Aerobic incubation	X	0.98	0.98	0.95	0.92	0.91	0.92	0.91
Anaerobic incubation	0.98	X	0.98	0.94	0.91	0.92	0.93	0.92
N by boiling water	0.98	0.98	X	0.96	0.92	0.92	0.95	0.94
Sugars by Ba(OH)$_2$	0.95	0.94	0.96	X	0.93	0.95	0.96	0.94
Sugars by K$_2$SO$_4$	0.92	0.91	0.92	0.93	X	0.79	0.83	0.81
Total-N	0.91	0.92	0.92	0.95	0.79	X	0.99	0.99
Organic-C Walkley-Black	0.92	0.93	0.95	0.96	0.83	0.99	X	0.99
Organic-C Tinsley	0.91	0.92	0.94	0.94	0.81	0.99	0.99	X

[1] All values of r significant at $P < 0.01$.

from several islands. Of the seven indexes tried (boiling water test was not used) total N and Cornfield's alkaline hydrolysable N were the best; but the correlations though significant were not high. The correlations were best on acid soils and soils with large exchange capacities. In East Africa, Robinson (1968a—d) tested several indexes and found that Bremner's (1965) aerobic incubation method appeared initially to be the best. However, use of the latter proved unsatisfactory because: (a) it required a long preincubation period, and (b) he obtained only fair correlation between this index and maize yields in the field. He was later able to show that the Keeney and Bremner (1966b) boiling water method which is much faster would work just as well as the aerobic incubation method. In Quebec, Canada, Kadirgamathaiyah and MacKenzie (1970) found that the boiling water and incubation methods were much better indexes of estimating soil available N or predicting yield response of sudangrass than was an acid hydrolysis method.

The most outstanding ongoing research in this field is that being carried on by Stanford and associates (Stanford, 1968a, b, 1969; Stanford and Legg, 1968; Stanford and De Mar, 1969, 1970; Stanford et al., 1970; Stanford and Smith, 1972, 1976; Stanford et al., 1973a, b, 1974; and others). They have succeeded in elucidating the nature of organic N in soil, the rate at which fertilizer N becomes incorporated into soil organic matter, and the rate of release.

These workers first defined the N mineralization potential (N_0) of a soil to be the quantity of soil organic N that is susceptible to mineralization according to first-order kinetics (Stanford and Smith, 1972). They presented a method for obtaining N_0 based on the equation:

$$\log(N_0 - N_t) = \log N_0 - kt/2.303 \qquad\qquad (30)$$

in which N_t denotes the cumulative amount of N mineralized during a specified period of incubation (t), and k is the rate constant. After several consecutive incubations (using Stanford's (1968a) method) at 35°C over a period of 30 weeks with intermittent leachings and determinations of N mineralized, N_0 was estimated from the regression of log ($N_0 - N_t$) on t for successive approximations of N_0. This regression method is cumbersome and can be avoided (Campbell et al., 1974) by using the least squares method of Hartley as adapted for computer use by Shih (1968). The foregoing study (Stanford and Smith, 1972) involved 39 diverse American soils. They found that values of N_0 ranged from 20 to >300 ppm N. The fraction of the total N comprising N_0 varied between 5 and 40% among soils, but the mineralization rate constant (k) did not differ significantly among most soils. The most reliable estimate of k = 0.054 ± 0.009 per week. Later, under greenhouse conditions, evidence was obtained that N_0 has intrinsic value in predicting the amounts of soil N mineralized under specified environmental conditions

(Stanford et al., 1973b). For example, they found that amounts of soil organic N mineralized during cropping plus the mineral N present initially in the soils correlated highly with amounts of soil N taken up by whole plants (Fig. 31). Moreover, A-values were similar to amounts of N mineralized before and during crop growth (r = 0.94**). This result is particularly significant, since amounts of N mineralized during crop growth were estimated from N mineralization potentials, taking into account the effects of temperature on the mineralization rate constant. Thus, the study provided preliminary evidence that the soil N mineralization potential offers a basis for reliably estimating amounts of soil N mineralized during selected periods of time under specified temperature regimes.

Stanford and coworkers had already demonstrated the possible significance of N_0 as a basis for predicting amounts of N mineralized under fluctuating temperatures and soil water contents (Stanford and Epstein, 1974; Stanford et al., 1973b). The basic concept is that the amount of N mineralized is proportional to N_0 (eq. 30), i.e., $-dN/dt = kN$. As shown earlier (p. 220) the rate constant, k, is influenced markedly by temperature and soil water content. The temperature coefficient, Q_{10}, is 2 (Stanford et al., 1973a). Relative N mineralization is a linear function of soil water content, expressed as percent of the optimum for biological activity (Stanford and Epstein, 1974). They used the following example to illustrate how N_0 may be used to estimate N mineralization while taking into account the effects of temperature and soil water content. For a given weekly period, it was assumed that antecedent potentially mineralizable N = 200 kg/ha; average temperature = 25°C; average soil water content = 75% of field capacity (field capacity is considered to be optimum for N mineralization).

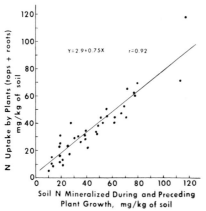

Fig. 31. N uptake by sudangrass (tops and roots) from zero N pots in relation to amounts of mineral N initially present in soils plus soil N mineralized during growth of the crop. (From Stanford et al., 1973.)

At 25°C, k = 0.027 weeks^{-1} (Stanford and Smith, 1972), and the amount of
N mineralized during 1 week, kN, at optimum soil water content is 200 ·
0.027, or 5.4 kg/ha. At 75% of field capacity, however, the amount mineralized is 0.75 kN, or 4.1 kg/ha. Although the validity of the foregoing views
has not yet been verified under field conditions, limited evidence from greenhouse studies involving fluctuating temperature and near-optimum soil water
tends to support the concept (Stanford et al., 1973b).

Carter et al. (1975, 1976) have recently demonstrated the importance of
measuring both residual mineral and mineralizable N in assessing N fertilizer
needs of sugarbeet. The relative contributions from these two sources of soil-
derived N vary widely among soils and crop management systems. Yearly
carry-over of mineral N is a function of water balance. In contrast, mineralization potential, N_0, is a definable and measurable quantity that reflects past
cropping history as well as inherent soil characteristics (Stanford and Smith,
1972). Under continuous corn, for example, with optimal fertilization, N_0
may not change perceptibly from one year to the next. Radical changes in
cultural practices, such as turning under established grass sod, however, can
markedly lower N_0 and N_t within a relatively short period of 1—2 years
Bennett et al. (1975).

One of the main impediments to Stanford's method has been the fact that
the determination was lengthy and laborious. Recently Stanford and Smith
(1976) were able to relate their boiling water ($CaCl_2$) method to the much
more rapid autoclaving method (Stanford and DeMar, 1970). The only soils
of the 54 soil types tested which were exceptions were some western U.S.A.
calcareous soils. It would appear then that if Stanford and associates can
demonstrate that this index works successfully in the field this will be a
momentous breakthrough in the study and practice of soil fertility.

FERTILITY RELATED FACTORS

The factors which will be discussed in this section might not be regarded
by some as qualifying under the definition of fertility (Introduction). However, they do modify the status of the soil such that the amounts and availability of nutrients are greatly affected. As such we felt that they deserved
some coverage in this chapter.

Soil aggregation and structure

Soil organic matter is responsible directly or indirectly, for making the
physical environment of soils suitable for the growth of crops. It exerts this
benefit largely through its effect on soil aggregation which in turn influences
soil incrustation, water infiltration, moisture content, drainage, tilth, aeration, temperature, microbial activities, and root penetration (Allison, 1973).

Soil structure is a function of aggregate size and distribution; it is of great importance in determining the fertility of soils. The best structure is one which has a fine-crumb macrostructure with aggregates 1—3 mm in size (Kononova, 1966). This type is found under perennial grass—legume vegetation. Root residues are superior to top residues in providing good soil aggregation (Allison, 1973) because roots are well distributed through soil thus the gums and polysaccharides arising from their decomposition are also well distributed and in a position to act as cement between soil particles as they are being formed into aggregates by the pressure of the roots. This type of fixation tends to protect the gums from decomposition by microbes. Top residues are also sources of gums but they are not intimately mixed with soil, thus the gums produced are rapidly decomposed and do not move far from the soil surface. Allison (1973) points out, and rightly so, that the emphasis commonly placed on the need to return residues to the soil to maintain good tilth and aggregation is surprising "in view of the often demonstrated fact that most of the aggregates so formed are non-stable and break down rapidly". Part of the reason for the transient effect of top residues on aggregation is the fact that the aggregation is promoted via fungi mycelia. Mould fungi appear in abundance on fresh residue and their mycelia enmesh soil particles and prevent their dispersion by water (Kononova, 1966). But, as the fungal mycelia decomposes, the aggregates break down. Soil aggregates formed in the gut of animals, especially earthworms are very stable.

Most Australian wheat soils have a high percentage of sand and silt. In such soils good structure is dependent on the organic materials added to the soil by the pasture in the rotation (Greenland, 1971). There is much more weight of roots in the top 15 cm of soil under pasture than under cereal crops. Furthermore, grasses have numerous fine roots and these are more effective in aggregate formation than fewer large roots. In a 4-year rotation of corn—oats—alfalfa—alfalfa, Rennie et al. (1954) found that soil structure began to deteriorate under corn, continued till the following crop of oats had attained considerable growth and then improved, reaching a maximum in the second year of alfalfa hay. Bolton and Webber (1952) working with a dark grey gleisolic clay soil in southern Ontario found that with respect to their effectiveness in promoting aggregation of the top 10 cm of the soil, bluegrass sod was > alfalfa-brome > oats > continuous corn.

Increased stability of aggregates usually accompanies increase in organic matter; this is primarily due to the polysaccharide content of the organic matter. Rennie et al. (1954) found that bacterial polysaccharides increased soil aggregates of two silt loams by as much as 50%; also that soil gums increased aggregation though not as much as bacterial gums. Martin and Richards (1969) found that some polysaccharides increased aggregation and hydraulic conductivity, others only increased aggregation, and still others increased conductivity but not aggregation.

Anderson and Peterson (1973) report that manure treatments had 7 times greater aggregate stability than non-manured soil. This resulted in an infiltration rate of 6.8 cm/h in manured plots compared to 0.5 cm/h in control plots. However, Kononova (1966) points out that because of the sparse manner in which manure is usually applied in the U.S.S.R. it has a weak effect in restoring soil structure.

In general it can be concluded that organic matter will improve the soil's crumb structure; it is important in its effect on the size, strength and mellowness of the crumb, and is therefore important to the farmer. It will influence seedbed preparation and it gives the farmer some degree of independence from the weather with regards to cultivation.

Colloidal properties

It is generally agreed that the colloidal complex of a soil plays a major role in supplying nutrients to plants grown thereon. Since humus is colloidal in nature its role in influencing soil fertility is considerable. As a colloid, humus is involved in three major functions: exchange capacity, buffer capacity and chelation of metals. We shall now examine these functions briefly.

Cation exchange capacity (CEC)

There is considerable uncertainty as to the explanation of the development of the negative charge on organic colloids. The charges may arise from the dissociation of the proton from carboxyl, phenolic-hydroxyl, enolic-hydroxyl, imide, and other groups (Russell, 1966; Allison, 1973). Russell (1966) points out that if the phenolic-hydroxyl groups are in the ortho position on a benzene ring the H will dissociate at the same pH as for carboxyl groups.

The organic matter of most mineral soils accounts for about 30—65% of the total CEC (Mitchell, 1932; Allison, 1973); in sandy and organic soils more than 50% of the CEC is likely due to the organic component of the soil. The organic matter of different soils vary greatly in their CEC; the more humified the organic matter the higher is its CEC (Allison, 1973). Allison (1973) reports values for the CEC of humus as 150—300 me./100 g. In comparison Grim (1953) gives values for kaolinite, illite, vermiculite and montmorillonite clays as 3—15, 10—40, 100—150 and 80—150 me./100 g, respectively.

Various factors affect the value obtained for CEC including pH, method of separating the contribution of clays from that of humus, the addition of manures, and so on. The CEC generally increases with increasing pH of the buffered saturated solution during its determination (Allison, 1973). McIntosh and Varney (1973) found that the CEC of a clay soil was increased by manure treatments which also increased the soil organic matter.

One method of determining the CEC of the organic component of a soil is by difference after determining the total CEC of the soil, destroying the organic matter by oxidation and redetermining the contribution of the mineral fraction only. Schnitzer (1965) used a different approach. He extracted and purified the organic matter of the Bh horizon of a podzol soil, determined its CEC directly and as expected found it to be much higher than when the determination was done by difference. A third approach used in segregating the contribution of the CEC due to organic matter from that due to the mineral colloids in mineral soils is multiple regression (Yuan et al., 1967; St. Arnaud and Sephton, 1972; Martel and Lavadière, 1976).

St. Arnaud and Sephton (1972) used the ammonium oxalate method together with regression analysis to develop predictive equations for the CEC of several chernozemic sub-group profiles occurring in the brown, dark brown and black soil zones of Saskatchewan. They obtained good predictive equations (Table XXI). The CEC of the organic matter in the A horizons increased from the Rego to the Orthic to the eluviated profile; that is, from the driest to the most moist position. In this experiment the average contribution of the organic matter was 250 me./100 g and for the clay 57. These average values can therefore be used as general values for chernozemic soils in western Canada while the equations can be used for more specific soil types. Martel and Lavadière (1976) carried out a similar study on podzolic and gleysolic Ap horizons in Quebec, Canada. They obtained average values of 161 ± 45 me./100 g for the organic matter and 29 ± 6 for the clay component. They suggested that these values were lower than St. Arnaud's because Quebec soils are less developed than are the chernozems of western Canada. In the work by Yuan et al. (1967) in Florida, U.S.A., 83 virgin surface soils which were mostly fine sands classified as inceptisols, spodosols, entisols, and mollisols were examined. They found that the CEC contributed by the organic matter was 138, 150, 187 and 197 me./100 g for inceptisols, spodosols, entisols and mollisols, respectively. The corresponding clay contributions were 90, 21, 58 and 128 me./100 g, respectively. The effect of humus in increasing CEC of soils is more important in clays than in sandy soils because in the latter it is difficult to maintain high humus under normal cropping practices.

Buffer capacity

A buffer solution is one which resists changes in pH on the addition of acid or base. Soils behave like buffers; most of this ability is due to their colloidal properties which are located in the humus and clays which hold hydrogen, aluminum, and other cations. Thus the resistance to pH change is low in sandy soils and high in fine textured and organic soils which have high CEC.

"The buffering capacity of soils is of great practical significance. If this did not exist it would be difficult to conceive of agriculture as we know it today. We would have a situation similar to that encountered in hydroponics.

TABLE XXI

Prediction equations for the CEC of five chernozemic soil horizons in terms of clay and organic matter contents (Adapted from St. Arnaud and Sephton, 1972)

Subgroup profiles	No. of samples	Without interaction	R^2 (%)	With interaction	R^2 (%)
Rego (A horizon)	21	$y = 2.14x_1 + 0.56x_2$	89	$y = 1.85x_1 + 0.60x_2 - 0.02x_3$	93
Orthic (A horizon)	111	$y = 2.23x_1 + 0.57x_2$	79	$y = 2.30x_1 + 0.57x_2 - 0.01x_3$	79
Orthic (B horizon)	99	$y = 2.86x_1 + 0.57x_2$	81	$y = 2.79x_1 + 0.59x_2 - 0.02x_3$	85
Eluviated (A horizon)	68	$y = 2.38x_1 + 0.55x_2$	54	$y = 2.56x_1 + 0.52x_2 - 0.02x_3$	57
Eluviated (B horizon)	45	$y = 2.97x_1 + 0.59x_2$	80	$y = 2.98x_1 + 0.59x_2 - 0.002x_3$	82

y = CEC, me/100 g; x_1 = % organic matter; x_2 = % clay; x_3 = interaction between organic matter and clay.

Plants could be grown but at great expense and attention to detail" (Allison, 1973). Acids and bases are continually being produced in field soils; without buffering capacity large pH and osmotic fluctuations would occur and crops and organisms would be injured because most of them grow well only within narrow pH and osmotic ranges.

Humus as a chelate

Much emphasis is presently being placed on the use of chelates in supplying plants with micro-nutrients. Humic colloids are believed to play an important role in controlling the availability of these metals through chelation. The mechanisms by which the chelates enter plants are not well understood. The metals can be held on the exchange complex of colloids but sometimes the bonds are too strong thus the metals are difficult to replace and the question arises as to whether they are held by very strong ionic bonds or by chelation.

In simple terms a metal chelate is defined as a complex formed when a metal ion is bound to an organic component by more than one bond to form a ring or cyclic structure. It is generally agreed that organic matter can hold metallic ions both by cation exchange and by chelation. Metals are held by chelates with varying degrees of stability; the stability is usually high but is a function of the number of atoms that form a ring with the cation, the number of participating rings, the nature of the ligands and the presence or absence of certain substituents in the chelate (Allison, 1973). Chelates are also water-soluble.

Duff et al. (1963) has demonstrated that some bacteria can produce 2-ketogluconic acid which chelates some metals. They also point out that micro-organisms play an important role in producing other acids (e.g., acetic, citric, lactic, formic, etc.) which act as chelates. Wright and Schnitzer (1963) have demonstrated quite clearly that the formation of a podzolic profile is largely dependent on chelation reactions. They suggest that under natural conditions fulvic acid is the chief ligand causing the movement of iron and aluminum in podzol development. Thus there is a strong possibility that soil organic matter is involved in chelation reactions with metals and that they play a very important role in making metal ions available to plants.

One other point should be mentioned here. It has been established that NH_3 (or NH_4) reacts with organic matter in such a way that the NH_3 becomes fixed or at least not easily recoverable (Coffee and Bartholomew, 1964; Broadbent and Thenabadu, 1967). The mechanism of this fixation reaction is not known but Broadbent and Thenabadu (1967) suggest that the NH_3 is fixed to the organic matter by chemical linkage. This phenomenon is of practical significance because of the extensive and even increasing use of ammoniacal fertilizer.

Acidity

The rate of mineralization is low but does occur in acid soils. As soils approach neutrality due to the addition of some alkaline amendment mineralization increases, particularly for a few months (Swaby, 1966; Black, 1968; Ishaque and Cornfield, 1972). This has mostly been demonstrated in laboratory experiments. Black (1968) states that the transitory nature of the effect of raising the pH on mineralization seems related to an increase in susceptibility of a fraction of the organic N to mineralization. After this portion is mineralized (perhaps 5 years) the pH effect diminishes. Transitory though it might be the effect is no doubt of great economic importance while it lasts because of the increased rate of nutrient release.

Moisture relationships

Organic matter may improve infiltration, decrease evaporation, improve drainage in fine textured soils, foster more extensive and deeper root systems which may make more moisture available to crops, and improve the efficiency of water use by the crop (Allison, 1973). These effects will depend on the crop, the soil type and the climate. If fertilizers are added the effect of organic matter might even be increased.

Organic matter increases infiltration by helping to hold water on the soil surface long enough for it to enter the soil and by improving the soil physical condition and thus reducing the time required for water to permeate the soil. In western Canada stubble mulch helps to trap snow, reduces icing of the soil surface during winter, and allows better infiltration of the spring melt. Organic mulches also protect the soil against crust formation and thus increase the rate of infiltration. Organic matter if mixed with the soil increases infiltration by improving aggregation and soil structure. However, this is only true under aerobic conditions; under anaerobic conditions gleization can occur and this forms a seal which prevents infiltration. This principle is currently used in western Canada and in the U.S.S.R. to seal dugouts against leakage of water (W. Nicholaichuk, Can. Dep. Agric. Res. Stn., Swift Current, Sask., personal communication, 1976). Some forms of organic matter actually decrease infiltration in soils. Such soils are non-wettable because they contain hydrophobic organic substances (De Bano, 1969; Fink, 1970). It has been suggested (Russell, 1966) that these substances are really large amounts of fungal cells, mainly Basidiomycetes. Even sandy soils may exhibit this feature if these substances are present.

Available water supply

Except for sandy soils organic matter does not increase available water. This is because the organic matter holds the water so tightly that it is not available to plants. This is especially true in peats. Even in sandy soils the effect on available water is small. Organic matter improves the structure of the soil; in this way the effective soil volume ramified by the roots may increase considerably and so the plants can secure more water than they would normally have obtained.

Drainage
To improve drainage, especially in low lying areas it is best to try to establish grasses and deep-rooted legumes since these will improve the aggregation and structure. Bolton and Webber (1952) report that increased aggregation may lower bulk density, increase percolation and improve drainage.

Erosion

There are few data that show accurately the magnitude of the total N lost by erosion. Furthermore, erosion losses are quite variable because the soils and topography are themselves variable.

In east Africa erosion control and the maintenance of permeability are closely linked and are dependent on the ability to prevent rain drops from actually hitting the soil surface (Russell, 1966). This problem is especially critical in Africa because a considerable proportion of the soils are of a fine sandy texture and they tend to seal very rapidly when subjected to heavy rainfall. When practical the use of a mulch is an ideal solution. Soil N is especially subject to water erosion because moving water loosens and floats away the surface soil where the organic matter is concentrated. Often the N in the eroded material is 3—5 times greater than that present in the soil proper (Allison, 1973).

Wind removes the lightest materials first and like water the eroded material is usually richer in nutrients than the soil left behind. Under western Canadian prairie conditions freeze-drying of soils can occur during cold periods when moist soils are bare of snow cover (Anderson and Bisal, 1969). This results in break down of clod structure and these soils become highly susceptible to wind erosion (Bisal and Nielsen, 1964). Trash cover can be used to trap snow and thus protect the soil against this effect (Anderson, 1968).

CONCLUSIONS

In this chapter we have discussed some of the numerous attributes of soil organic matter, the factors that affect its turnover, the quantification of its dynamics, and its potential for supplying N to plants. Steady progress has, and is being made in turnover studies; to this end the work of McGill, Paul and coworkers in western Canada and Jansson and Jenkinson in Europe, have been outstanding. In the quantification of the dynamics of soil organic matter, the recent work of Russell in Australia has been exemplary and will no doubt lead to renewed effort in the re-examination of long term crop sequence data which should result in more accurate estimates of the effect of management on organic matter losses and equilibrium levels. In the area of estimating availability of mineralizable N to plants, Stanford and his collaborators appear to be on the verge of a significant break-through in providing a precise, rapid, chemical N-availability index. Their method seems to be less empirical and is biologically more meaningful than any other chemical method; it might therefore prove to be versatile enough to provide satisfactory results under varied environmental conditions. The work of Laura, Broadbent, Sauerbeck, Jenkinson, and Sørenson, and others on "priming effect" and "salt (fertilizer) effect" have provided some stimulating theories. New and exciting approaches and ideas which are constantly being advanced by many imaginative scientists have opened up new avenues which must stimulate further efforts to move forward in search of solutions to other complex, unsolved problems.

Notwithstanding the progress that has been made, more accurate information on the contribution of added organic matter to the fertility of soils under various cultural and climatic conditions is required. We need to learn more about the long term effects of burning, and the effects of various kinds and amounts of manures and fertilizers on changes in the soil organic matter components. It is essential that more work be done on what the possible short and long term consequences of these practices will be on the productive capability of our soils, as well as on the possibilities of them causing pollution of streams and ground water, or increasing salinity or erosion.

There is a need for more tracer studies similar to those carried out by Jansson (1958), McGill et al. (1974), and Stanford (1960's to 1970's); these should be implemented on other soils and under other climatic conditions. Although tracers have not provided outstanding results from a practical agronomic standpoint, they have contributed significantly to the elucidation of the complex dynamics of soil organic matter and have provided some basic answers which would not have been possible with conventional techniques. As seen from the discussions in this chapter, tracers have assisted us greatly in understanding the basic nature and behavior of soil organic matter. It is imperative that we obtain such basic information if we are to make adequate forecasts and provide proper answers on the numerous soil, agronomic and pollution problems which are often associated with the many properties of soil organic matter.

Evidence as to the existence of several fractions of soil organic matter of varying stabilities is overwhelming; however, there is an urgent need for more investigations into the partitioning and characterization of these components. Any break-through in this area will greatly facilitate calculation of dynamics and thus provide a better basis for predictions of turnover rates of nutrients and equilibrium levels of organic matter.

Most of all we need to take the theory to the field; i.e., we need more testing of our dynamic models and theories. This is quite feasible in the present age of sophisticated high-speed computers (e.g., Russell's work). Laboratory derived coefficients, obtained under constant environmental conditions, on disturbed top soil (the common practice) provide precise results; but are they really applicable? In practice they usually fall short and at best they usually only provide us with trends. Thus the time has come for us to design more realistic multifactored experiments. Albeit, these experiments will likely be more complex, more time consuming and even more expensive; but we can't expect to obtain something for nothing! In such experiments several of the more important organic constituents should be labelled at the same time (e.g., ^{14}C, ^{35}S, ^{33}P, ^{15}N) and their concomitant transformations examined. Such studies will probably be more successful if research teams can be set up to include experts in soil microbiology, soil biochemistry, soil physics, humus chemistry, enzymology and soil fertility, all working together with well laid out and coordinated studies directed towards the achievement of predetermined objectives. The need for this type of multidisciplinary approach is obvious from the nature and complexity of organic matter as discussed throughout this chapter.

REFERENCES

Agarwal, A.S., Singh, B.R. and Kanehiro, Y., 1971a. Soil Sci. Soc. Am. Proc., 35: 96—100.

Agarwal, A.S., Singh, B.R. and Kanehiro, Y., 1971b. Soil Sci. Soc. Am. Proc., 35: 455—457.

Agarwal, A.S., Singh, B.R. and Kanehiro, Y., 1972. Plant Soil, 36: 529—537.

Ahmad, Z., Kai, H. and Harada, T., 1969. Soil Sci. Plant Nutr., 15: 252—258.

Ahmad, Z., Kai, H. and Harada, T., 1972. J. Fac. Agric. Kyushu Univ., 17: 49—65.

Aleksic, Z., Broeshart, H. and Middleboe, V., 1968. Plant Soil, 29: 474—478.

Alexander, M., 1961. Introduction to Soil Microbiology. Wiley, New York, N.Y., 472 pp.

Alexander, M., 1965. In: W.V. Bartholomew and F.E. Clark (Editors), Soil Nitrogen. Agronomy, 10, pp. 309—343.

Allison, F.E., 1966. Adv. Agron., 18: 219—258.

Allison, F.E., 1973. Soil Organic Matter and its Role in Crop Production. Developments in Soil Science, 3. Elsevier, Amsterdam, 637 pp.

Anderson, C.H., 1968. Can. J. Plant Sci., 48: 287—291.

Anderson, C.H., 1971. Can. J. Soil Sci., 51: 397—404.

Anderson, C.H. and Bisal, F., 1969. Can. J. Soil Sci., 49: 287—296.

Anderson, F.N. and Peterson, G.A., 1973. Agron. J., 65: 697—700.

Atkinson, H.J. and Wright, L.E., 1948. Sci. Agric., 28: 30—33.

Baeumer, K., 1970. Neth. J. Agric. Sci., 18: 283—292.

Bakermans, W.A.P. and De Wit, C.T., 1970. Neth. J. Agric. Sci., 18: 225—246.

Baldanzi, G., 1960. Int. Congr. Soil Sci., 7th, Madison, Wisc., 1960, Comm. III, pp. 523—530.

Bartholomew, W.V., 1965. In: W.V. Bartholomew and F.E. Clark (Editors), Soil Nitrogen. Agronomy, 10, pp. 287—302.

Bartholomew, W.V. and Kirkham, D., 1960. Int. Congr. Soil Sci., 7th, Madison, Wisc., 1960, Comm. III, pp. 471—477.

Bennett, O.L., Stanford, G., Matthias, E.L. and Lundberg, P.E., 1975. J. Environ. Qual., 4: 107—110.

Biederbeck, V.O. and Campbell, C.A., 1971. Soil Sci. Soc. Am. Proc., 35: 474—479.

Biederbeck, V.O. and Campbell, C.A., 1973. Can. J. Soil Sci., 53: 363—376.

Biederbeck, V.O. and Paul, E.A., 1973. Soil Sci., 115: 357—366.

Birch, H.F., 1960. Plant Soil, 12: 81—96.

Birch, H.F. and Friend, M.T., 1956. E. Afr. Agric. For. Res. Organ., Rep., 18.

Birch, H.F. and Friend, M.T., 1956. Nature, 178: 500—501.

Bisal, F. and Nielsen, K.F., 1964. Soil Sci., 98: 345—346.

Bishop, R.F., MacLeod, L.B. and Jackson, L.P., 1962. Can. J. Soil Sci., 42: 49—60.

Black, C.A., 1968. Soil—Plant Relationships (2nd ed.) Wiley New York, N.Y., 792 pp.

Bolton, E.F. and Webber, L.R., 1952. Sci. Agric., 32: 555—558.

Bremner, J.M., 1965. In: W.V. Bartholomew and F.E. Clark (Editors), Soil Nitrogen. Agronomy, 10, pp. 93—149.

Broadbent, F.E., 1970. Soil Sci., 110: 19—23.

Broadbent, F.E. and Nakashima, T., 1967. Soil Sci. Soc. Am. Proc., 31: 648—652.

Broadbent, F.E. and Nakashima, T., 1971. Soil Sci. Soc. Am. Proc. 35: 457—460.

Broadbent, F.E. and Tyler, K.B., 1962. Soil Sci. Soc. Am. Proc., 26: 459—462.

Broadbent, F.E.and Thenabadu, M.W., 1967. Soil Sci., 104: 283—288.

Broadbent, F.E., Jackman, R.H. and McNicoll, J., 1964. Soil Sci., 98: 118—128.

Broecker, W.S. and Olson, E.A., 1960. Science, 132: 712—718.

Brown, D.A. and Ferguson, W.S., 1956. Can. Dept. Agric. Res. Sta., Brandon, Man., Ann. Rep.

Brown, P.L. and Dickey, D.D., 1970. Soil Sci. Soc. Amer. Proc., 34: 118—121.

Calder, E.A., 1957. J. Soil Sci., 8: 60—72.

Caldwell, A.C., Wyatt, F.A. and Newton, J.D., 1939. Sci. Agric., 19: 258—270.

Cameron, D.R., Nyborg, M., Toogood, J.A. and Laverty, D.H., 1971. Can. J. Soil Sci., 51: 165—175.

Campbell, C.A., 1965. Ph.D. Thesis, Univ. of Sask., Saskatoon, Sask.

Campbell, C.A. and Biederbeck, V.O., 1972. Can. J. Soil Sci., 52: 323—336.

Campbell, C.A., Biederbeck, V.O. and Hinman, W.C., 1975. Can. J. Soil Sci., 55: 213—223.

Campbell, C.A., Biederbeck, V.O. and Warder, F.G., 1970. Can. J. Soil Sci., 50: 257—259.

Campbell, C.A., Biederbeck, V.O. and Warder, F.G., 1971. Soil Sci. Soc. Am. Proc., 35: 480—483.

Campbell, C.A., Paul, E.A. and McGill, W.B., 1976. In: Western Canada Nitrogen Symposium, Calgary, Alta., pp. 9—101.

Campbell, C.A., Paul, E.A., Rennie, D.A. and McCallum, K.J., 1967. Soil Sci., 104: 81—85.

Campbell, C.A., Paul, E.A., Rennie, D.A. and McCallum, K.J., 1967. Soil Sci., 104: 217—224.

Campbell, C.A., Stewart, D.W., Nicholaichuk, W. and Biederbeck, V.O., 1974. Can. J. Soil Sci., 54: 403—412.

Canada Department of Agriculture, 1972. Glossary of Terms in Soil Science. Publ. No. 1459, 66 pp.

Carter, J.N., Westerman, D.T. and Jensen, M.E., 1976. Agron. J., 68: 49—55.

Carter, J.N., Westerman, D.T., Jensen, M.E. and Bosma, S.M., 1975. J. Am. Soc. Sugar Beet Technol., 18: 232—244.

Chandra, P., 1964. Weed Res., 4: 54—63.

Cheng, H.H. and Kurtz, L.T., 1963. Soil Sci. Soc. Am. Proc., 27: 312—316.

Christenson, N.L., 1973. Science, 181: 66—68.

Chu, J.P.H. and Knowles, R., 1966. Soil Sci. Soc. Am. Proc., 30: 210—213.

Coffee, R.C. and Bartholomew, W.V., 1964. Soil Sci. Soc. Am. Proc., 28: 482—485.

Cordukes, W.E., MacLean, A.J. and Bishop, R.F., 1954. Can. J. Agric. Sci., 35: 229—237.

Cornfield, A.H., 1960. Nature, 187: 260—261.

Cornforth, I.S. and Walmsley, D., 1971. Plant Soil, 35: 389—399.

Craig, H., 1954. J. Geol., 62: 115—149.

Craswell, E.T. and Waring, S.A., 1972. Soil Biol. Biochem., 4: 427—433.

Date, R.A., 1973. Soil Biol. Biochem., 5: 5—18.

Daubenmire, R., 1968. Adv. Ecol. Res., 5: 209—266.

Dev, G., Sinha, B.K. and Narayan, K.G., 1970. J. Indian Soc. Soil Sci., 18: 221—225.

De Bano, L.F., 1969. Sci. Rev., 7: 11—18.

De Vries, H.I., 1958. Proc. K. Ned. Akad. Wetensch., B, 61: 1—9.

Doughty, J.L., 1939—1954. Ann. Rep. Soil Res. Lab., Agric. Canada, Swift Current, Sask.

Doughty, J.L., Cook, F.D. and Warder, F.G., 1954. Can. J. Agric. Sci., 34: 406—411.

Dowdell, R.J. and Cannell, R.Q., 1975. J. Soil Sci., 26: 53—61.

Dubetz, S. and Hill, K.W., 1964. Can. J. Plant Sci., 44: 139—144.

Dubetz, S., Kozub, G.C. and Dormaar, J.F., 1975. Can. J. Soil Sci., 55: 481—490.

Duff, R.B., Webley, D.M. and Scott, R.O., 1963. Soil Sci., 95: 105—114.

Dumanski, J., Kloosterman, B. and Brandon, S.E., 1975. Can. J. Soil Sci., 55: 181—187.

Durno, S.E., 1961. J. Ecol., 49: 347—351.

El-Shakweer, M.H., Gomah, A.M., Barakat, M.A. and Abdel-Ghaffar, A.S., 1976. In: IAEA/FAO-Agrochimica International Symposium on Soil Organic Matter, Brunswick, Germany, Sept., 1976. in press.

Ensminger, L.E. and Pearson, R.W., 1950. Adv. Agron., 2: 81—111.

Ferguson, W.S., 1967. Can. J. Soil Sci., 44: 286—291.

Ferguson, W.S. and Gorby, B.J., 1964. Can. J. Soil Sci., 44: 286—291.

Ferguson, W.S. and Gorby, B.J., 1971. Can. J. Soil Sci., 51: 65—73.

Fink, D.H., 1970. Soil Sci. Soc. Am. Proc., 34: 189—194.

Ford, G.W., Greenland, D.J. and Oades, J.M., 1969. J. Soil Sci., 20: 291—296.

Free, G.R., 1970. Cornell Univ. Agric. Exp. Sta., Bull. 1030.

Freney, J.R. and Simpson, J.R., 1969. Soil Biol. Biochem., 1: 241—251.

Friedlander, G. and Kennedy, J.W., 1960. Nuclear and Radiochemistry. Wiley, New York, N.Y., 468 pp.

Gasser, J.K.R., 1958. Nature, 181: 1334—1335.

Gasser, J.K. and Kalembasa, S.J., 1976. J. Soil Sci., 27: 237—249.

Geist, J.M., Reuss, J.O. and Johnson, D.D., 1970. Agron. J., 62: 385—389.

Goring, C.A. and Clark, F.E., 1948. Soil Sci. Soc. Am. Proc., 13: 261—266.

Greenland, D.J., 1971. Soils Fert., 34: 237—251.

Greenland, D.J. and Ford, G.W., 1964. Trans. Int. Congr. Soil Sci., 8th, Comm. III, pp. 137—148.

Greenland, D.J. and Nye, P.H., 1960. Intr. Congr. Soil Sci., 7th, Madison, Wisc., 1960, Comm. III, pp. 478—485.

Grier, C.C. and Cole, D.W., 1971. Northwest Sci., 45: 100—106.

Grim, R.E., 1953. Clay Mineralogy. McGraw Hill, New York, N.Y., 384 pp.

Gupta, U.C. and Reuszer, H.W., 1967. Soil Sci., 104: 395—400.

Haas, H.J., 1958. Agron. J., 50: 5—9.

Halstead, R.L. and Sowden, F.J., 1968. Can. J. Soil Sci., 48: 341—348.

Harmsen, G.W. and Kolenbrander, G.J., 1965. In: W.V. Bartholomew and F.E. Clark (Editors), Soil Nitrogen. Agronomy, 10, pp. 43—92.

Harmsen, G.W. and Van Schreven, D.A., 1955. Adv. Agron., 7: 299—398.

Hedlin, R.A., Smith, R.E. and Leclaire, F.P., 1957. Can. J. Soil Sci., 37: 34—40.

Heilman, P., 1975. Soil Sci. Soc. Am. Proc., 39: 778—782.

Hénin, S., Monnier, G. and Turc, L., 1959. In: Greenland, D.J., 1971. Soils Fert., 34: 237—251.

Hill, K.W., 1954. Soil Sci. Soc. Am. Proc., 18: 182—184.

Hobbs, J.A. and Brown, P.L., 1957. Kansas Agric. Exp. Stn. Tech. Bull., 89.

Hobbs, J.A. and Thompson, C.A., 1971. Agron. J., 63: 66—68.

Horton, H.A., 1942. Sci. Agric., 22: 546—551.

Iritani, W.M. and Arnold, C.Y., 1960. Soil Sci., 89: 74—82.

Ishaque, M. and Cornfield, A.H., 1972. Plant Soil, 37: 91—95.

Jackman, R.H., 1964. N. Z. J. Agr. Res., 7: 445—471.

Jansson, S.L., 1958. Ann. R. Agric. Coll. Swed., 24: 101—361.

Jansson, S.L., 1963. Soil Sci., 95: 31—37.

Jansson, S.L., 1971. In: A.D. McLaren and J. Skujins (Editors), Soil Biochemistry, 2. Dekker, New York, N.Y., pp. 129—166.

Jenkinson, D.S., 1966a. In: Rep. FAO/IAEA Tech. Meeting, Brunswick-Völkenrode, 1963. Pergamon, Toronto, pp. 187—197.

Jenkinson, D.S., 1966b. In: Rep. FAO/IAEA Tech. Meeting, Brunswick-Völkenrode, 1963. Pergamon, Toronto. pp. 199—208.

Jenkinson, D.S., 1968. J. Sci. Food Agric., 19: 160—168.

Jenkinson, D.S., 1971. Soil Sci., III: 64—70.

Jenny, H., 1930. Mo. Agric. Exp. Sta. Res. Bull., 152: 1—66.

Jenny, H., 1941. Factors of Soil Formation. McGraw-Hill, New York, N.Y., 281 pp.

Jenny, H., 1950. Soil Sci., 69: 63—69.

Jenny, H. and Raychaudhuri, S.P., 1960. Effect of Climate and Cultivation on Nitrogen and Organic Matter Reserves in Indian Soils. Indian Council of Agric. Res., New Delhi, 126 pp.

Jenny, H., Bingham, F. and Padilla-Saravia, B., 1948. Soil Sci., 66: 173—186.

Jenny, H., Gessel, S.P. and Bingham, F.T., 1949. Soil Sci., 68: 419—432.

Justice, J.K. and Smith, R.L., 1962. Soil Sci. Soc. Am. Proc., 26: 246—250.

Kadirgamathaiyah, S. and MacKenzie, A.F., 1970. Plant Soil, 33: 120—128.

Kai, H., Ahmed, Z. and Harada, T., 1969. Soil Sci. Plant Nutr., 15: 207—213.

Keeney, D.R. and Bremner, J.M., 1964. Soil Sci. Soc. Am. Proc., 28: 653—656.

Keeney, D.R. and Bremner, J.M., 1966a. Agron. J., 58: 498—503.

Keeney, D.R. and Bremner, J.M., 1966b. Nature, 211: 892—893.

Kononova, M.M., 1966. Soil Organic Matter (2nd ed.). Pergamon, Toronto, 544 pp.

Larson, W.E., Clapp, C.E., Pierre, W.H. and Morachan, Y.B., 1972. Agron. J., 64: 204—208.

Lathwell, D.J., Dubey, H.D. and Fox, R.H., 1972. Agron. J., 64: 763—766.

Laura, R.D., 1974. Plant Soil, 41: 113—127.

Laura, R.D., 1975. Plant Soil, 44: 463—465.

Laura, R.D., 1976. Plant Soil, 44: 587—596.

Legg, J.O. and Stanford, G., 1967. Soil Sci. Soc. Am. Proc., 31: 215—219.

Legg, J.O., Chichester, F.W., Stanford, G. and Demar, W.H., 1971. Soil Sci. Soc. Am. Proc., 35: 273—276.

Libby, W.F., 1955. Radiocarbon Dating (2nd ed.). Univ. Chicago Press, Chicago, 175 pp.

Livens, J., 1959. Agricultura, 7: 519—532.

Mack, A.R., 1962. Nature, 193: 803—804.

Mack, A.R., 1963. Can. J. Soil Sci., 43: 316—324.

Martel, Y.A. and Lavadière, M.R., 1976. Can. J. Soil Sci., 56: 213—221.

Martel, Y.A. and Paul, E.A., 1974. Can. J. Soil Sci., 54: 419—426.

Martin, J.P. and Richards, S.J., 1969. Soil Sci. Soc. Am. Proc., 33: 421—423.

Marumoto, T., Furukawa, K., Yoshida, T., Kai, H. and Harada, T., 1972. J. Fac. Agric. Kyushu Univ., 17: 37—47.

McCalla, T.M., Army, T.J. and Wiese, A.F., 1962. Agron. J., 54: 404—407.

McGill, W.B., Paul, E.A. and Sørenson, H.L., 1974. Matador Proj. Can. Int. Biol. Program, Saskatoon, Sask., Tech. Rep., 46.

McGill, W.B., Paul, E.A., Shields, J.A. and Lowe, W.E., 1973. In: T. Rosswall (Editor), Modern Methods in the Study of Microbial Ecology. Bull. Ecol. Res. Comm. (Stockholm), 17: 293—302.

McIntosh, J.L. and Varney, K.E., 1973. Agron. J., 65: 629—633.

Miller, R.D. and Johnson, D.D., 1964. Soil Sci. Soc. Am. Proc., 28: 644—647.

Minderman, G., 1968. J. Ecol., 56: 355—362.

Mitchell, J., 1932. J. Am. Soc. Agron., 24: 256—275.

Mitchell, J., Moss, H.C., Clayton, J.C. and Edmunds, F.H., 1944. Soil Surv. Rep., 12, Univ. of Sask., Saskatoon, Sask.

Newton, J.D., Wyatt, F.A. and Brown, A.L., 1945. Sci. Agric., 25: 718—737.

Newton, R., Young, R.S. and Malloch, J.G., 1939. Can. J. Res., Sect. C, 17: 212—231.

Nyborg, M., Neufeld, J. and Bertrand, R.A., 1976. In: Western Canada Nitrogen Symposium Proceedings, Calgary, Alta., pp. 102—127.

Nye, P.H., 1959. In: UNESCO Abidjan Symposium. Tropical Soils and Vegetation. pp. 59—63.

Overrein, L.N., 1971. Medd. Nor. Skogforsoeksves., 29(114): 261—280.

Parker, D.T., 1962. Soil Sci. Soc. Am. Proc., 26: 559—562.

Parsons, J.W. and Tinsley, J., 1975. In: J.E. Gieseking (Editor), Soil Components, 1. Organic Components. Springer-Verlag, New York, N.Y., pp. 263—304.

Paul, E.A., 1970. Recent Adv. Phytochem., 3: 59—104.

Porter, L.K., 1971. In: R.L. Dix and R.G. Beidleman (Editors), The Grassland Ecosystem. Range Science Dept., Sci. Ser. 2. Colorado State Univ., Fort Collins, Colo., pp. 377—402.

Porter, L.K., 1975. In: E.A. Paul and A. Douglas McLaren (Editors), Soil Biochemistry, 4. Dekker, New York, N.Y., pp. 1—30.

Porter, L.K., Stewart, B.A. and Haas, H.J., 1964. Soil Sci. Soc. Am. Proc., 28: 368—370.

Poyser, E.A., Hedlin, R.A. and Ridley, A.O., 1957. Can. J. Soil Sci., 37: 48—56.

Reichman, G.A., Grunes, D.L. and Viets, F.G. Jr., 1966. Soil Sci. Soc. Am. Proc., 30: 363—366.

Reid, A.S.J. and Miller, M.H., 1963. Can. J. Soil Sci., 43: 250—259.

Rennie, D.A., Truog, E. and Allen, O.N., 1954. Soil Sci. Soc. Am. Proc., 18: 399—403.

Ridley, A.O. and Hedlin, R.A., 1968. Can. J. Soil Sci., 48: 315—322.

Robinson, J.B.D., 1957. J. Agric. Sci., 49: 100—105.

Robinson, J.B.D., 1967. J. Soil Sci., 18: 109—117.

Robinson, J.B.D., 1968a. J. Soil Sci., 19: 269—279.

Robinson, J.B.D., 1968b. J. Soil Sci., 19: 280—290.

Robinson, J.B.D., 1968c. E. Afr. Agric. For. J., 33: 269—280.

Robinson, J.B.D., 1968d. E. Afr. Agric. For. J., 33: 299—301.

Rovira, A.D. and Greacen, E.L., 1957. Aust. J. Agric. Res., 8: 659—673.

Rowell, M.J., 1974. Production and Characterization of Enzymes Stabilized as Humic Acid Analogues. Ph.D. Thesis, Dept. Soil Sci., Univ. Sask., Saskatoon, Sask.

Russell, E.W., 1966. In: Rep. FAO/IAEA Tech. Meeting, Brunswick-Völkenrode, 1963. Pergamon, Toronto, pp. 3—19.

Russell, J.S., 1962. Trans. Int. Soil Conf., Comm. IV and V, Soc. Soil Sci. N.Z., pp. 191—197.

Russell, J.S., 1975. Soil Sci., 120: 37—44.

Ryan, J.A., Sims, J.L. and Peaslee, D.E., 1971. Agron. J., 63: 48—51.

Sabey, B.R., Bartholomew, W.V., Shaw, R. and Pesek, J., 1956. Soil Sci. Soc. Am. Proc., 20: 357—360.

Salisbury, H.F. and DeLong, W.A., 1940. Sci. Agric., 21: 121—132.

Salter, R.M. and Green, T.C., 1933. J. Am. Soc. Agron., 25: 622—629.

Sapozhnikov, N.A., Nesterova, E.I., Rusinova, I.P., Sirota, L.B. and Livandva, T.K., 1968. Int. Congr. Soil Sci., 9th, Adelaide, II: 467—474.

Sauerbeck, D., 1966. In: Rep. FAO/IAEA Tech. Meeting, Brunswick-Völkenrode, 1963. Pergamon, Toronto, pp. 209—221.

Scarsbrook, C.E., 1965. In: W.V. Bartholomew and F.E. Clark (Editors), Soil Nitrogen. Agronomy, 10, pp. 481—502.

Schnitzer, M., 1965. Nature, 207: 667—668.

Shields, J.A. and Paul, E.A., 1973. Can. J. Soil Sci., 53: 297—306.

Shields, J.A., Paul, E.A., Lowe, W.E. and Parkinson, D., 1973. Soil Biol. Biochem., 5: 753—764.

Shih, C.S., 1968. Computer Program S011 Nonlinear Statist. Res. Serv., Can. Dept. Agric., Ottawa, Ont.

Shreiner, O. and Brown, B.E., 1938. In: Soils and Men — The Yearbook of Agriculture, 1938. U.S. Dept. Agric., U.S. Govt. Printing Office, Washington, D.C., pp. 361—376.

Simonart, P. and Mayoudon, J., 1961. Pédologie, Gand, Symp. Int. 2 Appl. Sci. Nucl. Pedol., 91—103.

Sims, J.L., Wells, J.P. and Tackett, D.L., 1967. Soil Sci. Soc. Am. Proc., 31: 672—675.

Singh, B.R., Agarwal, A.S. and Kanehiro, Y., 1969. Soil Sci. Soc. Am. Proc., 33: 557—560.

Smith, D.W., 1970. Can. J. Soil Sci., 50: 17—29.

Smith, D.W. and Bowes, G.C., 1974. Can. J. Soil Sci., 54: 215—224.

Smith, J.H., 1966. In: Rep. FAO/IAEA Tech. Meeting, Brunswick-Völkenrode, 1963. Pergamon, Toronto, pp. 223—233.

Smith, R.M., Samuels, G. and Cernuda, C.F., 1951. Soil Sci., 72: 409—428.

Soil Survey Staff, 1960. Soil Classification, a Comprehensive System, 7th Approximation. Soil Conserv. Serv. U.S. Dept. Agric., Washington.

Sommerfeldt, T.G., Pittman, U.J. and Dubetz, S., 1974. In: P.J. Catania (Editor), Food, Fuel, Fertilizer. Proc. Symp. on Uses of Agric. Wastes, Regina, Sask., pp. 26—36.

Soper, R.J., Racz, G.J. and Fehr, P.I., 1971. Can. J. Soil Sci., 51: 45—49.

Sørenson, L.H., 1976. In: IAEA/FAO-Agrochimica Int. Symp. on Soil Organic Matter, Brunswick, Sept. 1976. in press.

Soulides, D.A. and Allison, F.E., 1961. Soil Sci., 91: 291—298.

Sowden, F.J. and Atkinson, H.J., 1968. Can. J. Soil Sci., 48: 323—330.

Stanford, G., 1968a. Soil Sci., 106: 345—351.

Stanford, G., 1968b. Soil Sci. Soc. Am. Proc., 32: 679—686.

Stanford, G., 1969. Soil Sci., 107: 323—328.

Stanford, G. and De Mar, W.H., 1969. Soil Sci., 107: 203—205.

Stanford, G. and De Mar, W.H., 1970. Soil Sci., 109: 190—196.

Stanford, G. and Epstein, E., 1974. Soil Sci. Soc. Am. Proc., 38: 103—107.

Stanford, G. and Legg, J.O., 1968. Soil Sci., 105: 320—326.

Stanford, G. and Smith, S.J., 1972. Soil Sci. Soc. Am. Proc., 36: 465—472.

Stanford, G. and Smith, S.J., 1976. Soil Sci., 122: 71—76.

Stanford, G., Legg, J.O. and Chichester, F.W., 1970. Plant Soil, 33: 425—435.

Stanford, G., Frere, M.H. and Schwaninger, D.H., 1973a. Soil Sci., 115: 321—323.

Stanford, G., Legg, J.O. and Smith, S.J., 1973b. Plant Soil, 39: 113—124.

Stanford, G., Carter, J.N. and Smith, S.J., 1974. Soil Sci. Soc. Am. Proc.. 38: 99—102.

Stanford, G., Frere, M.H. and Vanderpol, R.A., 1975. Soil Sci., 119: 222—226.

St. Arnaud, R.J. and Sephton, G.A., 1972. Can. J. Soil Sci., 52: 124—126.

Stevenson, F.J., 1965. In: W.V. Bartholomew and F.E. Clark (Editors), Soil Nitrogen. Agronomy, 10, pp. 1—42.

Stevenson, I.L. and Chase, F.E., 1953. Soil Sci., 76: 107—114.

Stevenson, J.L., 1956. Plant Soil, 8: 170—182.

Stewart, B.A., Johnson, D.D. and Porter, L.K., 1963a. Soil Sci. Soc. Am. Proc., 27: 656—659.

Stewart, B.A., Porter, L.K. and Johnson, D.D., 1963b. Soil Sci. Soc. Am. Proc., 27: 302—304.

Stone, E.L. and Fisher, R.F., 1969. Plant Sci., 30: 134—138.

Suess, H.E., 1955. Science, 122: 415—417.

Swaby, R.J., 1966. In: Rep. FAO/IAEA Tech. Meeting, Brunswick-Völkenrode, 1963. Pergamon, Toronto, pp. 21—31.

Takai, Y. and Harada, I., 1959. Soil Plant Food, 5: 46—50.

Tiedemann, A.R. and Klemmedson, J.O., 1973. Soil Sci. Soc. Am. Proc., 37: 107—111.

Tomlinson, T.E., 1973. In: Dowdell, R.J. and Cannell, R.Q., 1975. J. Soil Sci., 26: 53—61.

Truog, E., 1954. Commercial Fert., 88: 72—73.

Tusneem, M.E. and Patrick Jr., W.H., 1971. La. State Univ. Bull., 657.

Unger, P.W., 1968. Soil Sci. Soc. Am. Proc., 32: 427—429.

Vallis, I. and Jones, R.J., 1973. Soil Biol. Biochem., 5: 391—398.

Van Schreven, D.A., 1967. Plant Soil, 26: 14—31.

Van Schreven, D.A., 1968. Plant Soil, 28: 226—245.

Vlassak, K., 1970. Plant Soil, 32: 27—32.

Waksman, S.A. and Starkey, R.L., 1923. Soil Sci., 16: 343—357.

Walker, R.H. and Brown, P.E., 1936. In: P.E. Brown (Editor), Soils of Iowa. Iowa Agric. Exp. Sta. Spec. Rep., 3.

Warder, F.G. and Hinman, W.C., 1954—1964. Ann. Rep. Soils Sect. Agric. Canada, Res. Stn. Swift Current, Sask.

Waring, S.A. and Bremner, J.M., 1964. Nature, 201: 951—952.

Warren, R.G., 1956. Proc. Fertil. Soc., 37: 1—10.

Westerman, R.L. and Kurtz, L.T., 1973. Soil Sci. Soc. Am. Proc., 37: 725—727.

Westerman, R.L. and Tucker, T.C., 1974. Soil Sci. Soc. Am. Proc., 38: 602—605.

Wetselaar, R., 1968. Plant Soil, 29: 9—17.

Woodruff, C.M., 1949. Soil Sci. Soc. Am. Proc., 14: 208—212.

Wright, J.R. and Schnitzer, M., 1963. Soil Sci. Soc. Am. Proc., 27: 171—176.

Wright, L.E., Schurman, D.C. and Atkinson, H.J., 1950. Sci. Agric., 30: 447—455.

Yuan, T.L., Gammon, N. and Leighty, R.G., 1967. Soil Sci., 104: 123—128.

Chapter 6

SOIL ORGANIC SULFUR AND FERTILITY

V.O. BIEDERBECK

INTRODUCTION

Sulfur as a plant nutrient

Although S has long been known to be essential for plant growth it has received only recently the attention it deserves. Interest in fertility-related aspects of the S cycle is growing, as widespread crop deficiencies are reported with increasing frequency. The main reasons for this trend are: (a) increased use of high-analysis, S-free fertilizers; (b) decreased release of SO_2 by industrial and domestic fuel burning; (c) decreased use of S-containing pesticides; and (d) increased crop requirements and removal with more intensive farming. Deficiencies of S affect both yield and quality of crops.

Plants require S for the synthesis of some essential amino acids, protein, certain vitamins and coenzymes, glucoside oils, structurally and physiologically important disulfide linkages and sulfhydryl groups, and for the activation of certain enzymes (Coleman, 1966). Although the S content of plants varies depending on the supply available, some crops have definitely greater S requirements than others (Jordan and Ensminger, 1958; Whitehead, 1964) based on comparative crop uptake (Table I). Severalfold yield increases in response to S fertilization on deficient soils are not uncommon; this is particularly true with legumes (Coleman, 1966; Seim et al., 1969; McLachlan, 1975). S and N are closely associated in protein synthesis, thus S requirements vary according to the supply of N to crops. When S becomes limiting, additions of N do not change the yield or protein level of plants (Dijkshoorn and Van Wijk, 1967). Alfalfa requires one part S with every 11 to 12 parts N to ensure maximum protein production (Aulakh et al., 1976); with wheat, corn and beans one part S is required for every 12 to 15 parts N (Stewart and Porter, 1969). During early growth, cereals are also very sensitive to S deficiency and it has been suggested that many soils known to be deficient in S for alfalfa may also provide insufficient S for optimum cereal growth (Nyborg, 1968).

Sources of sulfur

In nature S occurs in a wide variety of organic and inorganic combinations, in various states of oxidation or reduction, and in solid, liquid or gaseous

TABLE I

Sulfur uptake and removal by certain crops
(Adapted from Whitehead, 1964; Coleman, 1966)

	Yield of product (kg D.M./ha/yr)	Sulfur in product (kg S/ha/yr)
Cereals *:		
Wheat	5,400	14—17
Corn	12,600	16—18
Vegetables:		
Cabbage	34,000	22—43
Onions	34,000	20—23
Potatoes	22,000	8—11
Sugar beet	33,000	21—31
Turnips	45,000	28—39
Forages:		
Lucerne	11,200	23—28
Clovers	9,000	17—22
Grasses	9,000	9—13

* D.M. yield includes straw.

form. However, plants obtain S primarily from soil as dissolved sulfate and, to a lesser extent, from the atmosphere by foliar absorption of SO_2. Thus, the S cycle resembles the N cycle in having an important atmospheric component and having its soil content associated with organic matter; but it differs in that the ultimate natural source of plant available S is the weathering of rock material.

The lithosphere contains about 0.06% S; this is present primarily as metallic sulfides in igneous rocks and as sulfates in sedimentary rocks (Jordan and Ensminger, 1958). The three principal sources of atmospheric S are (a) sulfates in sea spray precipitated onto coastal lands; (b) volatile sulfides released from the sea and marshlands; and (c) SO_2 released in industrial and urban areas by burning sulfurous fuels. The latter is considered to be the most important source (Coleman, 1966). Accession of S from the atmosphere can vary from <1 kg/ha/yr in remote rural areas to >130 kg/ha/yr near industrial centres. SO_2 is readily assimilated by S deficient plants (Ulrich et al., 1967) and even when sufficient sulfate is supplied some plants absorb as much as one third of their total S from the atmosphere (Olson, 1957). Whitehead (1964) states that S deficiency is unlikely to occur in areas where SO_2 enrichment is such that >11 kg S/ha/yr are carried down in precipitation.

Significant quantities of S can be added incidentally to soil by various agricultural practices such as application of superphosphate (12% S) or ammonium sulfate (24% S), manuring, use of sulfurous fungicides and irrigation (Mehring and Bennett, 1950; Jordan and Ensminger, 1958).

In the past, less intensive cropping and large S inputs from extraneous sources have made the indigenous soil S supply appear almost insignificant in many areas. However, with declining use of coal as fuel and with recent changes in agricultural practices the incidental accession of plant available S has sharply decreased while crop demands have steadily grown. Consequently, some Canadian soils, shown to be nondeficient 50 years ago (Wyatt and Doughty, 1928) have become highly S deficient (Nyborg, 1968). A similar shift toward S deficiency has occurred in many regions of the world (Coleman, 1966). More attention is now focused on the availability of soil S, which is governed primarily by the organic matter status. Thus interest in soil organic matter as a reserve of potential plant available S has been growing (Anderson, 1975). This chapter will, therefore, deal primarily with the nature, distribution and dynamics of soil organic S in relation to soil fertility.

NATURE AND DISTRIBUTION OF ORGANIC SULFUR IN SOIL

To evaluate the need for S fertilization and to assess the S supplying power of soils it is necessary to consider how S and other components of the soil organic reserve of nutrients are interrelated and to know the nature, amounts and distribution of S compounds in soil.

Numerous analyses by many workers have established that well over 90% of the total S in most noncalcareous surface soils is present in organic forms (Freney et al., 1962; Lowe, 1964; Rehm and Caldwell, 1968; Tabatabai and Bremner, 1972a; Jones et al., 1972; Bettany et al., 1973; Neptune et al., 1975; Scott and Anderson, 1976). The very close correlation between non-sulfate S, total N and organic C normally found in soils suggests that organic S forms a very consistent proportion of soil organic matter. Whitehead (1964) and Freney and Stevenson (1966) have reviewed some of the earlier studies on C/N/S ratios in soils throughout the world and have neglected to stress the differences by concluding that the mean ratios for different groups of soils are often remarkably similar in spite of great variations in climate and parent materials. However, the results of several American and Canadian studies suggest that climate, vegetation and other soil forming factors have produced marked differences between the mean C/N/S ratios of different soil types (Table II). The significance of these differences in terms of the S supply to crops is indicated by the general observation that S mineralization increases as the C/N/S ratio decreases and these relationships have recently been emphasized (Bettany et al., 1973, 1974; Kowalenko and Lowe, 1975b). The existence of a particularly close association between the N and S constituents of soil organic matter is indicated by the fact that the N/S ratios of most soils fall within the narrow range of 6.0—8.0 (Tabatabai and Bremner, 1972b; Scott and Anderson, 1976) and by reports of a closer correlation between total N and organic S than between organic C and organic S (Tabatabai

TABLE II

Mean ratios of carbon/nitrogen/sulfur in surface horizon of North and South American soils

Location	Description	C/N/S	Reference
Virgin soils:			
Alberta, Canada	Brown Chernozems (4) *	85 : 7.1 : 1	Lowe (1965)
	Black Chernozems (4)	96 : 7.7 : 1	Lowe (1965)
	Grey Wooded (7)	271 : 12.5 : 1	Lowe (1965)
	Gleysols (6)	78 : 5.0 : 1	Lowe (1969b)
Cultivated soils:			
Saskatchewan, Canada	Brown Chernozems (6)	57 : 6.1 : 1	Bettany et al. (1973)
	Dk. Brown Chernozems (13)	61 : 6.5 : 1	Bettany et al. (1973)
	Black Chernozems (9)	80 : 7.1 : 1	Bettany et al. (1973)
	Grey Black-Transitional (14)	95 : 7.6 : 1	Bettany et al. (1973)
	Grey Wooded (12)	124 : 10.4 : 1	Bettany et al. (1973)
Minnesota, U.S.A.	Brown Chernozems (6)	73 : 6.4 : 1	Evans and Rost (1945)
	Black Chernozems (9)	73 : 6.1 : 1	Evans and Rost (1945)
	Podzols (24)	116 : 8.5 : 1	Evans and Rost (1945)
Oregon, U.S.A.	Podzols (4)	141 : 9.3 : 1	Harward et al. (1962)
Mississippi, U.S.A.	Podzols (4)	139 : 13.1 : 1	Nelson (1964)
Iowa, U.S.A.	Soil type not reported (6)	85 : 7.7 : 1	Neptune et al. (1975)
Sao Paulo and Parana, Brazil	Soil type not reported (6)	121 : 6.3 : 1	Neptune et al. (1975)

* Figures in brackets refer to number of samples.

and Bremner, 1972a; Bettany et al., 1973; Neptune et al., 1975). This close N/S association is probably related to the general similarity in the cycling of these two elements within the soil—plant system.

Forms of organic sulfur

While much of the organic S in soils remains uncharacterized, some progress has been made in recent years in subdividing it into broad fractions which can serve as a useful preliminary for further identification (Anderson, 1975). The analytical approach currently in use, differentiates between three broad fractions of soil organic S: HI-reducible S, C-bonded S, and residual or inert S. The HI-S fraction contains S compounds that are not directly bonded to C and it is thought to consist primarily of sulfate esters and ethers in the form of phenolic sulfates, sulfated polysaccharides, choline sulfate and sulfated lipids (Freney, 1967; Tabatabai and Bremner, 1972b). As the S in this fraction can be readily hydrolyzed to inorganic sulfate by acid or alkali, HI-S is considered to be the most labile fraction of soil organic S (Spencer and Freney, 1960; Lowe, 1965; Freney et al., 1971; Cooper, 1972). This fraction is thought to be largely associated with active side chain components of fulvic and humic materials (Bettany et al., 1973). In most mineral soils HI-S is the dominant form of organic S, constituting between 33 and 78% of the total soil organic S (Table III).

The S directly bonded to C is determined by reduction to sulfide with Raney nickel (Lowe and DeLong, 1963). Although this method has been criticised for failing to reduce all C-bonded S compounds and for the possibility of producing artifacts (Freney et al., 1970; Tabatabai and Bremner, 1972b) it still is the best method available for assessing this component. Based on analyses of soil hydrolysates it was estimated that about half of the C-bonded S occurs as amino acids (Freney et al., 1972). The nature of the remainder of this fraction is unknown. In mineral soils C-bonded S accounts for between 5 and 35% of the total organic S (Table III). A much higher proportion of C-bonded S (47—58%) was reported for several organic soils (Lowe and DeLong, 1963). Based on its resistance to several extractants and its correlation with certain humus characteristics it has been postulated that C-bonded S is relatively stable and is primarily associated with the

TABLE III

Fractionation of organic sulfur in surface soils

Location	HI-reducible S as % of total		C-bonded S [1] as % of total		Residual, inert S as % of total		Reference
	range	mean	range	mean	range	mean	
Quebec, Canada (3) [2]	44—78	65	12—32	24	0—44	11	Lowe (1964)
Alberta, Canada (15)	25—74	49	12—32	21	7—45	30	Lowe (1965)
Saskatchewan, Canada (54)	28—59	45	n.d.		n.d.		Bettany et al. (1973)
Australia (24)	—	52	n.d.		n.d.		Freney (1967)
Australia (15)	32—63	47	22—54	30	3—31	23	Freney et al. (1970)
Iowa, U.S.A. (24)	36—66	52	5—20	11	21—53	37	Tabatabai and Bremmer (1972b)
Brazil (6)	36—70	51	5—12	7	24—59	42	Neptune et al. (1975)

[1] Determined by reduction with Raney nickel.
[2] Figures in brackets refer to number of samples.

strongly aromatic core of humic acids (Bettany et al., 1973).

It has been suggested that C-bonded S should be estimated as being the difference between total and HI-reducible S (Freney et al., 1970), and some workers have adopted this simple procedure. However, the assumption that all non-HI-reducible soil organic S consists of C-bonded (amino acid-) S is probably an oversimplification and is not justified in view of the existence in soils of sizeable amounts of inert S with characteristics differing from those of HI-S and C-S fractions (Lowe, 1965). The presence of this inert or residual fraction of soil organic S was first suggested by Lowe (1964) and later supported by results from other studies which indicate that this fraction accounts for between 3 and 59% of the total organic S in mineral soils (Table III). Whatever its origin, the organic S in this residual fraction is exceptionally stable since it resists degradation by drastic chemical treatments (Lowe, 1964, 1965). Thus this fraction is probably of little significance as a potential source of S for plants.

Most noncalcareous soils contain very little inorganic S, usually <5% of the total S (Freney et al., 1962; Tabatabai and Bremner, 1972b; Bettany et al., 1973). Under aerobic conditions the mineral S fraction consists almost exclusively of sulfates (Freney et al., 1962; Neptune et al., 1975). Only in soils developed under strongly reducing conditions and in poorly drained subsoils is sulfide accumulation a common occurrence (Whitehead, 1964). Some calcareous soils contain large amounts of sulfates, present as co-precipitated or co-crystallized impurities of $CaCO_3$ (Williams and Steinbergs, 1962; Scott and Anderson, 1976). However, these insoluble sulfates are considered to be of very low availability to plants (Williams and Steinbergs, 1964). In acid surface soils and in many subsoils with pH <6.5 soluble sulfates are retained by adsorption onto hydrated Fe and Al oxides, kaolinitic clays and organic colloids (Freney et al., 1962; Whitehead, 1964; Neptune et al., 1975; Scott, 1976). This adsorbed sulfate is readily available to plants (Williams and Steinbergs, 1964).

Amount and distribution of organic sulfur

The S content of soils can vary over an extremely wide range from 0.002 to 3.5% (Whitehead, 1964). In spite of this extreme variability certain trends and conclusions become apparent upon critical examination of the numerous analyses, that have been reported. The highest S contents are found in soils of tidal marshes and in swamps where sulfides have accumulated. Among agriculturally used soils, S contents are always considerably greater in peat, muck and other organic soils than in nearby mineral soils (Wyatt and Doughty, 1928; Lowe, 1964; Jones et al., 1972). In alkaline and saline soils of dry regions, large amounts of sulfates frequently accumulate near the surface and total S contents up to 2.24% were found in some of these soils in Alberta (Wyatt and Doughty, 1928; Lowe, 1969b). Calcareous soils, rich in insoluble sulfates, contain on an average two to three times more total S than non-calcareous soils developed under the same climate (Williams et al., 1960; Williams and Steinbergs, 1962).

In noncalcareous mineral soils total S contents usually range between 0.01 and 0.1% and are closely correlated with organic matter contents, emphasizing the dominance and importance of the organic S in these soils. Texture seems to have little or no influence on S contents since total S and clay contents are not significantly correlated (Rehm and Caldwell, 1968; Levesque, 1974). The three major factors affecting the amounts of total S in agricultural soils appear to be organic matter, pH and precipitation. This is indicated by many reports showing that soils with low organic matter contents, acidic soils, and soils subject to considerable leaching (e.g., Podzolic and Grey Wooded types) generally contain less S than do neutral, humus-rich soils and soils from drier regions (Wyatt and Doughty, 1928; Lowe, 1965, 1969b; Freney et al., 1970; Bettany et al., 1973; Levesque, 1974; Neptune et al., 1975; Scott and Anderson, 1976).

The distribution of S in soil profiles is subject to large quantitative and marked qualitative changes. Since the bulk of the S in noncalcareous soils is in organic forms total S usually decreases sharply with soil depth (Tabatabai and Bremner, 1972a; Levesque, 1974) as shown in Fig. 1. A tenfold decrease in S content from the A to the B horizon is not uncommon (Lowe, 1969b). However, while total and organic S decline, the sulfate content usually increases with depth. This may be due to leaching and to greater adsorption in acid subsoils under humid climates (Jordan and Ensminger, 1958; MacKenzie et al., 1967; Roberts and Koehler, 1968) or to accumulation of $CaSO_4$ in the C horizons of Chestnuts and Chernozems from drier regions (Whitehead, 1964; Lowe, 1965, 1969b). Analyses of profiles from a variety of soil types in Iowa (Tabatabai and Bremner, 1972b) and Alberta (Lowe, 1965, 1969b) show that the N/S ratio tends to decrease with depth and is <4 in several subsoils. In all soils, this narrowing of the N/S ratio was accompanied by a marked increase in the proportion of HI-reducible S and a decrease in the percentage of C-bonded S with increasing depth.

Effect of soil forming factors

There has been little research on the influence of soil forming factors on soil S. However, considering the very close correlation between total N and S in most surface soils, it can be assumed that the order of importance of these factors as they affect the S content is generally similar to that stated by Jenny (1930) with regard to the N content of American soils (i.e., climate > vegetation > topography = parent material > age).

The effects of climate and vegetation were indicated by Bettany et al. (1973) in a study of the distribution of S fractions in cultivated soils from grassland and forested areas developed under semiarid to subhumid condi-

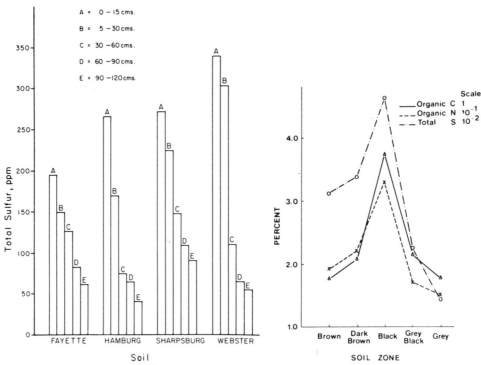

Fig. 1. Distribution of total S in profile samples of Iowa soils. (Adapted from Tabatabai and Bremner, 1972b.)

Fig. 2. Distribution of organic C, organic N, and total S among the soil zones in Saskatchewan. (From Bettany et al., 1973.)

tions. Across the five soil zones of Saskatchewan S contents were closely correlated with both organic C and organic N, with peak amounts occurring in the Chernozemic Black zone (Fig. 2). The Chernozemic soils contained more total S relative to C and N than was found in the Grey-Black and Grey Wooded soils. Studies of virgin Alberta soils by Lowe (1965) and early studies in Minnesota by Evans and Rost (1945) have also shown that Grey Wooded (Podzolic) soils have significantly wider C/N/S ratios than do Chernozems. Climate and vegetation also caused qualitative differences in soil organic S. With the transition from semiarid to subhumid conditions the proportion of total S in the HI-reducible fraction decreased from 52% in Brown Chernozems to 36% in Grey Wooded soils in Saskatchewan (Bettany et al., 1973) and from 63% in Chernozems to 33% in Grey Wooded soils in Alberta (Lowe, 1965).

The influence of restricted drainage on forms and distribution of S was examined by Lowe (1969b). In Gleysolic profiles he found considerably higher levels of total S than in Chernozems and Podzols. This was attributed to slower mineralization and less intensive leaching under the restricted drainage associated with the development of Gleysols. The most distinctive feature of these Gleysolic profiles was their much higher proportion of organic S in a reduced state (C-bonded S) than that found in better-drained profiles.

The nature of the parent material has very little influence on the organic S in surface soils (Scott and Anderson, 1976) but it can markedly affect the amounts and forms of inorganic S (Williams and Steinbergs, 1962; Neptune et al., 1975; Scott, 1976).

EFFECTS OF MANAGEMENT ON SOIL ORGANIC SULFUR

Effect of cropping

The breaking and cultivation of virgin soil has led invariably to large losses of soil organic matter (Allison, 1973; Campbell et al., 1976). Any change in the organic matter is likely to be reflected as a change in S status since >90% of the S in most noncalcareous soils is present in organic form. In western Canada, the surface horizons of virgin chernozemic and podzolic soils have much wider C/N/S ratios than those in the corresponding cultivated soils (Bettany et al., 1973). This narrowing of the C/N/S ratio upon cultivation indicates that losses of organic C and N are proportionately greater than losses of organic S during the initial period of rapid decomposition. A similar trend was observed under more humid conditions on a sandy loam in England (Table IV) where 100 years of cropping to a grass—arable rotation caused a marked reduction in C/N/S ratio and much greater losses of C and N than of

TABLE IV

Effect of 100 years of cultivation on carbon, nitrogen and sulfur contents in a sandy clay loam
(Adapted from Cowling and Jones, 1970)

Carbon, nitrogen, sulfur	Virgin soil (beech woodland)	Cropped * soil (grass/arable rotation)	Loss due to cultivation as % of total in virgin soil
Total C (%)	3.38	1.73	49
Total N (%)	0.233	0.154	34
Total S (%)	0.041	0.033	20
HCl-soluble SO_4-S (ppm)	42	22	48
C/N/S ratio	83 : 5.7 : 1	53 : 4.7 : 1	—

* During the last 20 yr this soil received average annual additions of 80 kg N and 30 kg S per ha.

S (Cowling and Jones, 1970). The reserve of potential plant available S was also depleted; the cropped soil contained less than half the amount of readily extractable S and about a third of the mineralizable S found in the corresponding virgin soil. Upon cultivation, the S mineralized becomes available for plant uptake, and if it is not adsorbed it may be lost by leaching (McLachlan, 1975). However, the losses of total S and the changes in organic S fractions due to cultivation depend primarily on the type of cropping system used. Twenty cycles of a wheat—fallow rotation on red-brown earths in Australia reduced the total S content by an average 29% and caused a marked net loss of S from the surface soils over and above that attributable to crop uptake. Surplus S released from the organic matter was lost from the surface soils by leaching of sulfates, and this was enhanced during fallow periods (Williams and Lipsett, 1961). Summerfallowing and long-term cultivation also increased the proportion and availability of a "heat-soluble", very labile fraction of soil S, probably as a result of some modification of the organic S. Continuous cropping to barley, for 50 years on a sandy loam near Rothamstead, resulted in comparatively smaller losses of total S, although some surfates were leached beyond the root zone of the grain crop (Mann, 1955). At the same time the proportion of humified S in the soil increased markedly.

The C and N (Chapter 5), high humus and S levels can be maintained when less intensive cropping systems such as forage leys and legume-based pastures are adopted. In eastern Canadian Podzols MacKenzie and co-workers (1967) were unable to detect any differences in total S contents and in extractable S between virgin profiles and profiles sampled under a variety of permanent pastures. When Australian Podzols were seeded to subterranean clover, fertilized regularly with superphosphates, and used as sheep pastures for 26 years there was a marked buildup of organic matter and a concomitant 58% increase in the S content of the surface soil (Donald and Williams, 1954). All of the fertilizer-applied S had become incorporated into the soil organic matter with 83% entering the C-bonded S fraction (Freney et al., 1975); thus leading to an eventual shortage of available S (Williams and Donald, 1957). This conservation of the fertilizer S agrees with other reports that leaching of radioactive fertilizer S could not be detected in experiments on grazed pastures (Till, 1975).

Although several workers have observed qualitative changes in soil organic S as a result of cultivation, no consistent trends with regard to the interchange between different organic forms of S are as yet apparent. The earlier mentioned narrowing of the C/N/S ratio together with some reported changes in S availability seems to suggest that the increased microbial activity, normally associated with cultivation of soil, could result in an accelerated transformation of the more resistant C-bonded S into the HI-reducible fraction. Studies with virgin profiles have indicated that soils with reduced microbial activity generally contain a greater proportion of C-bonded S (Lowe, 1965, 1969b).

However, more definitive studies, using modern methods of S fractionation together with tracer techniques are required to determine the long-term effect of cultivation and different cropping systems on the major organic fractions of soil S.

Effect of manures and residues

The benefits of regular manure applications to soil fertility, particularly under intensive cropping have been well documented (Allison, 1973). Unlike green manures, farmyard manures and composts mineralize very slowly in soil because they usually have already undergone considerable decomposition prior to application on the land.

The S content of animal manures can vary considerably depending on the type of feed and animal. S concentrations range from a low 0.04% in barnyard manure to a high of 0.50% in sheep and poultry manures (Mehring and Bennett, 1950). Some of the S originally contained in the faeces is likely to be lost by volatilization and leaching during the storage and handling of manure prior to application. In certain areas with intensive livestock operations where large quantities of feedlot wastes are regularly applied to small acreages, considerable amounts of organic S may be added to soils. However, under general agronomic practices manuring is considered to be a very minor source of S for crop production. In the U.S.A. S supplied with animal manures accounted for only 8% of the total applied to arable land as fertilizers, pesticides and soil amendments (Mehring and Bennett, 1950).

Green manures decompose rapidly and may increase the supply of plant available S, particularly if crucifers or legumes are used. In a greenhouse study with five different tagged sources of S, Jordan and Baker (1959) found that a green manure crop of peas supplied as much S to a growing crop of beans as most of the inorganic S sources and much more than equivalent additions of elemental S.

A considerable amount of the S taken up by crops may be returned to the soil in their residues and for some crops as much as 85% of the S in the above-ground parts is recycled in this manner (Nelson, 1973).

It is of considerable agronomic significance to know the extent to which S returned in crop residues becomes available to succeeding crops. The release of plant available S from decomposing crop residues depends primarily on the rate of decomposition and the S content of the residues. According to Stewart et al. (1966a), about 0.15% S in wheat straw is sufficient to ensure a maximum decomposition rate. For a wide variety of plant materials Barrow (1960) found only a broad relationship between S mineralized and the C/S ratio of these materials suggesting that initial C/S ratios of 200 and 420 are the lower and upper limits, respectively, of the range within which either mineralization or immobilization of S would occur. Below 200 he observed only mineralization and above 420 only immobilization. For wheat straw the C/S ratio for optimum decomposition was found to be less than 300 and additions of sulfate to soils containing low-S residues increased the rate of decomposition (Stewart et al., 1966a). Plant growth studies have led to conclusions similar to those derived from incubation studies; they show severe yield reductions on soils amended with low-S residues. When cotton residues with a C/S ratio of 148 and corn residues with a C/S ratio of 406 were added to a sandy loam and a clay loam with and without sulfate additions and subjected to cropping with turnips, Nelson (1973) observed severe yield depressions for the first crop but not for subsequent crops. The apparent recovery of residue-S in the absence of added sulfate was 12 and 35% on the two cotton residue treated soils, but due to complete immobilization no residue-S was recovered in any crop grown on soils amended with corn residues. There was a marked interaction between fertilizer-S and residue-S as the recovery from both S sources was reduced when they were added to the soil together. This effect was attributed to a lack of available N for both plant growth and microbial activity. The recovery of residue-S in crops and soils may also be decreased by the production of volatile S compounds during decomposition of some organic materials (Frederick et al., 1957; Nicolson, 1970). However, with the increased use of N and P fertilizers containing little or no S, immobilization of S and yield reductions accompanying decomposition of plant residues have become more prominent, even under field conditions. In humid climates S immobilization due to crop residue applications should aid the retention of fertilizer S in surface soil, but according to Jones et al. (1971) heavy additions of straw to soils in California failed to reduce leaching losses of S, but reduced leaching losses of nitrates.

In a field experiment where crop residues were added for 11 consecutive years to a silty clay loam, cropped to corn and well fertilized with N, Larson et al. (1972) found that soil organic S increased linearly in proportion to the amount of residues added. After 11 years organic S had increased by 16% at 2 tons/ha/yr and by 49% at 16 tons/ha/yr with alfalfa, and by 9% and 40% at the corresponding application rates with cornstalks. The increases in the organic S content were similar to those for organic C but greater than the gains in organic N and P. These results indicate that S levels can be maintained or increased, even under intensive cropping, when adequate quantities of crop residues are regularly returned to the soil.

Effect of liming

The addition of liming materials to arable land causes not only an increase in pH but also stimulates the production of sulfate from soil organic matter (Freney et al., 1962). Greenhouse studies have shown that liming greatly increased the S uptake by plants (Williams and Steinbergs, 1964; Probert, 1976). With a wide variety of noncalcareous soils from Australia, Williams (1967) found that the increase in the amounts of S mineralized during incu-

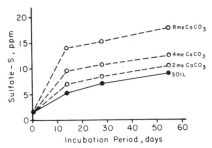

Fig. 3. Effect of CaCO$_3$ additions on S mineralization from a yellow podzolic soil during incubation. (Adapted from Williams, 1967.)

bation were generally large and depended primarily on the amount of CaCO$_3$ added (Fig. 3). The increases were apparently not related to soil type or any single soil property. In a tracer study with five different soils, Probert (1976) found that the increased S mineralization due to liming was a rather short lived effect. Although liming caused an initial increase in the phosphate extractable S fraction, most of the S released in response to liming originated in the bicarbonate extractable fraction which contains some labile organic S. As reported by Anderson (1975) there is also some evidence that liming may cause a relatively greater mineralization of S than of N, resulting in the formation of organic matter with a lower S content. Freney and Stevenson (1966) have concluded that the increase in soluble sulfates upon liming may be due to several factors; these are: (a) sulfate released from soil organic matter by bacteria growing better in a more favorable pH environment; (b) sulfate released from soil organic matter by chemical hydrolysis at alkaline pH; (c) adsorbed sulfate released from soil exchange sites because of an increase in pH; and (d) sulfate added with the liming materials.

SULFUR TRANSFORMATIONS IN SOIL

In many respects the transformations of S in the soil—plant system resemble those of the N cycle. Both cycles have an important atmospheric component and have most of their soil component associated with organic matter, while the very small but active inorganic fraction is subject to a variety of oxidation and reduction reactions. However, the S cycle differs from that of N in that the ultimate source of S under natural conditions is the weathering of the parent rock material. During the process of soil formation, the sulfides of primary minerals are converted to sulfates which may be transformed into a wide variety of organic compounds by the biosynthetic processes of microorganisms, plants, and animals. When plant and animal residues are returned to the soil they are attacked by microorganisms, thus releasing some of the S as sulfate, but most of it remains in organic form and eventually becomes part of the soil humus. The humus S is slowly remineralized and again recycled through microorganisms and plants. The balance between the various biosynthetic, catabolic, oxidation, and reduction processes illustrated in Fig. 4, depends primarily on climate, aeration status, soil type and agronomic practices.

Analogous to the role of nitrate in the N cycle, sulfate holds the key position in the natural cycling of S because under aerobic conditions sulfate is the usual inorganic end product of S metabolism in soils (Anderson, 1975). Purely chemical oxidation and reduction can occur in soils, but most of the S transformations are effected by microorganisms. The microbial transformations of S have been reviewed, in detail, by Alexander (1961), Starkey (1966), Freney and Stevenson (1966) and Freney (1967). In this section we shall deal primarily with processes involving organic forms of S (e.g., decomposition, mineralization and immobilization) rather than the various oxidative and reductive interchanges of inorganic S which are possible in soil.

Decomposition and stabilization of organic sulfur

For the purposes of this discussion decomposition is arbitrarily defined as the breakdown of large molecular organic S-containing materials to simple organic S compounds but not to inorganic S. Stabilization will refer to the processes of transformation and incorporation of relatively simple S compounds into the more complex, resistant, and humified fractions of soil organic matter.

As listed by Freney (1967), a great variety of S compounds have been isolated from plants, animals and microorganisms. In soil, there are three major sources of decomposable S: (a) fresh plant or animal residues; (b) the soil biomass including microbial cells and by-products of microbial synthesis; and

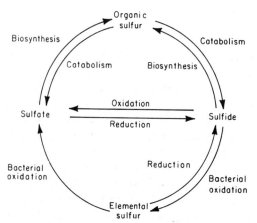

Fig. 4. Biological transformations of sulfur. (Adapted from Freney, 1967.)

(c) humus. Although the three fractions differ in biological stability, most of the S in these fractions occurs in the form of proteins or sulfate esters.

Enzymatic breakdown

Soil microorganisms produce a battery of extracellular proteolytic enzymes which can hydrolyze a great variety of peptide and ester bonds in molecules of varying sizes under a wide variety of physiological conditions. The multiplicity of different proteases synthesized by microbial cells and the low substrate specificity of these enzymes enable the soil microflora to be very active in the dissimilation of numerous types of proteinaceous materials (Biederbeck, 1969). The proteinaceous components of plant and microbial residues are sequentially broken down into polypeptides, peptides and eventually individual amino acids. Similarly, the sulfate esters occurring primarily as macromolecules in structural components of plant, animal and microbial tissues seem to require breakdown into low-molecular-weight sub-units before S hydrolysis by means of sulfohydrolases and arylsulfatases can occur (Houghton and Rose, 1976). The relative rates at which different synthetic sulfate esters, added to soils, are degraded suggest that some are hydrolyzed by enzymes already present in microbial cells or adsorbed onto soil colloids while others require enzyme induction in soil microorganisms. During the decomposition of large molecular S compounds there occurs not only an enzymatic breakdown into their basic organic components but also microbial assimilation of intermediates, as well as synthesis and excretion of a wide range of different organic S compounds. Starkey (1950) stressed this metabolic complexity when he wrote: "Consideration of the reactions whereby organic and inorganic sulfur compounds are transformed by microorganisms leads one to believe that many sulfur compounds are produced in soil, but because they are susceptible to decomposition, they do not accumulate and are not readily detected." Accordingly, the concentration of uncombined organic S compounds in soil has been found to be normally low and variable. Thus, naturally occurring low-molecular-weight sulfate esters do not accumulate in soils (Houghton and Rose, 1976) and only small amounts of the S-containing amino acids have been isolated from rhizosphere and non-rhizosphere soils (Paul and Schmidt, 1961). The amounts of free organic S compounds in soil are assumed to represent a balance between synthesis and destruction by microorganisms (Freney and Stevenson, 1966).

In contrast to the relatively rapid decomposition of fresh organic residues in soil the degradation and release of S from the large humus fraction is minimal and very low. Although soil humus contains considerable amounts of proteinaceous (Biederbeck and Paul, 1973) and large molecular sulfate ester materials, these humus components are very resistant to enzymatic hydrolysis (Paul, 1970; Houghton and Rose, 1976). This resistance to the effects of a wide array of microbial proteases and sulfatases can probably be attributed to a lack of flexibility of the substrate, and to mechanical as well as chemical shielding of localized areas of the substrate surface. However, the degradation of humic acid and other heteropolycondensates does occur in soil, albeit slowly, and is almost certainly carried out by microbial, multi-enzyme catalyzed sequences, producing low-molecular-weight aromatic and aliphatic intermediates (Mathur, 1971). Thus decomposition of humus-S could result from a sequential attack by depolymerizing and desulfating enzymes (Houghton and Rose, 1976).

Biomass

Simple, uncombined organic S compounds do not accumulate, as a result of protein and ester sulfate decomposition, in soil because these compounds are rapidly assimilated by microbial cells. Upon uptake by heterotrophic organisms the organic S in excess of that needed for microbial growth is partly mineralized by intracellular enzymes and released as sulfate, and partly excreted as organic metabolites produced by intracellular conversion. The S content of most microorganisms lies between 0.1 and 1.0% of the dry weight, and the most conspicuous S-containing cell constituents are the amino acids, cystine and methionine (Alexander, 1961). Based on microbial biomass data reported by Clark and Paul (1970), Kowalenko (Chapter 3, this book) has estimated that together the bacteria and fungi in the surface of a grassland soil account for 1.3% of the total soil organic S. This estimate agrees reasonably well with results of a tracer experiment by Van Praag (1973) which showed that not more than 2—3% of the total S in unamended surface horizons of Brown forest soils was in the fraction of "active organic phase S". It is noteworthy that these figures closely approach Jenkinson's (1966) estimates showing that biomass-C accounted for 2—3.5% of the total soil-C in unamended soils. Although the microbial mass contains only a very small proportion of soil organic S at any one time, this fraction is extremely labile and is the key or driving force for S turnover in soil.

Incorporation into humus

Decomposition and stabilization of organic S are two processes that proceed simultaneously and continuously in soil. The simple organic S compounds and microbial metabolites produced during decomposition may be stabilized initially by adsorption onto mineral and resistant organic colloids. This may be followed by a series of slow and poorly defined reactions such as complex formations and condensations of chemically dissimilar compounds, largely under the influence of various extracellular enzymes. This results in a very resistant, large molecular heteropolycondensate commonly referred to as humus. Very little is known about the types or sequences of reactions involved in the incorporation of S into humus but compounds whose reactions appear to be most important are those containing the —SH group (viz., cysteine, glutathione and methyl mercaptan). Since the reactions of —SH groups frequently resemble those of —NH₂ groups, S and N may be stabilized in humus in similar ways. Whitehead (1964) suggested that reactions between quinones and —SH groups may account for part of the C-bonded S in humus, because quinones react with —SH even more readily than with amino groups yielding condensation products containing the relatively stable thiazine ring. Thiol compounds can also react with aldehydes, including reducing sugars, to form condensation products in which S is bonded to C (Freney and Stevenson, 1966). Although there is no direct evidence for the occurrence of these thiol—quinone and thiol—aldehyde reactions in soils, they furnish a possible explanation for the content and stability of S in humus. However, there must also be other reactions involved in the stabilization of S-containing amino acids and the results of one tracer study suggest that amino acid-S from crop residues may be humified without complete prior decomposition to simple monomers. Scharpenseel and Krausse (1963) used ^{35}S-tagged sulfate, cystine, and methionine to follow the transfer of S compounds into humic acids. Very little sulfate, cystine, or methionine was incorporated directly into humic acid. However, when the labelled compounds were first assimilated by rye and the plants were allowed to decompose in soil, the hydrolysates of humic acid, derived from the decayed plants, showed high labelling in the methionine but not in the cystine fraction. Apparently, amino acids account only for part of the C-bonded S stabilized in humus since Lowe (1969a) was able to recover only an average of 39% of the C-bonded S as amino acid-S after hydrolysis of humic acids from different soils. The nature of the remaining C-bonded S remains unknown.

Another major fraction of humic S consists of the HI-reducible organic sulfates. These sulfates have been shown to be covalently bound and not just adsorbed to humic colloids (Freney, 1961; Houghton and Rose, 1976). However, nothing is known about the mode of incorporation of sulfate-esters and -ethers into soil humus. Humification must render the sulfate-esters very resistant to enzymatic hydrolysis since Houghton and Rose (1976) were unable to release any sulfate from humic acids during incubation with a variety of sulfohydrolases from bacteria and other organisms.

Fractionation of ^{35}S-labelled soil organic matter by Freney et al. (1971, 1975) has provided valuable information on the part played by the HI-reducible fraction in the stabilization and decomposition of soil organic S. When soils were incubated with labelled sulfate for periods up to 24 weeks, labelled S was incorporated into both HI-reducible and C-bonded organic S, but the former fraction was always more highly labelled. Furthermore, the initial rate of incorporation of labelled S into HI-reducible S was much greater than that into other forms of organic S. Marked increases in the HI-reducible fraction of the unlabelled S indicated the occurrence of conversion of C-bonded S to HI-reducible S during incubation. Sodium hydroxide extracted both labelled and unlabelled S in the same proportions as they were present in the organic matter; however, sodium bicarbonate preferentially extracted labelled S. In both extracts about 4/5 of the labelled S was present in the fulvic acid fraction; furthermore over 90% of the extracted labelled S was HI-reducible. Changes in soil organic S during plant growth showed that plants utilized some S from all organic S fractions (Freney et al., 1975). Much of the unlabelled S in the plants was derived from nonreducible fractions and approximately 40% came from the indigenous HI-reducible S. However, in the case of recently immobilized labelled S plants assimilated labelled S only from the HI-reducible fraction. In fallow soils, changes in organic S fractions were much smaller; however, they clearly showed that the bulk of the mineralized sulfate, both labelled and indigenous, was derived from the HI-reducible fraction. These tracer studies suggest that during biological transformations in soil, when S is simultaneously incorporated into and released from humus, the HI-reducible S in toto and especially that in fulvic acids functions as the major intermediate or active phase of soil organic S.

Mineralization—immobilization (turnover)

In the previous subsection decomposition was, for ease of discussion, arbitrarily defined as the breakdown of organic materials into small organic molecules. However, in nature the S metabolism is in continual motion with further conversion to mineral form and simultaneous resynthesis of organic S (turnover). Attention will now be focused on: (a) the conversion of simple organic S compounds to inorganic S by heterotrophic organisms in soil (i.e., S mineralization); and (b) the microbial uptake of mineral S and its conversion into various forms of organic S (i.e., S immobilization). The diverse microbial oxidations and reductions of inorganic S in soil have been reviewed excellently by Starkey (1966) and will not be discussed here.

Mineralization of amino acids

As mentioned earlier, uncombined organic S compounds do not accumulate in soil. They are rapidly assimilated by microorganisms and only the S in excess of microbial requirements is released in mineral form and thus rendered available for plant uptake. The principal end product of microbial decomposition of organic S in soil is sulfate under aerobic conditions, and sulfide under anaerobic conditions. Many reactions may be involved before these end products appear (Alexander, 1961).

The manner in which a wide variety of heterotrophic organisms influence the pathways and products of aerobic and anaerobic decomposition of various S-containing amino acids has been studied with pure cultures; the findings were reviewed by Freney (1967). However, relatively little is known about their decomposition in soil. The soil microflora is very heterogenous, consequently the reactions taking place in soil may be different from those with pure cultures because of interactions between species. Simple organic S compounds decompose at different rates under aerobic conditions. Frederick et al. (1957) observed that S was rapidly released from most S compounds of natural origin and that decomposition as determined by sulfate production was not related to any type of S linkage. Particular attention has been directed to the decomposition of S-containing amino acids because they constitute a significant fraction of the organic S in soils, plants, and microbial residues. Aerobic decomposition of cysteine and cystine in soils results in fairly complete transformation of the S to sulfate (Alexander, 1961). The conversion is rapid because many microorganisms attack the two compounds and their decomposition may proceed by any one of several competing pathways (Freney and Stevenson, 1966). There are also several pathways for methionine decomposition (Freney, 1967). In soil, there is sometimes complete conversion of methionine-S to sulfate; however, often little or no sulfate is released and the decomposition produces several volatile S compounds. It has been suggested that the mechanism of methionine decomposition varies with the soil type and conditions (Whitehead, 1964).

Formation of volatile S

The evolution of volatile S compounds during the decomposition of organic S in soil has long been recognized; but it was generally assumed that this was primarily associated with anaerobic decomposition processes. However, recent investigations using very sensitive gas chromatographic techniques have demonstrated the production of significant amounts of gaseous S even in aerobic soils (Lewis and Papavizas, 1970; Banwart and Bremner, 1975). During aerobic incubation of soils amended with residues from five cruciferous crops, Lewis and Papavizas (1970) observed abundant evolution of methyl mercaptan, dimethyl sulfide, and dimethyl disulfide; however, none of these volatiles were evolved from decomposing corn residues in soil. Banwart and Bremner (1975) studied the formation of volatile S compounds during the decomposition of 14, S-containing amino acids in three soils under aerobic and anaerobic conditions. Some S was volatilized from all but three (viz., cysteic acid, taurine and S-methyl methionine) of the added amino acids. The proportion of added amino acid S that was volatilized during incubation depended on the type of amino acid and varied from less than 0.1% to more than 50%. The amounts and types of volatile S formed were similar under aerobic and waterlogged conditions. The amounts measured may underestimate the actual production of volatile S compounds during decomposition because moist soils and soil microorganisms have the capacity to absorb many S gases (Bremner and Banwart, 1976).

Although there are reports that some microorganisms can produce the phytotoxic gas H_2S from S-containing amino acids (Freney, 1967; Swaby and Fedel, 1973); no trace of this gas could be detected during incubation of amended soils by several workers (Banwart and Bremner, 1975, 1976; Lewis and Papavizas, 1970). The volatile S compounds formed during decomposition may affect some N transformations in soil. Bremner and Bundy (1974) have shown that carbon disulfide is a potent inhibitor of nitrification and that this process is also retarded by methyl mercaptan, dimethyl sulfide and dimethyl disulfide. It could be speculated that this effect might produce anomalous rates of nitrification in soils amended with residues of cruciferous or other S-rich crops.

The occurrence of S volatilization could affect the results of mineralization studies with organic S amended soils because sulfate accumulation does not necessarily account for all of the S released during decomposition. As a possible explanation for substantial S losses observed in S balance studies with unamended and sulfate-treated sandy soil, Nicolson (1970) hypothesized that significant volatilization of S can occur even in unamended and sulfate-treated soils. He suggested that volatile S may be formed under aerobic conditions via the following pathways:

Nicolson's hypothesis was recently tested by Banwart and Bremner (1976). They incubated unamended and sulfate-treated samples of 25 different Iowa soils under aerobic and waterlogged conditions and found that volatile S compounds were released from 14 anaerobic soils but only from four of the aerobic soils. Where S was volatilized the amounts were so small ($<0.05\%$ of total soil-S) that they concluded that gaseous losses of S from soils will be insignificant under conditions likely to be encountered in the field.

Sulfatase activity in soil

Since much of the soil organic S is present in the form of sulfate esters, sulfatase enzymes may play an important role in the mineralization process. When a variety of synthetic sulfate esters were incubated with various soils most of these compounds were readily hydrolyzed by indigenous enzymes (Houghton and Rose, 1976). Apparently, many soils contain sulfohydrolases capable of releasing sulfate from alkyl-, aryl- and sugar-sulfates.

Arylsulfatases were the first sulfatases detected in soils (Tabatabai and Bremner, 1970a) and thus have received more attention than other S enzymes. Arylsulfatase activity has been reported for a variety of U.S., African and Canadian soils (Tabatabai and Bremner, 1970a; Cooper, 1972; Kowalenko and Lowe, 1975b). Irradiation treatment of soils has indicated that most of this activity is due to extracellular, particle-adsorbed arylsulfatases. The levels of arylsulfatase activity vary according to soil type, depth, and climate. Several workers have studied the relationships between the activity of this enzyme and various soil properties. In several profiles (Tabatabai and Bremner, 1970b), arylsulfatase decreased markedly with depth (Fig. 5) and the decline in activity was closely correlated with the decrease in soil organic

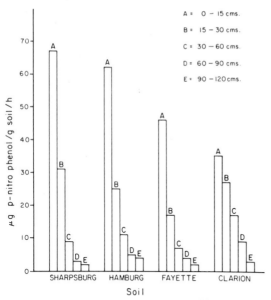

Fig. 5. Arylsulfatase activity in profile samples of Iowa soils. (Adapted from Tabatabai and Bremner, 1970b.)

matter (r = 0.783). In Iowa, Tabatabai and Bremner (1970b) examined 27 surface soils and found the activity of this enzyme was significantly correlated with organic C content (r = 0.506) but not with pH, S content, N content, or texture of the soils. Later (1972a) they also showed that mineralizable S was not significantly related to initial arylsulfatase activity. However, in a study with 20 Nigerian soils, Cooper (1972) found a significant correlation between arylsulfatase activity and total C (r = 0.691), organic S (r = 0.565), and HI-reducible S (r = 0.878). The latter correlation is most interesting since the HI-reducible ester sulfates are considered to be the natural substrates for sulfatase enzymes in soil. Under field conditions, arylsulfatase activity in some Nigerian soils was subject to wide seasonal variation as shown in Fig. 6. Thus, Cooper (1972) suggested that seasonal variation in weather conditions may influence arylsulfatase activity more than the various soil properties with which this enzyme was shown to be correlated. In four Canadian soils arylsulfatase activity declined sharply throughout a 14 week incubation period and although there was a slightly significant correlation between arylsulfatase activity and $CaCl_2$-extractable sulfate (r = 0.49) Kowalenko and Lowe (1975b) suggested that this enzyme was not a major factor in the release of sulfate from these soils. Considering the inconsistency of the data reported by various workers it must be concluded that the role of arylsulfatase in soil S mineralization remains uncertain and requires further investigation.

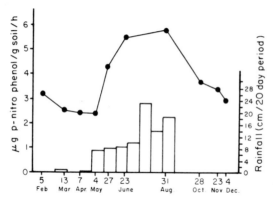

Fig. 6. Seasonal variation of arylsulfatase activity in the surface of a ferruginous tropical soil. (Adapted from Cooper, 1972.)

Relationship between N, C and S turnover

In view of the close resemblance between microbial conversions of N and S it has been assumed that the relative rates of mineralization of N and S from soil organic matter would be similar (Alexander, 1961), and therefore, that N and S will be mineralized in approximately the same ratio as they occur in soil organic matter (Walker, 1957; White, 1959). However, this hypothesis fails to explain the results of several studies where the ratio of N/S mineralized was much larger than that of the soil organic matter (Williams, 1967; Tabatabai and Bremner, 1972a; Haque and Walmsley, 1972). There have also been a few studies where the ratio of mineralized N/S was considerably smaller than that of the soil organic matter (Nelson, 1964; Simon-Sylvestre, 1965), and other studies in which N was mineralized without any release of S during incubation (Hesse, 1957; Barrow, 1961). The difference between N and S mineralization in soils is emphasized by the fact that, although the amounts of N and organic S are highly correlated in most soils, most workers have reported that the amounts of S mineralized during incubation were not significantly correlated with total N or mineralizable N (Harward et al., 1962; Williams, 1967; Haque and Walmsley, 1972; Tabatabai and Bremner, 1972a). Under field conditions Simon-Sylvestre (1965) found a very different mineralization pattern for S than for N with sulfate levels in an uncropped soil undergoing much greater seasonal variations than levels of mineral N (Fig. 7). However, a few workers have been able to show a close relationship between N and S mineralization (Nelson, 1964; Kowalenko and Lowe, 1975b).

In an attempt to explain the different behaviour of organic N and S during mineralization, Freney et al. (1962) have suggested: (1) that the N and the S mineralized could originate from different fractions of soil organic matter which decompose at different rates; and (2) that the ratio of N/S mineralized

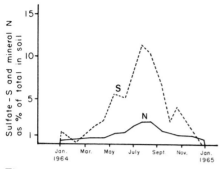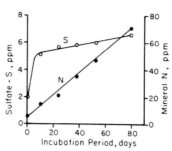

Fig. 7. Seasonal variation of sulfur and nitrogen mineralization in the surface of a summerfallowed loam. (Adapted from Simon-Sylvestre, 1965.)

Fig. 8. Sulfur and nitrogen mineralized from a calcareous sand, maintained moist after field sampling and during incubation at 30°C. (Adapted from Williams, 1967.)

depends more on the N/S ratio of recently added organic matter than on the ratio in the bulk of the soil organic matter. It is also possible that N and S mineralization processes may differ in type and efficiency of microorganisms involved as well as in the stability of extracellular enzymes. Williams (1967) observed considerable differences between the degree of suppression of S and of N mineralization in response to soil treatment with toluene and formaldehyde. The results of several recent studies (Williams, 1967; Haque and Walmsley, 1972; Kowalenko and Lowe, 1975b) seem to indicate that the main reason for the lack of correlation between S and N mineralization in most soils is the frequently observed, rapid initial release of S without a corresponding release of N from soil organic matter (Fig. 8). This initial flush in S mineralization appears to be an inherent characteristic of the S metabolism in soils, because it occurs not only upon air drying but also in soils which are maintained moist after field sampling (Williams, 1967). The very transient nature of the observed flush in S release, and Nelson's (1964) observation that S and N mineralization were not correlated after one month of incubation but were highly correlated after two and six months, indicates the need for a re-evaluation of S mineralization studies with more emphasis on length of incubation so as to properly assess the N/S relationships.

Attempts to relate S mineralization to other soil properties have also been quite variable. In Mississippi, Nelson (1964) found significant relationships between S mineralized from native soil organic matter during six months incubation and their organic S ($r = 0.929$) and organic C contents ($r = 0.918$). However, other workers have been unable to demonstrate such relationships in soils from Oregon (Harward et al., 1962), Iowa (Tabatabai and Bremner, 1972a), Australia (Williams, 1967), the West Indies (Haque and Walmsley, 1972) and Canada (Bettany et al., 1974). In fact, Williams (1967) reported that the amounts of S mineralized in 17 Australian soils were not closely related to any of the many soil properties investigated. Nevertheless, the results of most incubation studies seem to suggest that larger amounts of S are generally mineralized from soils with relatively high organic matter contents and particularly from soils with low C/N/S ratios (Harward et al., 1962; Nelson, 1964; Haque and Walmsley, 1972; Bettany et al., 1974; Kowalenko and Lowe, 1975b). Thus the existence of a rather complex interrelationship among soil organic C, N and S seems evident. However, in view of the many inconsistencies observed it can be concluded that the relationship between the amounts of S mineralized and the C/N/S ratio and other characteristics of soil organic matter cannot be clarified without further improvement in the fractionation and identification of soil organic S. In this respect organic S is no different from organic N (Chapter 5, this book).

The results of incubation studies should be interpreted with caution, because (as is the case with N) the mineralization of S from soil organic matter may be obscured by fixation, immobilization or volatilization of some of the released S (Chandra and Bollen, 1960; Freney, 1967; Nicolson, 1970). Definite relationships between microbial S requirements, the extent of S immobilization, and the S content of substrates have been reported for the decomposition of various pure organic substances (Stotzky and Norman, 1961; Stewart et al., 1966b) and crop and animal residues (Barrow, 1960; Stewart et al., 1966a; Freney, 1967) in soil, but not for the slow release of S from native soil organic matter. This is indicated by the very wide variations in rates and amounts of net mineralization of S that have been reported for unamended soils. In Australia, net immobilization of considerable amounts of added sulfate has been shown to occur during incubation even in the absence of organic amendments or growing plants (Freney and Spencer, 1960; Freney et al., 1971).

Effect of plants on S turnover

The mineralization of S from soil organic matter is markedly influenced by the presence of growing plants. In a pot experiment Freney and Spencer (1960) found that more than twice as much S was mineralized during 14 months in soils planted to *Phalaris* as in the corresponding uncropped soils. They also reported that increasing rates of S fertilization decreased the net mineralization from soil organic S, but the cropped soils still mineralized considerably more S than the uncropped soils. In another Australian study, the addition of 30 ppm sulfate-S also caused a greater depression of S mineralization in bare than in planted pots (Nicolson, 1970). However, Bettany et al. (1974) recently reported that S mineralization in Saskatchewan soils cropped to alfalfa was not affected by addition of 25 ppm tagged sulfate-S. This discrepancy in the effect of S fertilization on soil organic S mineralization may be due to greater S deficiency in the Australian soils. In unfertilized soil with and without plants cropping to clover caused not only a marked increase in S mineralization but it also decreased total S losses (Nicolson,

1970). Similarly, a recent tracer study indicated that S mineralization in soils cropped to *Sorghum* was severalfold greater than in uncropped soils (Freney et al., 1975). The stimulating effect of growing plants is obviously of great significance in the cycling of S in the soil-plant system and the increased mineralization may be due to greater microbial activity in the rhizosphere of the plants and/or the excretion by plant roots of enzymes which catalyze the decomposition of soil organic matter (Freney, 1967).

In conclusion it would appear that in contrast to the behaviour of N, the rate and extent of the release of plant available S from soil organic matter is not closely governed by the major soil characteristics such as organic C, N and S content.

ENVIRONMENTAL FACTORS AFFECTING MINERALIZATION

In the previous section the microbiological processes affecting S mineralization and the relationship of the latter to N and C turnover were discussed. Since S mineralization in soils is catalyzed by a wide variety of heterotrophic microorganisms, it can be expected that environmental factors which influence microbial growth will also affect the rate of S mineralization. Sudden changes in weather conditions are known to cause physical, chemical, and biological changes in soil, and these will affect soil fertility by modifying the turnover of nutrients.

Moisture

Soil moisture affects the activity of sulfatases, the rate of S mineralization, the form in which S is released from organic matter, and the movement of sulfates in soil. In tropical soils the level of activity of the arylsulfatase enzyme was shown to be closely related to the moisture status (Cooper, 1972); under field conditions, seasonal changes in moisture were more important than various other soil properties in controlling the activity of this enzyme. When soils were incubated at varying moisture levels, S mineralization was sharply reduced at moisture contents appreciably above or below field capacity (Williams, 1967). The formation of sulfate followed a pattern similar to that of nitrate (Fig. 9), suggesting that the apparent decline in S mineralization at moisture levels close to saturation was probably due to poor aeration. Under such high moisture conditions the measurement of sulfate production tends to underestimate the total S mineralization in soils because reduced forms of inorganic S gradually accumulate as their oxidation is inhibited. When soils containing considerable organic matter become waterlogged sulfate is reduced to sulfide by sulfate-reducing bacteria growing anaerobically on organic compounds (Starkey, 1966). The S mineralized in saturated soils is less mobile because under reducing conditions iron is rendered more solu-

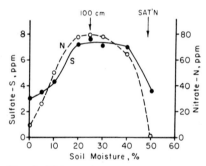

Fig. 9. Effect of moisture content on sulfate and nitrate production in a Yellow podzolic soil during two weeks of incubation at 30°C. (Adapted from Williams, 1967.)

ble and reacts readily with sulfides to form precipitates of iron sulfides. Thus Jones et al. (1971) found that temporary waterlogging decreased leaching losses of S from grassland soils by an average of 40%, and Haque and Walmsley (1974) reported considerably greater retention of tagged sulfate in a sandy loam with high organic matter content than in a clay soil containing less organic matter when they were leached. When the moisture content decreases and the soils become aerobic again, the sulfides are readily reoxidized to sulfate by thiobacilli and other microorganisms.

Since the moisture effects reported by Williams (1967) are based on laboratory studies, where soil moisture and other interacting factors such as temperature were kept at constant levels, it is difficult to assess the contribution of moisture per se to the mineralization of S under natural conditions.

Temperature

Although the influence of temperature on the mineralization of N and C has been thoroughly investigated very little is known about its effect on S mineralization because sulfate release from soil organic matter is usually assessed by incubating soils at a constant temperature of 30°C. It can be assumed that S mineralization will decrease with temperature and be minimal at low temperatures because of sharply reduced intra- and extra-cellular enzymatic activity. Such a temperature-induced slowdown of mineralization would explain the relatively greater S contents found in soils formed under northern climates. When moist soils were incubated for 64 days at different temperatures, Williams (1967) found that no sulfate was released from soil organic matter at 10°C or below. As expected, the rate of S mineralization increased with increasing temperatures up to 35°C. However, as shown in Fig. 10 a 10°C rise in soil temperature stimulated ammonification disproportionately more than S mineralization. By contrast, Nicolson (1970) observed that increasing soil temperatures between 10° and 20°C increased

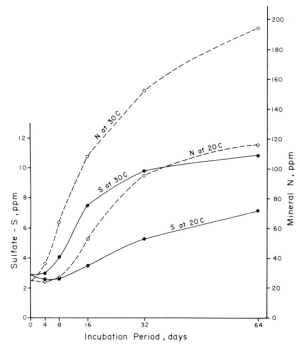

Fig. 10. Effect of temperature on S and N mineralization from a Yellow podzolic soil during incubation. (Adapted from Williams, 1967.)

plant growth but had very little effect on S mineralization in cropped and uncropped soil. The strong inhibition of S mineralization at medium to low soil temperatures is likely to prevent excessive leaching losses of soil S in temperate and subtropical climates during seasons that are both wet and cool.

Wide temperature fluctuations and periodic soil freezing and thawing are known to cause a flush of N mineralization (Chapter 5, this book).

Considering the somewhat contradictory findings mentioned above and the lack of information on the influence of fluctuating vs constant temperatures and on the effect of freeze-thaw cycles on S mineralization, it is evident that there is a need for research designed to elucidate such relationships.

Drying, and wetting and drying

Drying effects

Part of the organic S in soil appears to be very unstable and is readily converted to sulfate by physical treatments such as heating, air drying, or grinding (Freney, 1967). Frequently, soils are subjected to some or all of these treatments during preparation for laboratory or greenhouse studies on S mineralization. Numerous reports (Freney, 1958; Barrow, 1961; Williams and Steinbergs, 1964; Williams, 1967; Tabatabai and Bremner, 1972a; Walker and Doornenbal, 1972; Kowalenko and Lowe, 1975a) show that air drying per se releases considerable amounts of sulfate from a wide variety of soils. Little is known about the mechanisms involved, but Williams (1967) has suggested that they are associated with the same processes that bring about the increased solubility of organic matter after drying or heating (Birch, 1959). Similarly, Barrow (1961) suggested that the mineral S released during drying was split from organic sulfates present in soil. The drying process may also cause the release of some simple organic S compounds. Some of the released S appears to be readily adsorbed by soil colloids, because phosphate and bicarbonate solutions, which are known to extract variable proportions of organic and adsorbed sulfate, generally extract more S from dried soils than do calcium chloride solutions (Williams and Steinbergs, 1964; Williams, 1967). The release of S during soil drying appears to be an abiological phenomenon since it occurs immediately, and since the amounts released during air drying are frequently larger than the amounts released during subsequent moist incubation of the same soils (Barrow, 1961; Freney et al., 1962; Williams, 1967). It has also been shown that the rapid S transformations occurring during air drying are completely unrelated to soil microbiological activity as measured by CO_2 evolution (Kowalenko and Lowe 1975a). The amounts of S released from soils by drying are markedly affected by the manner and frequency of drying; they increase with increasing drying temperatures and very large quantities are released by ovendrying at $105°C$ (Barrow, 1961; Williams, 1967). The increases in extractable sulfate upon air drying at room temperatures vary widely according to soil type and extractant used, but based on data reported by various workers, average increases range from 20 to 80% (Freney, 1958; Barrow, 1961; Williams, 1967; Tabatabai and Bremner, 1972a).

Drying has a noticeably different effect on S than on C and N mineralization. While drying prior to incubation enhances the mineralization of both C and N for several weeks, drying causes an immediate release of S but has little effect on subsequent S mineralization (Barrow, 1961). This immediate release of sulfate is proportionately much greater than the change in mineral N during drying. Lengthening the period of air drying prior to rewetting and incubation is known to increase the amount of inorganic N produced (Birch, 1959) but there is no parallel effect on S mineralization (Williams, 1967). It is unlikely that these divergent tendencies can be attributed to enzymatic changes because Tabatabai and Bremner (1970b) found air drying of field moist soils caused marked increases in arylsulfatase activity (average, 43%). This activity did not decrease significantly when air dried soils were stored at $22°-24°C$ for three months.

The sulfate released as a result of air drying is readily available to plants. Barrow (1961) found that even where fertilizer S had been added plant growth was markedly increased by prior soil drying. The increased availability of S caused by soil drying can seriously affect the interpretation of pot experiments designed to compare the S status of soils because even if all soils are dried at the same temperature the relative availability of S may be no indication of the relative availability when the soils were fresh. Similarly, in estimating S mineralized during laboratory incubation, Kowalenko and Lowe (1975a) recommend that soil drying after incubation be avoided since analysis of moist samples will give a more reliable index of microbial S mineralization.

Wetting and drying

Cycles of wetting and drying are known to cause flushes in CO_2 and mineral N with each successive cycle producing a slightly smaller flush (Birch, 1960). A similar cumulative effect of wetting and drying on S mineralization was observed by Williams (1967). However, intermittent moisture stress treatments over a period of 27 weeks were found to have little effect on S uptake by plants and did not stimulate the release of mineral S from soil organic matter in pot experiments (Freney et al., 1975). These results suggest that gradual moisture fluctuations in the range between field capacity and wilting point have little influence on S mineralization and that rather drastic changes in soil moisture conditions are required to produce a flush in S. Rewetting of soils stored air dry in the laboratory led to enhanced S mineralization in some, but not in all soils examined by Williams (1967). This flush of mineralization occurred immediately after rewetting and thereafter the rate of mineralization was similar to that in the original undried soils. During this flush, the enhancement of S mineralization was relatively greater than that of N. It has been suggested by some workers (Barrow, 1961; Till, 1975) that the mineral S released as a result of repeated wetting and drying is derived primarily from ester sulfates in the soil organic matter. This theory is supported by Cooper's (1972) finding that the initial rapid release of sulfate upon wetting of some Nigerian soils is associated with a decrease in the size of the HI-reducible organic S fraction.

Based on results of laboratory and greenhouse studies it can be concluded that the greatly increased availability of S caused by soil wetting and drying would also be important under field conditions and Barrow (1961) claims this is the major factor contributing to the flush of growth commonly observed on S-deficient soils in Australia after dry periods. It appears that in regions where drastic soil moisture changes periodically occur the abiological rapid release of sulfate during soil wetting and drying may be more significant in supplying S to crops than the slow release by microbial decomposition. However, extensive field experimentation making use of tagged S plus meaningful fractionation, is required so as to assess the practical significance of abiological versus biological phenomena in the release of S from soil organic matter under changing soil conditions.

SULFUR AVAILABILITY AND AVAILABILITY INDEXES

In considering sulfur-soil fertility interactions the availability of soil S to the crop is of major concern. As discussed earlier (Introduction) this concern has increased because S accessions from extraneous sources have decreased considerably and the incidence of S-deficiency has become a worldwide phenomenon. It is important to know how availability can be measured and related to crop yield. Consequently, considerable effort has been expended in devising suitable methods for the diagnosis of S deficiency in plants and soils and for the prediction of fertilizer requirements. This subject has been reviewed by Ensminger and Freney (1966).

The methods that have been proposed are largely empirical and those commonly used measure only the S available at a particular time. However, to predict accurately the amount of fertilizer required it is necessary to know not only the amount of soil S that is available at the time of planting but also the contribution that will be made during crop growth through the mineralization of soil organic S. Ideally, a soil test method should also estimate the possible incorporation of fertilizer S into unavailable forms in soil and it should take into account the residual effects of previous fertilizer applications as well as the possibility of leaching losses.

Although most surface soils contain several hundred kilograms S per hectare only a very small amount is directly available to plants and very little additional S, usually 4—14 kg S/ha (Alexander, 1961), is made available to crops each year by mineralization of soil organic S. The directly available soil S is defined as S in a chemical form that can be readily absorbed by plant roots. Roots can assimilate some simple organic S compounds, such as S-containing amino acids, but these are very minor sources of plant available S because they do not persist in soil. Thus the most important forms of directly available S in aerobic soils are dissolved and adsorbed sulfates.

Based on the soil and plant-nutrition aspects discussed in the preceding sections it is suggested that the main factors affecting the availability of S in soils are: organic matter content, C/N/S ratio, mineralization rate, leaching losses, and N and P fertilization practices.

The development of diagnostic methods for estimating the S status or S supplying power of soils has generally followed the approaches used for other elements such as N and P. These approaches are based on chemical and/or biological assays, and include plant tissue analyses, extraction of various soil S fractions, soil incubation, microbiological assays, and plant growth experiments.

Plant analyses

The use of plant tissue analysis for assessing S deficiencies in soils is based on the premise that an essential element should be present in the plant at a concentration just sufficient for unrestricted growth. This concentration is known as the "critical percentage". The main advantage of this techinique is that the critical percentage should be the same for each plant species over a wide variety of soil types and climatic conditions (Ensminger and Freney, 1966). This is true for the critical S concentration in alfalfa (0.20% S), but for other species there have not been sufficient data published (Martin and Walker, 1966). In addition to total S, sulfate-S and the S/N ratio in plant tissues have also been used for detection of S deficiency in soils. This type of analysis will only be useful if these plant consitituents vary only with the S-supplying power of soils. Unfortunately the concentration of S and sulfate in plants is affected by various factors such as other nutrients, particularly N supply (Jones et al., 1972), the part and age of the plant analyzed (Martin and Walker, 1966), and several environmental variables. Although tissue analyses are generally closely correlated with the S status of soils (Ensminger and Freney, 1966) the major disadvantage of these techniques is that they often do not reveal a deficiency until it is too late for S fertilization to affect crop yields.

Soil analyses

Since the nature of the mineralizable or readily available fraction of soil S is not yet known, workers in this field have relied on arbitrary analytical methods. Thus numerous extractants have been suggested for estimating available soil S (Beaton et al., 1968) but there is no general agreement as to the estimate which best defines a soil's S supply. Most of the soil tests developed up to 1966 were reviewed and discussed by Ensminger and Freney (1966), thus they will only be summarized briefly here and more attention will be given to those methods which attempt to estimate the likely contribution from soil organic S.

Extraction of sulfate S

A wide variety of extractants has been used for measuring extractable sulfate. Commonly used extractants include water (Spencer and Freney, 1960; Fox et al., 1964; Walker and Doornenbal, 1972), neutral salt solutions such as $CaCl_2$ (Barrow, 1961; Bettany et al., 1974; Kowalenko and Lowe, 1975a) LiCl, and $MgCl_2$ (Roberts and Koehler, 1968; Tabatabai and Bremner, 1972a), phosphate solutions (Fox et al., 1964; Jones et al., 1972; Probert, 1976), and acidic solutions such as various acetates and Bray P_1 (Rehm and Caldwell, 1968; Hoeft et al., 1973). Phosphate solutions extract more sulfate from most soils than can be extracted with water or chloride solutions because phosphate ions displace the adsorbed sulfate which is known to be readily available to plants. Tracer experiments have shown that the S extracted with calcium phosphate comes from the same pool of soil S as that used by plants; Probert (1976) suggests that calcium phosphate is the best extractant for characterizing the immediate S supply of soils. Close correlations between phosphate extractable S and plant uptake have been reported from pot and field studies with a wide range of soils (Fox et al., 1964; Barrow, 1969; Jones et al., 1972; Hoeft et al., 1973). The results of field experiments with alfalfa suggest that the critical level of phosphate extractable S below which crop response to S is assured on most soils is about 6 ppm. On Alberta soils, Walker and Doornenbal (1972) found the same critical level based on water extractable S and obtained a high level of prediction (86%) of S deficiency when compared with the field growth response of alfalfa to S fertilizer.

The good correlations frequently observed between many of the above-mentioned extraction methods and plant growth are surprising because these methods are assumed to extract only soil mineral S and because the S uptake during plant growth is usually considerably greater than the concomitant decline of extractable S in the soil (Spencer and Freney, 1960; Williams and Steinbergs, 1964; Jones et al., 1972; Bettany et al., 1974). However, a critical examination of these methods suggests that they invariably include some estimate of the readily soluble soil organic S. Water as well as chloride and phosphate solutions can extract variable amounts of labile organic S (Williams and Steinbergs, 1959; Spencer and Freney, 1960; Freney et al., 1971; Kowalenko and Lowe, 1975a). The Johnson and Nishita (1952) reduction method, which measures organic as well as inorganic sulfates, is normally used for the final sulfate determination in the extracts, thus the resultant estimates of extractable S will include some soil organic S. Furthermore, soils are usually dried prior to potting or extraction of sulfates and since air-drying is known (previous section) to render some organic S (primarily ester sulfates) extractable and plant available, this type of treatment will boost the apparent level of available S.

The basic weakness of the methods used to measure extractable sulfate is that they are only correlated with S uptake on S-deficient soils and fail to provide an estimate of the reserve forms of S which is required for an assessment of the long-term S supply in non-deficient soils (Freney et al., 1962; Rehm and Caldwell, 1968).

Extraction of organic S

Since the total S content of soils is not related to the mineralizable or plant available S (Ensminger and Freney, 1966), some workers have devised special methods in an attempt to extract the labile fraction of soil organic S. To be useful as a soil test for S availability these methods must provide a reasonable estimate of that part of the soil organic S which is mineralized during the growing season. When soils contain very low amounts of sulfate, the amounts mineralized may be critical in preventing S deficiency in plants. Thus, a survey by Hamm et al. (1973) has shown that if the contributions of mineralized S are neglected 53% of the soils tested in the Grey soil zone and

18% of those from the Black in Saskatchewan can be considered as potentially S deficient for alfalfa and rapeseed. The importance of the mineralized S was also emphasized by Bettany et al. (1974) who found that most of the S taken up by alfalfa from several Saskatchewan soils was obtained from S mineralized during plant growth.

The methods proposed for extraction and estimation of labile soil organic S do include the readily available sulfate S and although they are generally based on arbitrary procedures some of these attempts have proven rather successful.

(a) Heat-soluble S. This method (Williams and Steinbergs, 1959) measures the amount of water soluble S released during gentle hydrolysis as a result of sequential wet and dry heating of soils. It extracts amounts of S similar to those obtained by direct extraction with hot water (Spencer and Freney, 1960). Fox et al. (1964) reported that heat-soluble, and autoclave methods extracted > twice and four times as much soil S, respectively, as that extracted by phosphate solutions. Pot experiments in Australia (Williams and Steinbergs, 1959; Spencer and Freney, 1960) and field calibrations in Nebraska (Fox et al., 1964) have demonstrated that heat-soluble S is highly correlated with S uptake by plants. This very labile fraction of soil organic S appears to provide a satisfactory index of S availability in a variety of soils.

(b) Acetate-extractable S. McClung et al. (1959) used neutral ammonium acetate to extract a S fraction from Brazilian soils which correlated well with plant response. This method generally extracts amounts of S similar to those extracted by phosphate solutions, although it does not displace adsorbed sulfates. Apparently this fraction contains relatively little organic S, but it was found to provide a reasonable index of the S status in some Australian (Spencer and Freney, 1960) and Minnesota soils (Rehm and Caldwell, 1968).

(c) Bicarbonate-extractable S. Extraction of S with 0.5 M NaHCO$_3$ was first suggested by Kilmer and Nearpass (1960). It has become the most widely used soil test for S that measures mineral plus some organic S. The pH adjustment of this extractant is important because at pH 10 bicarbonate solutions extract considerably more S than at the normally used pH 8.5. This method measures adsorbed sulfate together with free sulfate and some labile organic S (Williams and Steinbergs, 1964; Freney et al., 1971) and the amounts extracted are generally double those extracted with phosphate solutions (Rehm and Caldwell, 1968; Probert, 1976). The bicarbonate-extractable S fraction appears to provide a reliable index of S availability. For example, it was shown to be very well correlated with S "A" values ($r = 0.89$) for 30 soils from the southeastern U.S.A. (Kilmer and Nearpass, 1960). On some Minnesota soils plant uptake of S was more closely correlated with this fraction ($r = 0.95$) than with acetate- or phosphate-extractable S (Rehm and Caldwell, 1968). Decreases in bicarbonate-extractable S during plant growth accounted for most of the plant uptake and were more closely correlated with the latter than were the decreases measured by four other extrac-

tants (Williams and Steinbergs, 1964). Bicarbonate was also found to extract preferentially the recently immobilized organic S which consists mainly of HI-reducible S (Freney et al., 1971). However, in a recent study of available and isotopically exchangeable S in several Australian soils Probert (1976) found that the bicarbonate extractant measured a much larger pool of soil S than that which is available to plants. He concluded that this fraction does not represent the potentially mineralizable S since it was not sufficiently correlated with the "L" values of the soils examined.

(d) Reducible S. This method measures the HI-reducible S in soils and as mentioned earlier (see the section on p. 275) it accounts for 1/3 to 2/3 of the total organic S in most mineral soils. Although this fraction is generally thought to be more labile or degradable than the C-bonded and the residual fraction of soil organic S (see the section on p. 285), it is much too large a fraction to provide a suitable index of plant available S. Nevertheless, Spencer and Freney (1960) found that reducible S was significantly correlated with both S uptake and yield on several Australian soils. In an earlier study Williams and Steinbergs (1959) showed very close correlations between reducible S and total S, organic S and NaOH-soluble S, but reducible S was not significantly correlated with plant uptake.

(e) Reserve S. Bardsley and Lancaster (1960) suggested that a fraction designated as "reserve S" could be used as an indicator of the soil S status. This method involves the ignition of a soil—sodium bicarbonate mixture followed by the extraction of sulfates. Although this fraction contains essentially the total organic soil S and some inorganic S it was found to be reasonably well correlated with S uptake by clover ($r = 0.79$).

Incubation techniques

Incubation procedures of the type commonly used for assessing the availability of N in soils (Chapter 5) appear to offer little promise for estimating the S availability status of soils. This is because only a few ppm of sulfate S are normally released during laboratory incubation of soils, particularly in the absence of plants (Nicolson, 1970). Some success was reported by Harward et al. (1962); the correlation of S released on incubation plus extractable S versus percentage yield was $r = 0.75$. However, this correlation was considerably weaker than that obtained for "A" values or % S in the plants.

Microbial assays

Aspergillus niger has been used by several workers as a test organism for available soil S (Ensminger and Freney, 1966). This fungus can obtain similar amounts of S from soils as can be extracted with phosphate solutions. Spencer and Freney (1960) found that the growth of *Aspergillus* on several Australian soils was closely correlated with S uptake by plants ($r = 0.83$).

In assessing the aforementioned tests it becomes apparent that some degree of correlation with plant uptake has been found for every method suggested. However, these correlations with S tests, particularly with those which include some measure of labile soil organic S, have been largely restricted to pot experiments. In such tests the extractability and availability of soil S can be markedly influenced by the pretreatment of the soil; consequently this artifact will contribute significantly to the observed correlations. More emphasis should therefore be placed on field testing of these availability indexes. A real need still exists for a quick and simple laboratory procedure which will adequately estimate the S supplying power of soils. To develop such a soil test it is necessary to identify those fractions of the soil organic S that decompose during the growing season and provide plant available S. It will be difficult to attain this goal with existing fractionation procedures, because Freney et al. (1975) have established that all of the major soil organic S fractions contribute available S for plant uptake and that no single fraction is likely to be of much value for assessing S availability. Until fractionation techniques yielding biologically more meaningful fractions can be developed it might be useful to re-examine, modify and improve some of the more promising soil tests for labile S (e.g., heat-soluble and bicarbonate extractable S).

CONCLUSIONS

In this chapter we have discussed the nature, distribution and transformations of soil organic S, its relationship to other soil organic constituents, the factors that affect its turnover, and its function as a reserve for the supply of S to crops. During the last two decades steady progress has been made toward a better understanding of S dynamics in the soil—plant system. Thus S can no longer be viewed as the "neglected nutrient" or the "step child of soil fertility research", a concern frequently expressed during the early 1950's. The fractionation techniques developed by Australian and Canadian workers have provided a useful basis for further partitioning and characterization of soil organic S. Future progress in attempts to identify the source, nature and dynamics of the most labile components of organic S will depend largely on the partitioning into fractions of greater biological significance. This information is without doubt of great agronomic importance because it provides the means for assessing more accurately the availability of S in soils. The research to date has indicated that, contrary to earlier assumptions, there are marked differences between the dynamics of S and N as exemplified by the dissimilar mineralization pattern in response to plant growth, fertilization and changes in environmental conditions. Consequently, experimental techniques and approaches that have been used successfully to characterize the N turnover are not necessarily applicable to the study of S dynamics. However, it should be possible to delineate more clearly the differences and similarities between the cycling of these two nutrients by simultaneous labelling of soil organic matter with ^{35}S and ^{15}N and subsequent tracing of their concomitant transformations. In the area of S immobilization and mineralization, the imaginative and comprehensive tracer studies conducted by Freney et al. (1971, 1975) represent a significant advance by providing some insight into the quantitative transfer of S into and out of major soil organic fractions. It is hoped that their results will stimulate further efforts to elucidate the complex dynamics of soil organic S.

From an agronomic standpoint more accurate information is required on the long-term effects of cultivation and different cropping systems on the quantity and nature of the organic S reserve in soils. We need to learn more about the factors that govern the release of plant available S from this reserve such as the role of arylsulfatase and other S-hydrolysing soil enzymes, the rhizosphere effect, and the effects of freeze–thaw cycles and other drastic environmental changes. It is essential that the theories developed on the basis of trends observed in laboratory and greenhouse experiments be taken to the field and tested under more realistic conditions. The development of sound models for the turnover of S requires a more effective use of tracers in well co-ordinated multifactored laboratory and field experiments which are designed to quantify the various biological interconversions of S within the complex soil–plant system.

REFERENCES

Alexander, M., 1961. Introduction to Soil Microbiology. Wiley, New York, N.Y. 472 pp.
Allison, F.E., 1973. Soil Organic Matter and its Role in Crop Production. Developments in Soil Science, 3. Elsevier, Amsterdam, 637 pp.
Anderson, G., 1975. In: J.E. Gieseking (Editor), Soil Components, 1. Organic Components. Springer, New York, N.Y., pp. 333—341.
Aulakh, M.S., Dev, G. and Arora, B.R., 1976. Plant Soil, 45: 75—80.
Banwart, W.L. and Bremner, J.M., 1975. Soil Biol. Biochem., 7: 359—364.
Banwart, W.L. and Bremner, J.M., 1976. Soil Biol. Biochem., 8: 19—22.
Bardsley, C.E. and Lancaster, J.D., 1960. Soil Sci. Soc. Am. Proc., 24: 265—268.
Barrow, N.J., 1960. Aust. J. Agric. Res., 11: 960—969.
Barrow, N.J., 1961. Aust. J. Agric. Res., 12: 306—319.
Barrow, N.J., 1969. Soil Sci., 108: 193—201.
Beaton, J.D., Burns, G.R. and Platou, J., 1968. Sulphur Inst. Tech. Bull. 14: 1—56.
Bettany, J.R., Stewart, J.W.B. and Halstead, E.H., 1973. Soil Sci. Soc. Am. Proc., 37: 915—918.
Bettany, J.R., Stewart, J.W.B. and Halstead, E.H., 1974. Can. J. Soil Sci., 54: 309—315.
Biederbeck, V.O., 1969. Microbial Degradation and Chemical Characterization of Soil Humic Nitrogen. Ph.D. Thesis, Dept. Soil Sci., Univ. of Sask., Saskatoon, Sask.
Biederbeck, V.O. and Paul, E.A., 1973. Soil Sci., 115: 357—366.
Birch, H.F., 1959. Plant Soil, 11: 262—286.
Birch, H.F., 1960. Plant Soil, 12: 81—96.
Bremner, J.M. and Banwart, W.L., 1976. Soil Biol. Biochem., 8: 79—83.
Bremner, J.M. and Bundy, L.G., 1974. Soil Biol. Biochem., 6: 161—165.
Campbell, C.A., Paul., E.A. and McGill, W.B., 1976. Canada Nitrogen Symp., Calgary, Alta, pp. 9—101.

Chandra, P. and Bollen, W.B., 1960. Appl. Microbiol., 8: 31—38.
Clark, F.E. and Paul, E.A., 1970. Advan. Agron., 22: 375—435.
Coleman, R., 1966. Soil Sci., 101: 230—239.
Cooper, P.J.M., 1972. Soil Biol. Biochem., 4: 333—337.
Cowling, D.W. and Jones, L.H.P., 1970. Soil Sci., 110: 346—354.
Dijkshoorn, W. and Van Wijk, A.L., 1967. Plant Soil, 26: 129—157.
Donald, C.M. and Williams, C.H., 1954. Aust. J. Agric. Res., 5: 664—687.
Ensminger, L.E. and Freney, J.R., 1966. Soil Sci., 101: 283—290.
Evans, C.A. and Rost, C.O., 1945. Soil Sci., 59: 125—137.
Fox, R.L., Olson, R.A. and Rhoades, H.F., 1964. Soil Sci. Soc. Am. Proc., 28: 243—246.
Frederick, L.R., Starkey, R.L. and Segal, W., 1957. Soil Sci. Soc. Am. Proc., 21: 287—
    292.
Freney, J.R., 1958. Soil Sci., 86: 241—244.
Freney, J.R., 1961. Aust. J. Agric. Res., 12: 424—432.
Freney, J.R., 1967. In: A.D. McLaren and G.H. Peterson (Editors), Soil Biochemistry, 1.
    Dekker, New York, N.Y., pp. 229—259.
Freney, J.R. and Spencer, K., 1960. Aust. J. Agric. Res., 11: 339—345.
Freney, J.R. and Stevenson, F.J., 1966. Soil Sci., 101: 307—316.
Freney, J.R., Barrow, N.J. and Spencer, K., 1962. Plant Soil, 27: 295—308.
Freney, J.R., Melville, G.E. and Williams, C.H., 1970. Soil Sci., 109: 310—318.
Freney, J.R., Melville, G.E. and Williams, C.H., 1971. Soil Biol. Biochem., 3: 133—141.
Freney, J.R., Stevenson, F.J. and Beavers, A.H., 1972. Soil Sci., 114: 468—476.
Freney, J.R., Melville, G.E. and Williams, C.H., 1975. Soil Biol. Biochem., 7: 217—221.
Hamm, J.M., Bettany, J.R. and Halstead, E.H., 1973. Comm. Soil Sci. Plant Anal., 4:
    219—231.
Haque, I. and Walmsley, D., 1972. Plant Soil, 37: 255—264.
Haque, I. and Walmsley, D., 1974. Plant Soil, 40: 145—152.
Harward, M.E., Chao, T.T. and Fang, S.C., 1962. Agron. J., 54: 101—106.
Hesse, P.R., 1957. Plant Soil, 9: 86—96.
Hoeft, R.G., Walsh, L.M. and Keeney, D.R., 1973. Soil Sci. Soc. Am. Proc., 37: 401—
    404.
Houghton, C and Rose, F.A., 1976. Appl. Environ. Microbiol., 31: 969—976.
Jenkinson, D.S., 1966. J. Soil Sci., 17: 280—302.
Jenny, H., 1930. Missouri Agr. Exp. Sta. Res. Bul., 152: 1—66.
Johnson, C.M. and Nishita, H., 1952. Anal. Chem., 24: 736—742.
Jones, L.H.P., Cowling, D.W. and Lockyer, D.R., 1972. Soil Sci., 114: 104—114.
Jones, M.B., Williams, W.A. and Martin, W.E., 1971. Soil Sci. Soc. Am. Proc., 35: 542—
    546.
Jordan, H.V. and Ensminger, L.E., 1958. Advan. Agron., 10: 407—434.
Jordan, J.V. and Baker, G.O., 1959. Soil Sci., 88: 1—6.
Kilmer, V.J. and Nearpass, D.C., 1960. Soil Sci. Soc. Am. Proc., 24: 337—340.
Kowalenko, C.G. and Lowe, L.E., 1975a. Can. J. Soil Sci., 55: 1—8.
Kowalenko, C.G. and Lowe, L.E., 1975b. Can. J. Soil Sci., 55: 9—14.
Larson, W.E., Clapp, C.E., Pierre, W.H. and Morachan, Y.B., 1972. Agron. J., 64: 204—
    208.
Levesque, M., 1974. Can. J. Soil Sci., 54: 333—335.
Lewis, J.A. and Papavizas, G.C., 1970. Soil Biol. Biochem., 2: 239—246.
Lowe, L.E., 1964. Can. J. Soil Sci., 44: 176—179.
Lowe, L.E., 1965. Can. J. Soil Sci., 45: 297—303.
Lowe, L.E., 1969a. Can. J. Soil Sci., 49: 129—141.
Lowe, L.E., 1969b. Can. J. Soil Sci., 49: 375—381.
Lowe, L.E. and DeLong, W.A., 1963. Can. J. Soil Sci., 43: 151—155.

MacKenzie, A.F., DeLong, W.A. and Ghanem, I.S., 1967. Plant Soil, 27: 408—414.
Mann, H.H., 1955. J. Soil Sci., 6: 241—247.
Martin, W.E. and Walker, T.W., 1966. Soil Sci., 101: 248—257.
Mathur, S.P., 1971. Soil Sci., 111: 147—157.
McClung, A.C., DeFreitas, L.M.M. and Lott, W.L., 1959. Soil Sci. Soc. Am. Proc., 23: 221—224.
McLachlan, K.D., 1975. In: K.D. McLachlan (Editor), Sulphur in Australasian Agriculture. Sidney Univ. Press, Sidney, pp. 58—67.
Mehring, A.L. and Bennett, G.A., 1950. Soil Sci., 70: 73—81.
Nelson, L.E., 1964. Soil Sci., 97: 300—306.
Nelson, L.E., 1973. Soil Sci., 115: 447—454.
Neptune, A.M.L., Tabatabai, M.A. and Hanway, J.J., 1975. Soil Sci. Soc. Am. Proc., 39: 51—55.
Nicolson, A.J., 1970. Soil Sci., 109: 345—350.
Nyborg, M., 1968. Can. J. Soil Sci., 48: 37—41.
Olson, R.A., 1957. Soil Sci., 84: 107—111.
Paul, E.A., 1970. Recent Advan. Phytochem., 3: 59—104.
Paul, E.A. and Schmidt, E.L., 1961. Soil Sci. Soc. Am. Proc., 25: 359—362.
Probert, M.E., 1976. Plant Soil, 45: 461—475.
Rehm, G.W. and Caldwell, A.C., 1968. Soil Sci., 105: 355—361.
Roberts, S. and Koehler, F.E., 1968. Soil Sci., 106: 53—59.
Scharpenseel, H.W. and Krausse, R., 1963. Z. Pfl. Ernähr. Düng. Bodenk., 101: 11—23.
Scott, N.M., 1976. J. Sci. Food Agric., 27: 367—372.
Scott, N.M. and Anderson, G., 1976. J. Sci. Food Agric., 27: 358—366.
Seim, E.C., Caldwell, A.C. and Rehm, G.W., 1969. Agron. J., 61: 368—371.
Simon-Sylvestre, G., 1965. C.R. Hebd. Seances Acad. Agric. Fr., 51: 426—431.
Spencer, K. and Freney, J.R., 1960. Aust. J. Agric. Res., 11: 948—959.
Starkey, R.L., 1950. Soil Sci., 70: 55—65.
Starkey, R.L., 1966. Soil Sci., 101: 297—306.
Stewart, B.A. and Porter, L.K., 1969. Agron. J., 61: 267—271.
Stewart, B.A., Porter, L.K. and Viets Jr., F.G., 1966a. Soil Sci. Soc. Am. Proc., 30: 355—358.
Stewart, B.A., Porter, L.K. and Viets Jr., F.G., 1966b. Soil Sci. Soc. Am. Proc., 30: 453—456.
Stotsky, G. and Norman, A.G., 1961. Arch. Mikrobiol., 40: 370—382.
Swaby, R.J. and Fedel, R., 1973. Soil Biol. Biochem., 5: 773—781.
Tabatabai, M.A. and Bremner, J.M., 1970a. Soil Sci. Soc. Am. Proc., 34: 225—229.
Tabatabai, M.A. and Bremner, J.M., 1970b. Soil Sci. Soc. Am. Proc., 34: 427—429.
Tabatabai, M.A. and Bremner, J.M., 1972a. Agron. J., 64: 40—44.
Tabatabai, M.A. and Bremner, J.M., 1972b. Soil Sci., 114: 380—386.
Till, A.R., 1975. In: K.D. McLachlan (Editor), Sulphur in Australasian Agriculture. Sidney Univ. Press, Sidney, pp. 68—75.
Ulrich, A., Tabatabai, M.A., Ohki, K. and Johnson, C.M., 1967. Plant Soil, 26: 235—252.
Van Praag, H.J., 1973. Plant Soil, 39: 61—69.
Walker, D.R. and Doornenbal, G., 1972. Can. J. Soil Sci., 52: 261—266.
Walker, T.W., 1957. J. Brit. Grassl. Soc., 12: 10—18.
White, J.G., 1959. New Zealand, J. Agr. Res., 2: 255—258.
Whitehead, D.C., 1964. Soils Fert., 27: 1—8.
Williams, C.H., 1967. Plant Soil, 26: 205—223.
Williams, C.H. and Donald, C.M., 1957. Aust. J. Agric. Res., 8: 179—189.
Williams, C.H. and Lipsett, J., 1961. Aust. J. Agric. Res., 12: 612—629.
Williams, C.H. and Steinbergs, A., 1959. Aust. J. Agric. Res., 10; 340—352.
Williams, C.H. and Steinbergs, A., 1962. Plant Soil, 17: 279—294.
Williams, C.H. and Steinbergs, A., 1964. Plant Soil, 21: 50—62.
Williams, C.H., Williams, E.G. and Scott, N.M., 1960. J. Soil Sci., 11: 334—346.
Wyatt, F.A. and Doughty, J.L., 1928. Sci. Agric., 8: 549—555.

SUBJECT INDEX